Bacterial Toxins

METHODS IN MOLECULAR BIOLOGY™

John M. Walker, SERIES EDITOR

METHODS IN MOLECULAR BIOLOGY™

Bacterial Toxins

Methods and Protocols

Edited by

Otto Holst

Research Center Borstel, Germany

Humana Press ✳ **Totowa, New Jersey**

Cover design by Patricia Cleary.
Cover illustration from: Chapter 6, Structure–Function Analysis of Cysteine-Engineered Entomopathogenic Toxins, by J.-L. Schwartz and L. Masson.

For additional copies, pricing for bulk purchases, and/or information about other Humana titles, contact Humana at the above address or at any of the following numbers: Tel.: 973-256-1699; Fax: 973-256-8341; E-mail: humana@humanapr.com; or visit our Website: http://humanapress.com

Printed in the United States of America. 10 9 8 7 6 5 4 3 2 1

Library of Congress Cataloging in Publication Data

Bacterial toxins : methods and protocols / edited by Otto Holst.
        p.   cm. -- (Methods in molecular biology ; 145)
     Includes bibliographical references and index.
     ISBN 0-89603-604-9 (alk. paper)
        1. Bacterial toxins--Research--Methodology. I. Holst, Otto.
     II. Series: Methods in molecular biology (Clifton, N.J.) ; v. 145.
     QP632.B3B334 2000
     579.3'165--dc21   99-41237
                                                                                            CIP

Preface

The interest of investigators across a broad spectrum of scientific disciplines has been steadily stimulated by the field of bacterial toxin research, an area that makes use of a large variety of biological, chemical, physicochemical, and medically oriented approaches. Researchers studying bacterial toxins need to be acquainted with all these disciplines in order to work effectively in the field. To date, there has been no published collection offering detailed descriptions of the techniques and methods needed by researchers operating across the field's diverse areas. The present volume *Bacterial Toxins: Methods and Protocols,* is intended to fill this gap.

Bacterial Toxins: Methods and Protocols consists of two sections: one on protein toxins (15 chapters) and one on endotoxins (5 chapters). Each section is introduced by an overview article (Chapters 1 and 16). The protocols collected represent state-of-the-art techniques that each have high impact on future bacterial toxin research. All methods are described by authors who have regularly been using the protocol in their own laboratories. Included in each chapter is a brief introduction to the method being described.

Since the goal of the book this to outline the practical steps necessary for successful application of the methods, the major part of each chapter provides a step-by-step description of the method treated. Each chapter also possesses a Notes section, which deals with difficulties that may arise when using the method, and with the modifications and limitations of the technique. In sum, our volume, *Bacterial Toxins: Methods and Protocols* should prove useful to a broad spectrum of researchers, including those without any previous experience with a particular technique.

Otto Holst

Contents

Contributors

DAVID W. K. ACHESON • *Division of Geographic Medicine and Infectious Diseases, New England Medical Center Hospital, Boston, MA*

JOSEPH E. ALOUF • *Institut Pasteur de Lille, Lille, France*

CHRISTIAN ALTENBACH • *Jules Stein Eye Institute and Department of Chemistry and Biochemistry, University of California, Los Angeles, CA*

PETER AMERSDORFER • *Phylos Inc., Lexington, MA*

MARIE-FRANCE BADER • *INSERM U338, Strasbourg, France*

R. JOHN COLLIER • *Department of Microbiology and Molecular Genetics, Harvard Medical School, Boston, MA*

GREGORY DE CRESCENZO • *Biotechnology Research Institute, National Research Council Canada, Montréal, Canada*

MAURO DALLA SERRA • *CNR-ITC, Centro di Fisica Degli Stati Aggregati, Povo, Italy*

MIKAEL DOHLSTEN • *AstraZeneca, Lund, Sweden*

ALI EL BAYÂ • *Institut für Infektiologie, Westfälische Wilhelms-Universität Münster, Münster, Germany*

VOLKER T. EL-SAMALOUTI • *Department of Immunology and Cell Biology, Research Center Borstel, Borstel, Germany*

HANS-DIETER FLAD • *Department of Immunology and Cell Biology, Research Center Borstel, Borstel, Germany*

MICHAEL C. GOODNOUGH • *Department of Food Microbiology and Toxicology, University of Wisconsin, Madison, WI*

LUTZ HAMANN • *Department of Immunology and Cell Biology, Research Center Borstel, Borstel, Germany*

KEN-ICHI HARADA • *Laboratory of Instrumental Analytical Chemistry, Faculty of Pharmacy, Meijo University, Nagoya, Japan*

OTTO HOLST • *Division of Medical and Biochemical Microbiology, Research Center Borstel, Borstel, Germany*

WAYNE L. HUBBELL • *Jules Stein Eye Institute and Department of Chemistry and Biochemistry, University of California, Los Angeles, CA*

ERIC A. JOHNSON • *Department of Food Microbiology and Toxicology, University of Wisconsin, Madison, WI*

ANNE V. KANE • *Center for Gastroenterology Research on Absorptive and Secretory Processes, New England Medical Center Hospital, Boston, MA*

GERALD T. KEUSCH • *Division of Geographic Medicine and Infectious Diseases, New England Medical Center Hospital, Boston, MA*

KARIN KRISTENSSON • *Active Biotech, Lund, Sweden*

LINDA LAWTON • *School of Applied Sciences, Robert Gordon University, Aberdeen, UK*

BUKO LINDNER • *Division of Biophysics, Research Center Borstel, Borstel, Germany*

CARL J. MALIZIO • *Department of Food Microbiology and Toxicology, University of Wisconsin, Madison, WI*

JAMES D. MARKS • *SFGH-Department of Anesthesia, University of California, San Francisco, CA*

LUKE MASSON • *Biotechnology Research Institute, National Research Council Canada, Montréal, Canada*

ALBERTO MAZZA• *Biotechnology Research Institute, National Research Council Canada, Montréal, Canada*

GIANFRANCO MENESTRINA • *CNR-ITC, Centro di Fisica Degli Stati Aggregati, Povo, Italy*

JUSSI MERILUOTO • *Department of Biochemistry and Pharmacy, Åbo Akademi University, Turku, Finland*

JORDI MOLGÓ • *Laboratoire de Neurobiologie Cellulaire et Moléculaire, Gif-sur-Yvette, France*

JOHN R. MURPHY • *Department of Medicine, Boston University School of Medicine, Boston, MA*

KYOUNG JOON OH • *Department of Microbiology and Molecular Genetics, Harvard Medical School, Boston, MA*

WILLIAM D. PICKING • *Department of Molecular Biosciences, University of Kansas, Lawrence, KS*

BERNARD POULAIN • *Laboratoire de Neurobiologie Cellulaire, Strasbourg, France*

KRISTIAN RIESBECK • *Active Biotech, Lund, Sweden, and Department of Medical Microbiology, University Hospital MAS, Malmoe, Sweden*

ALEXANDER ROSENDAHL • *AstraZeneca, Lund, Sweden*

JAMES C. RICHARDS • *Institute for Biological Sciences, National Research Council of Canada, Ottawa, Canada*

M. ALEXANDER SCHMIDT • *Institut für Infektiologie, Westfälische Wilhelms-Universität Münster, Münster, Germany*

JEAN-LOUIS SCHWARTZ • *Biotechnology Research Institute, National Research Council,and Groupe de Recherche en Transport Membranaire, Université de Montréal, Montréal, Canada*

ULRICH SEYDEL • *Division of Biophysics, Research Center Borstel, Borstel, Germany*

PIERRE THIBAULT • *Institute for Biological Sciences, National Research Council of Canada, Ottawa, Canada*

ARTUR J. ULMER • *Department of Immunology and Cell Biology, Research Center Borstel, Borstel, Germany*

ANDRE WIESE • *Division of Biophysics, Research Center Borstel, Borstel, Germany*

JOHANNA C. VANDERSPEK • *Department of Medicine, Boston University School of Medicine, Boston, MA*

LARS VON OLLESCHIK-ELBHEIM • *Institut für Infektiologie, Westfälische Wilhelms-Universität Münster, Münster, Germany*

1

Bacterial Protein Toxins

An Overview

Joseph E. Alouf

> *To the physiologist the poison becomes an instrument which dissociates and analyzes the most delicate phenomenon of living structures and by attending carefully to their mechanism in causing death, he can learn indirectly much about the physiological processes of life.*
>
> Claude Bernard, *La Science Experimentale*
> Paris, 1878

1. Introduction

This short overview attempts to highlight the current state of the art relevant to bacterial protein toxins. In particular we outline the major achievements in this field during the past decade and briefly describe some significant hallmarks of toxinological research since the advent of modern methodologies elucidating the biochemistry, genetics, and cell biology of these fascinating bacterial effectors.

Valuable information on the progress of our knowledge during the past 15 yr can be found in recently published books *(1–13)* and in the series (eight to date) *Bacterial Protein Toxins* (European Workshops Books) published by Academic Press (London, 1983) and thereafter every other year by Gustav Fischer Verlag (Stuttgart/Jena).

Valuable general reviews have also been published *(14–37)* that may help the reader to find specific information and appropriate bibliography.

2. What are Bacterial Toxins?

In microbiology, the term bacterial toxin, coined 110 yr ago by Roux and Yersin *(38)*, designates exclusively the special class of bacterial macromolecu-

From: *Methods in Molecular Biology, vol. 145: Bacterial Toxins: Methods and Protocols*
Edited by: O. Holst © Humana Press Inc., Totowa, NJ

Table 1
Repertoire of Bacterial Protein/Peptide Toxins
(as of June, 1999: 323)

148 (46%) from Gram-positive bacteria
175 (54%) from Gram-negative bacteria
Extracellular toxins: 75%
Intracellular toxins : 25%
Membrane damaging/pore-forming cytolysins: 110
(approx 35% of protein toxins)

lar substances that when produced during natural or experimental infection of the host or introduced parenterally, orally (bacterial food poisoning), or by any other route in the organism results in the impairment of physiological functions or in overt damage to tissues. These unfavorable effects may lead to disease and even to the death of the individual.

Bacterial toxins are differentiated into two major classes on the basis of their chemical nature, regardless of their cellular location and the staining features of the bacteria that produce them: bacterial protein toxins and the toxic lipopolysaccharide complexes present at the surface of the outer membrane of the cell walls of Gram-negative bacteria.

The protein toxins that are the subject of this chapter constitute a wide collection of more than 300 distinct entities (**Table 1**) which are mostly released from bacterial cells during growth and therefore are considered as exotoxins. However, ca. 25% of protein toxins remain either intracytoplasmic or more or less firmly associated to the cell surface. Their eventual release outside the bacterial cell takes place during the decline of bacterial growth or after cell death, generally through autolytic processes.

2.1. Historical Background

The concept that pathogenic bacteria might elaborate harmful substances to the infected host emerged shortly after the discovery of these microorganisms as etiological agents of human diseases. In 1884, Robert Koch suggested that cholera was elicited by a bacterial component released by *Vibrio cholerae*, but parenteral injection of bacterial culture filtrates in animals did not produce any toxic effect and the idea of an extracellular cholera poison was abandoned. Then 4 yr later, Roux and Yersin (Institute Pasteur, Paris) discovered the first bacterial toxin (diphtheria toxin) in the culture filtrate of *Corynebacterium diphtheriae (38)* after the failure of Loeffer 1 yr earlier to prove the release of this toxin. The Institute Pasteur investigators thus brought up the prototype of a new class of extraordinarily toxic, pharmacologically and physiologically active factors of immense potential for medicine, microbiology, immunology, molecular and cellular biology, and the neurosciences.

Two other major toxins were soon to follow: tetanus toxin discovered in 1890 independently by Faber, Briedel and Frankel, Tizzoni and Cattani, and botulinum toxin discovered in 1896 by Van Ermengem *(38)*. From that time through the end of the 1940s, about 60 toxins were identified, among them clostridial toxins as a result of the experience gained during World War I gas gangrenes. The onset of World War II stimulated further research into toxinogenic anaerobes. A milestone was the discovery by Macfarlane and Knight that *C. perfringens* α-toxin was a phospholipase C. This toxin became the first for which a biochemical mode of action was recognized at the molecular level. It is now the prototype of the group of at least 14 cytolytic toxins that disrupt eucaryotic cell membranes by hydrolysis of their constitutive phospholipids *(23)*. The beginning of the 1950s witnessed the discovery of *Bacillus anthracis* toxin (anthrax toxin) by H. Smith and his co-workers *(40)*. A great advance in our understanding was the observation in India in 1953 by De and co-workers that the injection of living *V. cholerae* or cell-free filtrates into the lumen of a ligated loop of rabbit ileum caused accumulation of a large amount of fluid having gross similarity to cholera. This led to the discovery of cholera toxin which revived and experimentally confirmed the old Koch prediction of the reality of an enteropathogenic cholera exotoxin. However, it was some 17 yr after De's initial work that the putative enterotoxin was isolated and purified in 1969 by Finkelstein and his co-workers *(29,38)*. Cholera toxin, a 84-kDa oligomeric protein is the prototype of a wide family of biochemically, immunologically, and pharmacologically related toxins found in human and porcine *E. coli* strains, *V. cholerae*, *V. mimicus*, non-01 *Aeromonas hydrophila*, *Campylobacter jejuni*, *Salmonella enterica* sv. *typhi* and sv. *typhimurium*, and *Plesiomonas shigelloides (29,34)*. From 1970 to 1983 the bacterial toxin repertoire encompassed about 220 proteins and peptides. At present (1999) it comprises 323 different members (**Table 1**).

One notes that the discovery of this considerable class of toxins over more than one century was the combined fruit of rational design based on technological advances and also chance or serendipity to a certain degree. This was particularly the case for diphtheria and cholera toxins.

2.2. Structural and Genetic Aspects

2.2.1. Molecular Topology

A striking feature of bacterial protein toxins is the broad variety of molecular size and topological features in contrast to the more homogeneous structures of protein effectors of eucaryotic origin (e.g., hormones, neuropeptides, cytokines, growth factors).

2.2.1.1. SINGLE-CHAIN MOLECULES

Most protein toxins occur as single-chain holoproteins varying from approx 2–3 kDa such as the *E. coli* 17/18-amino-acid residue thermostable enterotoxin *(29)* and the *S. aureus* 26-amino-acid residue δ-toxin *(39)*, to the 150-kDa tetanus and botulinum neurotoxins *(26)* up to the approx 300-kDa *Clostridium difficile* toxins A (308 kDa) and B (270 kDa), which are the largest single-chain bacterial proteins hitherto identified *(35)*.

2.2.1.2. OLIGOMERIC MOLECULES

Several toxins occur as multimolecular complexes comprising two or more noncovalently bonded different subunits. Cholera toxin, *E. coli* thermolabile enterotoxins I and II (LT-I, LT-II), and other related enterotoxins form an heterohexamer AB5 composed of one 28-kDa A-subunit (the ADP-ribosylating moiety of the toxin) and five identical 11.8 kDa B-subunits *(29,34)*. Shiga toxin and *E. coli* shiga-like toxins (verotoxins) are also composed of a single 32-kDa A-subunit in association with a pentamer of 7.7-kDa B-subunits. The A-subunit is the holotoxin component that acts as an *N*-glycosidase to cleave a single adenine residue from the 28S rRNA component of the eucaryotic ribosomal complex *(18)*. Pertussis toxin has the most complex structure known so far among protein toxins *(41)*. The toxin is an A–B type hexamer composed of five dissimilar subunits, S1–S5. S1 is the enzymatic ADP-ribosylating A-subunit ($M_r = 26220$) and the B oligomeric moiety contains the S2 (21.9 kDa), S3 (21.8 kDa), S4 (12 kDa), and S5 (11.7 kDa) protomers complexed in a 1:1:2:1 molar ratio.

2.2.1.3. MACROMOLECULAR COMPLEXES OF TOXINS ASSOCIATED
        TO NONTOXIC MOIETIES

This situation is known for the 150-kDa botulinum neurotoxins (BoNT) found in bacterial cultures and contaminated foodstuffs. These complexes, referred to as progenitor toxins, comprise three different forms: M toxin (300 kDa), L toxin (500 kDa) and LL toxin (900 kDa). The smaller M toxin is composed of a BoNT molecule (150 kDa) in association with a similarly sized nontoxic protein (150 kDa). The larger L and LL progenitor toxins additionally contain an undefined number of proteins with hemagglutinin (HA) activity. The form of progenitor toxin found varies between the different toxinogenic types, and more than one form may be produced by a single strain. All three forms have been found in type A *Clostridium botulinum* strains. *C. botulinum* type G strains produce the L toxin. The botulinum toxin of type E and F strains is composed exclusively of M progenitor toxin *(8)*.

2.2.1.4. MULTIFACTORIAL TOXINS

A number of toxins designated *binary toxins* are composed of two independent single chains not joined by either covalent or noncovalent bonds. In this respect, they differ from oligomeric toxins, the protomers of which are assembled in a defined structure (holotoxin). The moieties of binary toxins should act in concert to be efficient. Each individual protein separately expresses little or no toxicity.

Binary toxins are produced by a variety of Gram-positive bacteria, for example, *S. aureus* leucocidin and γ-toxin *(42,43)*, *Enterococcus faecalis* hemolysin/bacteriocin *(44)*, *Clostridium botulinum* C2 toxin, *Cl. perfringens* iota-toxin, and *Cl. spiroforme* iota-like toxin *(35,45)*. *Bacillus anthracis* three-component toxin is more complex. Two different sets of active toxin result from the combination of either the lethal factor (LF) and protective antigen (PA) leading to the metalloprotease lethal toxin, or the combination of PA and edema factor (EF) which constitutes the calmodulin-dependent adenyl cyclase *(46–48)*.

2.2.1.5. PROTOXIN FORMS

Several protein toxins are secreted in their mature form into the culture medium as inactive protoxins similarly to several proenzymes (zymogens). These protoxins are converted to active toxins by proteolytic enzymes present in the medium or by treatment with proteases that split off small fragments from the precursor, for example, *C. perfringens* ε- and iota-toxins, the C2-toxin of *C. botulinum* (component C2-II), and the membrane damaging toxin aerolysin of *Aeromonas hydrophila*.

2.2.1.6. THREE-DIMENSIONAL CRYSTAL STRUCTURE

Since the elucidation in 1986 of the three-dimensional structure of *P. aeruginosa* exotoxin A, that of 28 other toxins (among them 10 of the family of Gram-positive cocci superantigens) has been established so far (**Table 2**).

2.2.2. Molecular Genetics

The past 15 yr (1983–1998) may be considered as the golden age of the molecular genetics of bacterial protein. More than 150 structural genes have been cloned and sequenced (vs only 10 by the end of 1982). About 85% of the genes are chromosomal. The other genes are located on mobile genetic elements: bacteriophages, plasmids, and transposons. Bacteriophagic genes were found to encode, among other toxins: diphtheria toxin *(38,49)*, cholera toxin *(50)*, *S. pyogenes* erythrogenic toxins A and C and *S. aureus* enterotoxins A and E (*16* and **Table 3**), *C. botulinum* toxins C1 and D *(8,26)*, and *E. coli*

Table 2
Three-Dimensional Structure of Crystallized Toxins[a]

1. *P. aeruginosa* exotoxin A (**ref. 76**)
2. *E. coli* LT-1 toxin (**refs. 77,78**)
3. *Bacillus thuringiensis* δ-toxin (**ref. 79**)
4. Oligomer B of *E. coli* shiga-like toxin (verotoxin) (**ref. 80**)
5. Diphtheria toxin (**refs. 81,82**)
6. *Aeromonas hydrophila* proaerolysin (**ref. 83**)
7. *Pertussis* toxin (**refs. 84,85**)
8. *Shigella dysenteriae* toxin (**ref. 86**)
9. Oligomer B of the choleragenoid form of cholera toxin (**ref. 87**)
10. Cholera toxin (holotoxin) (**ref. 88**)
11. *S. aureus* α-toxin (**ref. 89**)
12. *S. aureus* exfoliative toxin (**refs. 90,91**)
13. Anthrax toxin P component (protective antigen) (**ref. 92**)
14. Hc fragment of tetanus neurotoxin (**ref. 93**)
15. Perfringolysin O (**ref. 94**)
16. *Clostridium perfringens* α-toxin (**ref. 95**)
17. *Clostridium botulinum* neurotoxin A (**ref. 96**)
18. *S. aureus* leucocidin (LᴜᴋE-PV) (**ref. 96a**)
19. *S. aureus* leukocidin (LᴜᴋE) (**ref. 96b**)

Superantigens

20. *S. aureus* enterotoxin B (**ref. 97**) Enterotoxin B—CMH of class II complex (**ref. 98**)
21. *S. aureus* enterotoxin C1 (**ref. 99**)
22. *S. aureus* enterotoxin C2 (**ref. 100**)
23. *S. aureus* enterotoxin A (**refs. 101,102**)
24. *S. aureus* enterotoxin D (**ref. 103**)
25. *S. aureus* toxic-shock syndrome toxin-1 (TSST-1) (**refs. 104,105**)
26. TSST-1—CMH class II complex (**ref. 106**)
27. *S. pyogenes* erythrogenic (pyrogenic) exotoxin C (**ref. 107**)
28. *S. pyogenes* erythrogenic (pyrogenic) exotoxin A (**ref. 107a**)
29. Streptococcal super antigen (SSA) (**ref. 107b**)

[a]For general references see (*10,11*).

shiga-like toxins I and II (*18*). Plasmid-borne genes encode tetanus toxin (*8,26*); anthrax toxin complex PA, EF and LF (*46–48*); *S. aureus* enterotoxin D (*16*); and *E. coli* heat-labile and heat-stable enterotoxins (*29,34*). Heat-stable enterotoxin was also shown to be encoded by the transposon gene. The determination of the nucleotide sequences of toxin genes made it possible to deduce the primary structure of the relevant encoded proteins, thereby paving the way for

Table 3
Molecular Characteristics of Superantigenic Toxins
from *Staphylococcus aureus* and *Streptococcus pyogenes*
(Refs. *16,108–111e* and Table 2)

Toxins (Acronymes)	Gene[a] localization	AA[b]	Mol. mass (Da)	Human T-cell receptor Vβ motif(s) recognized
Staphylococcus aureus				
Enterotoxin A (SEA)[c]	B	233	27078	1.1, 5.3, 6.3, 6.4, 6.9, 7.3,7.4, 9.1, 18
Enterotoxin B (SEB)[c]	C	239	28336	3, 12, 14, 15, 17, 20
Enterotoxin C1 (SEC)[c]	C	239	27496	12
Enterotoxin C2 (SEC)[c]	C	239	27531	12, 13.2, 14, 15, 17, 20
Enterotoxin C3 (SEC)	C	239	27563	5, 12, 13.2
Enterotoxin D (SED)[c]	P	228	26360	5, 12
Enterotoxin E (SEE)	B	230	26425	5.1, 6.3, 6.4, 6.9, 8.1, 18
Enterotoxin G (SEG)	ND[d]	233	27023	ND
Enterotoxin H (SEH)	ND	218	25145	ND
Enterotoxin I (SEI)	ND	218	24928	ND
Toxic shock syndrome toxin-1 (TSST-1)[d]	C	194	22049	2
Streptococcus pyogenes				
Erythrogenic (pyrogenic) exotoxin A	B	222	27787	2, 12, 14, 15
Erythrogenic (pyrogenic) exotoxin C	B	208	24354	1, 2, 5.1, 10
Streptococcal superantigen (SSA)	ND	234	26892	1, 3, 15
Mitogen F (SPE F DNase B)	ND	228	25363	2, 4, 8, 15, 19
Mitogen M (SPM)	ND	ND	28000	13
Mitogen SPM-2	ND	ND	29000	4, 7, 8
Exotoxin MZ (SMEZ)	ND	234	25524	2, 8 (rabbit)

[a]C, Chromosomal; B, bacteriophage; P, plasmid.
[b]AA, Number of amino acid residues of mature form.
[c]Three-dimensional crystal structure determined.
[d]ND, not determined.

considerable progress in the elucidation of the mechanism of action of the toxins at the molecular level and for a deep understanding of structure–activity relationships. This achievement was particularly facilitated by the modification of the structural genes by site-directed mutagenesis, deletion, gene fusion, transposon insertion, and gene expression in appropriate bacteria followed by the purification of the recombinant proteins and the evaluation of their biological activity.

Table 4
Toxins Exhibiting Enzyme Activity

1. ADP-ribosylating toxins (*see* **Table 8**) Transferase and NADase activity

2. Phospholipases: (*see* **refs.** *14,23,28,33,51*) Cytolytic phospholipases C: *C. perfringens* α-toxin; *C. sordellii* γ-toxin; *C. novyi* β- and γ-toxins, *C. haemolyticus, P. aeruginosa, P. aureofaciens, A. hydrophila, B. cereus* hemolysins; *Acinetobacter calcoaceticus* phospholipase *Vibrio damsela* cytolytic phospholipase D, *Rickettsia prowazeki* hemolytic phospholipase, A *S. aureus* β-toxin, (hemolysin, sphingomyelinase), *Corynebacterium ovis* lethal toxin (phospholipase D)

3. Adenylcyclases: (*see* **refs.** *46–48,55,76*) *Bordetella pertussis* adenylate cyclase, *Bacillus anthracis* bifactorial edema toxin, *P. aeruginosa* exoenzyme y

4. Metalloproteases: (*see* **refs.** *8,26,36,46,59*) *Tetanus* and *botulinum* A, B, C, D, E, F, G, (light chain) zinc-dependent neurotoxins, *Bacillus anthracis* bifactorial lethal toxin, *Bacteroides fragilis* zinc-dependent enterotoxin

5. Ribonucleases: (*see* **ref.** *18*) *S. dysenteriae* shiga toxin, Shiga-like SLT-I and SLT-II toxins (verotoxins)

6. Glucosyltransferases: (*see* **refs.** *30,60,63*) *C. difficile* A, B toxins, *C. sordellii* and *C. novyi* lethal α-toxin

7. Deamidase activity: (*see* **refs.** *61,62*) *E. coli* cytotoxic necrotizing factor-1

8. Protease activity: (*see* **ref.** *58*) *S. aureus* epidermolytic toxin (exfoliatin) A, B, cysteine proteinase streptococcal pyrogenic exotoxin B (*see* **ref.** *13*, Chapter 32)

3. Mechanisms of Action of Protein Toxins on Eucaryotic Target Cells

Toxin effects on target cells may be classified operationally into two types: *type I effects* (toxins strictly acting at the surface of the cell [cytoplasmic membrane] without penetrating into the cell cytosol) and *type II effects* (toxins ultimately acting on specific molecular targets into the cytosol after binding on specific site[s] on cell surface and translocation across the membrane bilayer). A large number of both type I and type II toxins exhibit identified enzymatic activities (**Tables 4** and **5**).

3.1. Toxins Strictly Acting on the Cell Surface

This process concerns two distinct types of toxins with totally different molecular modes of action: damage to the cytoplasmic membrane or induction of biological effects through signal transduction processes.

Table 5
Functional typology of ADP-ribosylating toxins

Group I: Toxins causing inhibition of protein synthesis: diphtheria toxin and pseudo-
monas exotoxin A *(49,64)*
Acceptor cell protein: elongation factor-2 (EF-2)
EF–2 AA target: Diphthamide-715 (posttranslational histidine)
Cell killing/host death (MLD: 0.1 µg/kg)

Group II: Toxins causing alteration of transmembrane signal transduction
  A. *Cholera, E. coli* LT-I, LT-II/and related Gram-negative bacteria enterotoxins
(29,34)
            Acceptor protein(s): αs subunit of heterotrimeric GTP-binding proteins
            (G-protein), Target AA: Arg-201
  B. *Pertussis* toxin: *(41)*
            Acceptor protein(s): G-proteins, αi subunit of heterotrimeric GTP-
            binding proteins (G-protein); target amino acid: Cys-351
            Dysregulation of adenylate cyclase activation and accumulation of
            cyclic AMP leading to massive intestinal fluid secretion (diarrhea) induced
            by cholera toxin or impairment of various signal transduction processes
            induced by pertussis toxin

Group III: Toxins causing disorganization of cytoskeleton actin *(35)*
Subgroup IIIA
  1. *Clostridium botulinum* C2 toxin
  2. *Clostridium spiroforme* toxin
  3. *Clostridium perfringens* iota-toxin
  4. *Clostridium difficile* C2-like toxin (CDT)
            Acceptor protein: actin, target amino acid: Arg-177
            Depolymerization of actin F filaments (massive edema of intestinal
            endothelium and increase of fluid secretion)
Subgroup IIIB
  1. *Clostridium botulinum* C3 enzyme
  2. *Clostridium limosum* C3-like enzyme
  3. *Bacillus cereus* C3-like enzyme
  4. *Legionella pneumophila* C3-like enzyme
  5. *Staphylococcus aureus* EDIN (epidermal cell differentiation inhibition,
      C3-like enzyme)
            Acceptor protein: Rho (small G-protein), target amino acid: Asn-41
            Disruption of actin filaments

Group IV: *Bacillus sphaericus* mosquitocidal toxin *(112)*
Acceptor proteins: 38- and 42-kDa proteins cell extract
Target amino acid: unknown
Larval death

Group V: *Pseudomonas aeruginosa* exoenzymes S and T *(73,74)*
            Acceptor protein: Rho, Vimentin, target amino acid: unknown
            Increased *P. aeruginosa* virulence

3.1.1. Toxins Eliciting Cell Damage (Lysis) by Disruption of the Cytoplasmic Membrane

These toxins, also known as *cytolysins* (*hemolysins* when acting on erythrocytes and *leukotoxins* when acting on leukocytes) constitute 35% of the entire bacterial toxin repertoire (**Table 1**). They are produced by both Gram-positive and Gram-negative bacteria. Two classes of toxins could be differentiated within this group.

3.1.1.1. ENZYMATICALLY ACTIVE TOXINS

These toxins belong to the group of the phospholipases A, C, and D (*23,28,51* and **Table 4**) that hydrolyze the phospholipids of cell membranes, leading to the destabilization of the lipid bilayer, which may result in cell lysis.

3.1.1.2. PORE-FORMING TOXINS

These toxins which have so far no known enzymatic activity, impair cell membranes by insertion into the cytoplasmic phospholipid–cholesterol bilayer, creating stable defined pores/channels of various sizes. This effect is often accompanied by the oligomerization of toxin molecules. Staphylococcal α, γ, and δ toxins create such pores (*20,28,39,42,43,51*). Among this group of toxins two important families have been widely investigated: the RTX toxins from Gram-negative bacteria and the cholesterol-binding (also known as sulfhydryl-activated) toxins from Gram-positive bacteria.

3.1.1.2.1. The Family of the RTX Pore-Forming Toxins. The members of these closely structurally and functionally related cytolysins are produced by a variety of Gram-negative bacteria associated with diseases in humans and animals (**Table 6**). These cytolytic proteins have been designated *repeats in toxins* (RTX) on the basis of a common series of glycine- and aspartate-rich nonapeptide tandem repeat units (L/I/F-X-G-G-G-X-N/D-D-X) clustered near the C-terminal ends of protein molecules (*20–22,31*). The prototypical member of this family is the widely investigated 110-kDa *E. coli* hemolysin (*52,53*).

All of the RTX toxins are either secreted into the culture medium by a unique leader peptide-independent process or are localized to the cell surface. Interestingly, as first shown for *E. coli* hemolysin and thereafter for the other toxins of the family, the genes for toxin synthesis and secretion are for the most part grouped together on bacterial chromosome to form an operon comprising four determinants arranged in the order *hlyC*, *hlyA*, *hlyB*, and *hlyD*. Gene *hlyA* encodes the 110-kDa hemolysin protein (pro-HlyA) which represents an inactive precursor of the mature toxin. The conversion of pro-HlyA to the hemolytically active hemolysin (HlyA) takes place in the cytoplasm of *E. coli* and is mediated by HlyC. The physiologically active form of HlyC seems to be a homodimer of the 20-kDa gene product of *hlyC*. The proteins encoded by *hlyB*

Table 6
The Family of RTX Cytolysins from Gram-Negative Bacteria (20–22,31)

Bacterium	Toxin	Disease association	Target cell specificity
Escherichia coli	Hemolysin (HlyA)	Urinary tract infections, septicemia in humans	Wide
Proteus mirabilis	Hemolysin (HlyA)	Urinary tract infections, septicemia in humans	Wide
Proteus vulgaris	Hemolysin (HlyA)	Urinary tract infections, septicemia in humans	Wide
Proteus penneri	Hemolysin (HlyA)	Urinary tract infections, septicemia in humans	Wide
Morganella morganii	Hemolysin (HlyA)	Urinary tract infections, septicemia in humans	Wide
Pasteurella haemolytica	Leukotoxin (LktA)	Pasteurellosis in ruminants	Narrow
Actinobacillus actinomycetemcomitans	Leukotoxin (AktA)	Periodontitis, endocarditis, and abscesses in humans	Narrow
Actinobacillus pleuropneumoniae	Hemolysin (ApxlA) Hemolysin (ApxllA) Cytolysin (ApxlllA)	Porcine pleuropneumonia	Wide Wide Narrow
Bordetella pertussis	Adenylate cyclase/ hemolysin (cyclo-lysin) (CyaA)	Whooping cough in humans	Wide

Apart from *B. pertussis* hemolysin, the molecular masses of the other RTX cytolysins range from 103 to 120 kDa and the number of glycine-rich repeats varies from 13 to 17 (*54*). The molecular mass of pertussis toxin is 177 kDa and the number of glycine-rich repeats is 41 (*55*).

and *hlyD* are localized in the cytoplasmic membrane of *E. coli* and involved in the secretion of hemolysin into the medium.

Hemolysin adsorption to erythrocyte membranes depends on the binding of Ca^{2+} to the set of glycine- and aspartate-rich nonapeptide repeats. In addition, the interaction of hemolysin with erythrocytes also requires the activation of pro-HlyA by HlyC. Under in vitro conditions, HlyC activates pro-HlyA by directing the acyl-acyl carrier protein (acyl-ACP)-dependent covalent fatty acid acylation of the hemolysin protein *(54)*.

3.1.1.2.2. The Family of the Cholesterol-Binding ("Sulfhydryl-Activated" Cytolytic Toxins). This family of 50- to 60-kDa single-chain amphipathic proteins is the largest group of bacterial toxins known to date. It has been the subject of a number of reviews *(15,56,57)*. The toxins are produced by 23 taxonomically different species of Gram-positive bacteria from the genera *Streptococcus, Bacillus, Paenibacillus, Brevibacillus, Clostridium, Listeria,* and *Arcanobacterium* (**Table 7**). New members may be discovered.

The cytolysins of this family share the following features: (1) they are antigenically related, lethal to animals, and highly lytic toward eucaryotic cells, including erythrocytes (hence the name "hemolysins" is often used); (2) their lytic and lethal properties are suppressed by sulfydryl-group blocking agents and reversibly restored by thiols or other reducing agents except for pyolysin and intermedilysin. These properties are irreversibly abrogated in the presence of very low concentrations of cholesterol and other 3β-hydroxysterols; (3) membrane cholesterol is thought to be the toxin-binding site at the surface of eucaryotic cells; (iv) toxin molecules bind as monomers to membrane surface with subsequent oligomerization into arc- and ring-shaped structures surrounding large pores generated by this process; (5) the 13 structural genes of the toxins cloned and sequenced to date (**Table 7**) are all chromosomal; (6) the primary structure of the proteins deduced from the nucleotide sequence of encoding genes shows obvious sequence homology, particularly in the C-terminal part and a characteristic consensus common sequence (undecapeptide) near the C-terminus of the molecules, containing in most toxins the unique Cys residue of the protein critical for biological activity.

Apart from pneumolysin, which is an intracytoplasmic toxin, all the other toxins are secreted in the extracellular medium. Among the species producing this group of toxins, only *L. monocytogenes* and *L. ivanovii* are intracellular pathogens that grow and release their toxins in the phagocytic cells of the host.

3.1.2. Receptor-Targeted Toxins

These types of toxins bind to appropriate cell surface receptor(s) with subsequent triggering of intracellular processes via transmembrane signaling without being internalized *(28)*. Two main classes of toxins exhibit these properties.

Table 7
The Family of the Cholesterol-Binding
"Sulfhydryl-Activated" Cytolytic Toxins (15,56,57)

Bacterial genus	Species	Toxin name[a]	Gene acronym[a]
Streptococcus	*S. pyogenes*	Streptolysin O (SLO)	*slo*
	S. equisimilis b	Streptolysin O (SLO)	*slo*
	S. canis c	Streptolysin O (SLO)[d]	*slo*
	S. pneumoniae	Pneumolysin (PLY)	*ply*
	S. suis	Suilysin (SLY)	*sly*
	S. intermedius	Intermedilysin (ILY)	*ily*
Bacillus	*B. cereus*	Cereolysin O (CLO)	*clo*
	B. thuringiensis	Thuringiolysin O (TLO)	*tlo*
Brevibacillus	*B. laterosporus*	Laterosporolysin (LSL)	*lsl*
Paenibacillus	*P. alvei*	Alveolysin (ALV)	*alv*
Clostridium	*C. tetani*	Tetanolysin (TLY)	*tly*
	C. botulinum	Botulinolysin (BLY)	*bly*
	C. perfringens	Perfringolysin O (PFO)	*pfo*
	C. septicum	Septicolysin O (SPL)	*spl*
	C. histolyticum	Histolyticolysin O (HLO)	*hlo*
	C. novyi A (oedematiens)	Novyilysin (NVL)	*nvl*
	C. chauvoei	Chauveolysin (CVL)	*cvl*
	C. bifermentans	Bifermentolysin (BFL)	*bfl*
	C. sordellii	Sordellilysin (SDL)	*sdl*
Listeria	*L. monocytogenes*	Listeriolysin O (LLO)	*llo*
	L. ivanovii	Ivanolysin (ILO)	*ilo*
	L. seeligeri	Seeligerolysin (LSO)	*lso*
Arcanobacterium (*Actinomyces*)	*A. pyogenes*	Pyolysin (PLO)	*plo*

[a]The abbreviation of toxin names (in parentheses) and gene acronyms are those reported in the literature or suggested by the author of the chapter.
[b]Group C streptococcus.
[c]Group G streptococcus.
[d]Also called canilysin.
Underlined: toxins purified to apparent homogeneity and cloned and sequenced genes.

3.1.2.1. *E. COLI* HEAT-STABLE ENTEROTOXIN I

This toxins binds to the guanylate cyclase receptor (a membrane-spanning enzyme) of target cells, leading to the activation of cellular guanylate cyclase, which provokes an elevation of cyclic GMP that stimulates chloride secretion and/or inhibition of NaCl absorption and results in net fluid secretion (diarrhea) (29,34). A similar effect is observed with the hormone guanylin, which shares 50% homology with the toxin and activates the same receptor.

3.1.2.2. SUPERANTIGENIC TOXINS

This group of T-cell stimulating proteins particularly comprises *S. aureus* enterotoxins and toxic-shock toxin-1 and *S. pyogenes* pyrogenic exotoxins A and C (**Table 3** and **ref.** *16*) and other similar molecules which are responsible for severe diseases. The superantigens possess the remarkable property of triggering the proliferation (mitogenic effect) of a large proportion (2–20%) of T lymphocytes (human and other animal species). This process involves, unlike for conventional antigens, the binding of superantigen molecules in their native state (unprocessed) to nonpolymorphic regions of major histocompatibility (MHC) class II molecules on antigen-presenting cells and concomitantly to certain specific motifs on the Vβ chain of the T-cell receptor. This linkage triggers lymphocyte proliferation and subsequently elicits the release of massive amounts of cytokines and other effectors. In diseases involving in vivo superantigen production, particularly during streptococcal and staphylococcal toxic shocks and other severe disorders, the in vivo release of pro-inflammatory cytokines is considered as the major pathogenic process in toxic shock (**Table 8**).

3.2. Toxins Acting Ultimately in the Cytosol (The Conundrum of the Molecular of Action of A–B Type Toxins)

The toxic activity of a great number of bacterial (and certain plant) toxins takes place in target cell cytosol or on the inner leaflet (cytosolic site) of the cytoplasmic membrane. The toxins called A–B type toxins damage or kill the cells by various mechanisms, most of them enzymatic (**Tables 4** and **8**) as a result of the inactivation of molecular targets essential for cell functions. However, these toxins share common operational features that involve initial binding of toxin molecule to cell surface receptors followed by endocytosis and translocation into the cytosol through the endocytic vesicular membrane.

All toxins acting according to this process consist of two functionally different domains designated A and B, which can be two different proteins (in the case of oligomeric or binary toxins) or two regions of a single polypeptide chain (e.g., diphtheria toxin, *P. aeruginosa* exotoxin A). The A domain is the active catalytic part of toxin molecule which acts inside the cytosol and the B domain is the part that binds to the appropriate receptors at the cell surface and allows toxin translocation across the membrane. To reach the cytosol, a proteolytic cleavage ("nicking") of toxin molecule is required (*25*). The cleaved ("activated") fragment A is the effective enzymatic moiety, acting depending on the nature of the toxin involved, as ADP-ribosyltransferases, adenylcyclases, metalloproteases, ribonucleases (*N*-glycosidases), glucosyl transferases, and deamidases (**Tables 3** and **8**, and **refs.** *18,19,30,35,36,41, 46,49,60–64,73,74*).

Table 8
Examples of Toxins Known to Be or Likely to be Involved in Disease Causation

Disease	Microorganism	Toxin(s)	Role in disease	References
Diphtheria	C. diphtheriae	Diphtheria toxin	Cardiac cells, epithelial damage, necrosis, nerve paralysis, coma, death	(3,38,64)
Tetanus	C. tetani	Tetanus toxin	Muscle spastic paralysis, death	(8,26,36,37)
Botulism	C. botulinum	Botulinal toxins	Muscle flaccid paralysis, death	(8,26,27,36,37)
Whooping cough	B. pertussis	Pertussis toxin, adenylate cyclase	Paroxysmal cough, convulsions, death	(31,41,55)
Diarrhea, bloody diarrhea, hemolytic uremic syndrome	E. coli	Shiga-like toxin, LT and ST enterotoxins	Traveler's diarrhea, bladder/kidney infections	(18,29,32, 34,35,60)
Shigellosis	Shigella spp.	Shiga toxin	Excerbate diarrhea, dysentery, neurological effects	(18,29,35)
Cholera	V. cholerae	Cholera, toxins	Profuse diarrhea, dehydration	(32,35,50)
Scarlet fever	S. pyogenes	Erythrogenic (pyrogenic) toxins	Cutaneous rash, death	(16,67,68) (58)
Scalded skin syndrome	S. aureus	Exfoliative/Epidermolytic	Whole body erythema, massive scaling toxins A, B	
Toxic shock syndrome	S. aureus	Enterotoxins, toxic shock syndrome toxin-1	Vomiting, diarrhea fever, shock, death	(16,27,67,68)
	S. pyogenes	Erythrogenic (pyrogenic) toxins		(67–71)
Gas gangrene	Clostridium spp.	Certain clostridial toxins	Tissue damage, gangrene, death	(1,4,17,33,60)
B. fragilis diarrhea	B. fragilis	Metalloprotease toxin	Diarrhea	(29,59)
Anthrax	B. anthracis	Anthrax toxin complex	Pulmonary edema circulatory collapse, death	(46–48)
Pseudomembranous colitis	C. difficile	A, B toxins	Enterocolitis, diarrhea	(35)
Gastric/duodenal ulcer	H. pylori	Vacuolating toxin, other toxins?	Mucosal destruction Gastric cancer?	(65–67)

For general texbook references see (2,4,6–9).

A number of toxins have been shown to be nicked by furin (25) a newly recognized eucaryotic protease (75), including the protective antigen of anthrax toxin, diphtheria, pseudomonas, shiga and shiga-like toxins, and botulinum toxin C2. In contrast, cholera, tetanus, botulinum neurotoxin A, and *C. perfringens* iota-toxins are not recognized by furin but are cleaved by other proteases.

4. Implication of Bacterial Protein Toxin in Infectious and Foodborne Diseases

A certain number of criteria are required for the assessment of toxin-associated natural bacterial infectious disease(s) (adaptation to toxins of the classical Koch's postulates for bacterial pathogens). Ideally, the following five criteria should be fulfilled: (1) The bacterial microorganism clearly identified as the pathogenic agent of the disease produces component(s) characterized as a toxin or an array of toxins. (2) The administration to appropriate animal(s) of the toxin(s) separated from the relevant bacteria or produced by genetic engineering from a heterologous tox⁻ recombinant bacterial strain produces symptoms and/or pathophysiological disorders that mimic those observed in the natural disease or at least those elicited in experimental animals by the cognate toxin-producing bacteria. (3) The in vitro challenge of appropriate animal organs, tissues, or cells with the isolated toxin(s) elicits certain pathophysiological, biochemical, or metabolic manifestions observed in the host infected with the relevant toxinogenic bacteria (e.g., hemolysis, histological lesions, production of effectors involved in the host such as cytokines, inflammatory factors, and other mediators of pathological significance). (4) Toxin concentration in the organism of the host infected by the toxinogenic bacteria should be compatible with the characteristics of the natural disease. (5) The disease can be prevented by immunization against the toxin(s).

However, a toxin may play a role in bacterial pathogenesis without fulfilling all these criteria. For example, a failure to demonstrate toxin production in vitro does not necessarily mean that the bacterial strain is unable to produce toxin(s) in the host (in vivo toxin expression and release). Historically this was the case in the attempts of Koch to demonstrate toxin release in *Vibrio cholerea* cultures.

4.1. Toxin-Mediated or Associated Bacterial Diseases

The toxins of pathogenic interest exhibit a variety of effects in bacterial diseases. Bacteria that colonize a wound or mucosal surface but do not invade target cells can produce toxins that act locally or enter the bloodstream and attack internal organs (e.g., *Corynebacterium diphtheriae, Vibrio cholerae*).

Bacteria growing in a wound (e.g., streptococci, staphylococci) can produce toxins that destroy host tissue and kill phagocytes in the immediate vicinity of the bacteria, thus facilitating bacterial growth and spread.

On the basis of the above-mentioned criteria, the bacterial diseases among many others listed in **Tables 8** and **9** are toxin-associated (toxinoses). These toxins belong to various functional classes such as the ADP-ribosylating toxins, superantigens, pore-forming (hemolytic, membrane-damaging) toxins, neurotoxic and other metalloproteases, glucosyl transferases, and RNA–*N*-glycosidases.

4.2. Food Poisoning

On the other hand, toxins formed in food and then ingested along with the food can be the source of pathological symptoms *(27)*. Most bacterial toxins involved in food poisoning are enterotoxins (toxins acting on mucosal cells of the intestinal tract). However, there are also other foodborne toxins, such as botulinal neurotoxins, that are not enterotoxic.

The toxinogenic bacteria involved in food poisoning can be divided into four classes on the basis of clinical effects:

1. Bacteria causing infection (*Salmonella enterica*, *Shigella* spp., *Campylobacter* spp., *Yersinia enterocolitica*, *Listeria monocytogenes*, some *E. coli* and some *Aeromonas* spp.).
2. Bacteria interacting (colonization) with the host before toxin production (*Vibrio cholerae*, *Vibrio parahaemolyticus*, *E. coli* [ETEC] and some *Aeromonas* spp.).
3. Bacteria releasing enterotoxin in the host intestine without any direct tissue interaction (*Bacillus cereus*, *Clostridium perfringens*).
4. Intoxication (preformed toxins in food before ingestion): *Clostridium botulinum*, *Bacillus cereus* (emetic type), *Staphylococcus aureus*.

5. Concluding Remarks

As briefly scanned in this chapter, a wealth of outstanding achievements in our knowledge of the structure, genetics, and molecular mechanisms of action of bacterial protein toxins was gathered in the past decade, providing a basis for a deeper understanding of bacterial pathogenesis and thereby allowing the development of prophylactic and therapeutic strategies to combat toxin-induced diseases. The great strides that have been made in the field of bacterial toxinology also led to exciting developments in cell biology, neurobiology, and immunology and to promising applications of genetically engineered toxins (novel vaccines, pharmacologic agents) not only in the domain of infections diseases but also in that of oncology, immunopathology, and nervous system disorders.

Table 9
Main Toxin-Producing Bacteria Involved in Human or Animal Digestive Tract Diseases (27–30,32,34,35)

Bacteria	Enterotoxins	Mechanisms of action	Diseases
V. cholerae	Cholera toxin (cytotonic enterotoxin)	cAMP activation	Cholera
	Cofactors: Zot toxin[a], Ace toxin[a]		
	ST (thermostable toxin)		
V. parahaemolyticus	TDH (thermostable direct hemolysin)		Gastroenteritis
	TDH (thermostable direct hemolysin)		
E. coli (ETEC)	Heat-labile (LT-I/LT-II) enterotoxins	cAMP activation	
Salmonella enterica	Heat-stable (ST-I/ST-II) enterotoxins	cGMP activation	Diarrhea, traveller's diarrhea
Aeromonas hydrophila	(both cytotonic)		
Yersinia enterocolitica			
Campylobacter jejuni			
Shigella sp.	Shiga toxins I/II: cytotoxic	Inhibition of protein synthesis	Diarrhea, dysentery, hemorrhagic colitis
E. coli (EPEC, EHEC)	Shiga-like toxins or verotoxins I/II (cytotoxic)		Hemolytic–uremic syndrome
E. coli (EAggEC)	East I (heat stable enterotoxin)	cGMP activation	Diarrhea
S. aureus	A, B, C1, C2, C3, D, E, G, H, I enterotoxins	Effect on vagus nerve	Food poisoning
B. cereus	Nonhemolytic enterotoxin, lethal toxin	Cytotoxic	Food poisoning
	Hemolytic HBL enterotoxin	Cytotoxic	
Clost. perfringens A	Enterotoxin	Pore formation	Food poisoning
Clost. perfringens B and C	β and β2 toxin (cytotoxic)	?	Necrotizing enteritis
Clost. difficile	Toxins A and B	Modification of cellular actin	Pseudomembranous colitis
			Antibiotic-associated diarrhea
Clost. septicum	α-Toxin	Cytolytic	Necrotizing enterocolitis
Clost. sordellii	Lethal toxin	Glucosyltransferase on Rho	Hemorrhagic enteritis
Clost. perfringens E	Iota-toxin	ADP-ribosylation of cellular actin	Necrotizing/hemorrhagic enteritis
Clost. botulinum C & D	C2 toxin		
Clost. spiriforme	Iota-like toxin		Enteritis (animal)
Bacteroides fragilis	Enterotoxin	Zn protease: metzincine	Diarrhea

[a]Zot, Zonula occludens toxin; Ace, accessory cholera enterotoxin.

References

1. Harshman, S., ed. (1988) Microbial toxins: tools in enzymology. *Methods Enzymol.* **Series 165**. Academic Press, San Diego.
2. Hardegree, C. and Tu, A. T. (1988) *Bacterial Toxins. Handbook of Natural Toxins*, Vol. 4. Marcel Dekker, New York.
3. Moss, J. and Vaughan, M., eds. (1990) *ADP-Ribosylating Toxins and G Proteins*. American Society for Microbiology Press, Washington, DC.
4. Dorner, F. and Drews, J., eds. (1990) *Pharmacology of Bacterial Toxins*. Pergamon Press, Oxford, England.
5. Alouf, J. E. and Freer, J. H., eds. (1991) *Sourcebook of Bacterial Protein Toxins*. Academic Press, London, England.
6. Salyers, A. A. and Whitt, D. D., eds. (1994) *Bacterial Pathogenesis. A Molecular Approach*. American Society for Microbiology Press, Washington, DC.
7. Mims, C., Dimmock, N., Nash, A., and Stephen, J., eds. (1995) *Mims's Pathogenesis of Infections Diseases* 4th ed., Academic Press, London, England.
8. Montecucco, C., ed. (1995) *Clostridial Neurotoxins*. Springer-Verlag, Berlin.
9. Moss, J., Iglewski, B., Vaughan, M., and Tu, A. T., eds. (1995) *Bacterial Toxins and Virulence Factors in Diseases* (*Handbook of Natural Toxins,* Vol. 8), Marcel Dekker, New York.
10. Parker, M. M. (1996) *Protein Toxin Structure*. R. G. Landes, Austin, TX.
11. Montecucco, C. and Rappuoli, R., eds. (1997) *Guidebook to Protein Toxins and their Use in Cell Biology*. Oxford University Press, Oxford, England.
12. Aktories, K., ed. (1997) *Bacterial Toxins*, Chapman and Hall, Weinheim.
13. Alouf, J. E. and Freer, J. H., eds. (1999) *Bacterial Protein Toxins: A Sourcebook*. Academic Press, London, England.
14. Hatheway, C. L. (1990) Toxigenic clostridia. *Clin. Microbiol. Rev.* **3,** 66–98.
15. Alouf, J. E. (1999) The family of the structurally-related, cholesterol-binding ('sulf-hydryl-activated') toxins, in *Bacterial Protein Toxins:* A Sourcebook (Alouf, J. E. and Freer, J. H., eds.), Academic Press, London, England, pp.443–456.
16. Alouf, J. E., Müler-Alouf, H., and Köhler, W. (1999) Superantigenic *Streptococus pyogenes* erythrogenic/pyrogenic exotoxins, in *Bacterial Protein Toxins: A Sourcebook* (Alouf, J. E. and Freer, J. H., eds.), Academic Press, London, England, pp. 567–588.
17. Rood, J. I. and Cole, S. T. (1991) Molecular genetics and pathogenesis of *Clostridium perfringens. Microbiol. Rev.* **55,** 621–648.
18. Tesh, V. L. and O'Brien, A. D. (1991) The pathogenic mechanisms of Shiga toxin and the Shiga-like toxins. *Mol. Microbiol.* **5,** 1817–1822.
19. Considine, R. V. and Simpson, L. L. (1991) Cellular and molecular actions of binary toxins possessing ADP-ribosyltransferase activity. *Toxicon* **29,** 913–936.
20. Braun, V. and Focareta, T. (1991) Pore-forming bacterial toxins (cytolysins). *Crit. Rev. Microbiol.* **18,** 115–158.
21. Welch, R. A. (1991) Pore-forming cytolysins of Gram-negative bacteria. *Mol. Microbiol.* **5,** 521–528.

22. Coote, J. (1992) Structural and functional relationships among the RTX toxin determinants of Gram-negative bacteria. *FEMS Microbiol. Rev.* **88,** 137–162.

23. Titball, R. W. (1999) Membrane-damaging and cytotoxic phospholipases, in *Bacterial Protein Toxins: A Sourcebook* (Alouf, J. E. and Freer, J. H., eds.) Academic Press, London, England, pp. 310–329.

24. Alouf, J. E. (1994) A decade of progress in toxin research, in *Bacterial Protein Toxins* (Freer, J. H., Aitken, R., Alouf, J. E., Boulnois, G., Falmagne, P., Fehrenbach, F., Montecucco, C., Piemont, Y., Rappuoli, R., Wadström, T., and Witholt, B., eds.), Gustav Fischer Verlag, Stuttgart, pp. 3–29.

25. Gordon, V. M. and Leppla, S. H. (1994) Proteolytic activation of bacterial toxins: role of bacterial and host cell proteases. *Infect. Immun.* **62,** 333–340.

26. Montecucco, C. and Schiavo, G. (1995) Structure and function of tetanus and botulinum neurotoxins. *Quart. Rev. Biophysics* **28,** 423–472.

27. Granum, P. E., Tomas, J. M., and Alouf, J. E. (1995) A survey of bacterial toxins involved in food poisoning: a suggestion for bacterial food poisoning toxin nomenclature. *Int. J. Food Microbiol.* **28,** 129–144.

28. Balfanz, J., Rautenberg, P., and Ullmann, U. (1996) Molecular mechanisms of action of bacterial exotoxins. *Zbl. Bakt.* **284,** 170–206.

29. Sears, C. L. and Kaper, J. B. (1996) Enteric bacterial toxins: mechanisms of action and linkage to intestinal secretion. *Microbiol. Rev.* **60,** 167–215.

30. Von Eichel-Streiber, C., Boquet, P., Sauerborn, M., and Thelestam, M. (1996) Large clostridial cytotoxins—a family of glycosyltransferases modifying small GTP-binding proteins. *Trends Microbiol.* **4,** 375–382.

31. Coote, J. (1996) The RTX toxins of Gram-negative bacterial pathogens: modulators of the host immune system. *Rev. Med. Microbiol.* **7,** 53–62.

32. Fasano, A. (1997) Cellular microbiology: how enteric pathogens socialize with their intestinal host. *ASM News* **63,** 259–265.

33. Songer, J.G. (1997) Bacterial phospholipases and their role in virulence. *Trends Microbiol.* **5,** 156–161.

34. Nataro, J. P. and Kaper, J. B. (1998) Diarrheagenic *Escherichia coli. Clin. Microbiol. Rev.* **11,** 142–201.

35. Popoff, M. R. (1998) Interactions between bacterial toxins and intestinal cells. *Toxicon* **6,** 665–685.

36. Montecucco, C., Schiavo, G., Tugnoli, V., and de Grandis, D. (1996) Botulinum neurotoxins: mechanisms of action and therapeutic applications. *Mol. Med. Today* **2,** 418–424.

37. Schantz, E. J. and Johnson, E. A. (1992) Neurotoxins in medicine. *Microbiol. Rev.* **56,** 80–99.

38. Alouf, J. E. (1988) From 'diphtheritic' poison to molecular toxinology. *Bull. Inst. Pasteur* **86,** 127–144.

39. Smith, H., Keppie, J., and Stanley, J. J. (1955) The chemical basis of the virulence of *Bacillus anthracis* in vivo. *Br. J. Exp. Pathol.* **36,** 460–472.

40. Alouf, J. E., Dufourcq, J., Siffert, O., Thiaudiere, E., and Geoffroy, C. (1989) Interaction of staphylococcal delta toxin and synthetic analogues with erythro-

cytes and phospholipid vesicles. Biological and physical properties of the amphipathic peptides. *Eur. J. Biochem.* **183**, 381–390.

41. Locht, C. and Antoine, R. (1995) A proposed mechanism of ADP-ribosylation catalyzed by the pertussis toxin S1 subunit. *Biochimie* (Paris) **77**, 333–340.

42. Guidi-Rontani, C., Fouque, F., and Alouf, J. E. (1994) Bifactorial versus monofactorial molecular status of *Staphylococcus aureus* γ-toxin. *Microbial Pathogen.* **16**, 1–14.

43. Supersac, G., Prevost, G., and Piemont, Y. (1993) Sequencing of leucocidin R from *Staphylococcus aureus* P83 suggests that staphylococcal leucocidins and gamma-hemolysin are members of a single, two-component family of toxins. *Infect. Immun.* **61**, 580–587.

44. Segarra, R. A., Booth, M., Morales, D., Huycke, M. M., and Gilmore, M. S. (1991) Molecular characterization of the *Enterococcus faecalis* cytolysin activator. *Infect. Immun.* **59**, 1230–1246.

45. Perelle, S., Gibert, M., Boquet, P., and Popoff, M. R. (1993) Characterization of *Clostridium perfringens* iota-toxin genes and expression in *Escherichia coli*. *Infect. Immun.* **61**, 5147–5156.

46. Klimpel, K. R., Molloy, S. S., Thomas, G., and Leppla, S. (1993) Anthrax toxin protective antigen is activated by a cell surface protease with the sequence specificity and catalytic properties of furino. *Proc. Nat. Acad. Sci. USA* **89**, 10,277–10,283.

47. Mock, M. and Ullmann, A. (1993) Calmodulin activated bacterial adenylate cyclase as virulence factors. *Trends Microbiol.* **1**, 187–191.

48. Dai, Z., Sirard, J. C., Mock, M., and Koehler, T. (1995) The *atxA* gene product activated transcription of the anthrax toxin genes is essential for virulence. *Mol. Microbiol.* **16**, 1171–1181.

49. Collier, R. J. (1994) Diphtheria toxin: crystallographic structure and function, in *Bacterial Protein Toxins, FEMS Symposium 73* (Freer, J. H., Aitken, R., Alouf, J. E., Boulnois, G., Falmagne, P., Fehrenbach, F., Montecucco, C., Piemont, Y., Rappuoli, R., Wadström, T., and Witholt, B., eds.), Gustav Fischer Verlag, Stuttgart, pp. 217–230.

50. Waldor, M. K. and Mekalanos, J. J. (1996) Lysogenic conversion by a filamentous phage encoding cholera toxin. *Science* **272**, 1910–1914.

51. Bernheimer, A. W. and Rudy, B. (1986) Interaction between membranes and cytolytic peptides. *Biochim. Biophys. Acta* **864**, 123–141.

52. Koronakis, V. and Hughes, C. (1996) Synthesis, maturation and export of the *E. coli* hemolysin. *Med. Microbiol. Immunol.* **185**, 65–71.

53. Braun, V., Schönherr, R., and Hobbie, S. (1993) Enterobacterial hemolysins: activation section and pore formation. *Trends Microbiol.* **3**, 211–216.

54. Ludwig, A., Garcia, F., Bauer, S., Jarchau, J., Beuz, R., Hoppe, J., and Goebel, W. (1996) Analysis of the in vivo activation of hemolysin (HlyA) from *Escherichia coli*. *J. Bacteriol.* **178**, 5428–5430.

55. Betsou, F., Sebo, P., and Guiso, N. (1995) The C-terminal domain is essential for protective activity of the *Bordetella pertussis* adenylate cyclase-hemolysin. *Infect. Immun.* **63**, 3309–3315.

56. Morgan, P. J., Andrew, P. W., and Mitchell, T. J. (1996) Thiol-activated cytolysins. *Rev. Med. Microbiol.* **7,** 221–229.
57. Alouf, J. E. and Palmer, M. (1999) Streptolysin O, in *Bacterial Protein Toxins: A Sourcebook* (Alouf, J. E. and Freer, J. H., eds.), Academic Press, London, England.
58. Bailey, C. J., Lockhart, B. L., Redpath, M. B., and Smith, T. B. (1995) The epidermolytic (exfoliative) toxins of *Staphylococcus aureus. Med. Microbiol. Immunol.* **184,** 53–61.
59. Franco, A. A., Mundy, L. M., Trucksis, M., Wu, S., Kaper, J. B., and Sears, C. L. (1997) Cloning and characterization of the *Bacteroides fragilis* metalloprotease toxin gene. *Infect. Immun.* **65,** 1007–1013.
60. Popoff, M., Chaves-Olarte, E., Lemichez, E., von Eichel-Streiber, C., Thelestam, M., Chardin, P., Cussac, D., Antonny, B., Chavrier, P., Flatau, G., Giry, M., de Gunzburg, J., and Boquet, P. (1996) Ras, rap and rac small GTP-binding proteins are targets for *Clostridium sordellii* lethal toxin glucosylation. *J. Biol. Chem.* **271,** 10,217–10,224.
61. Flatau, G., Lemichez, E., Gauthier, M., Chardin, P., Paris, S., Fiorentini, C., and Boquet, P. (1997) Toxin-induced activation of the G protein p21 Rho by deamidation of glutamine. *Nature* **387,** 729–733.
62. Schmidt, G., Sehr, P., Wilm, M., Selzer, J., Mann, M., and Aktories, K. (1997) Gln 63 of Rho is deamidated by *Escherichia coli* cytotoxic necrotizing factor-1. *Nature* **387,** 725–729.
63. Selzer, J., Hofmann, F., Rex, G., Wilm, M., Mann, M., Just, I., and Aktories, K. (1996) *Clostridium novyi* α-toxin-catalyzed incorporation of GlcNac into Rho subfamily proteins. *J. Biol. Chem.* **271,** 25,173–25,177.
64. Lemichez, E., Bomsel, M., Devilliers, G., Vanderspek, J., Murphy, J. R., Lukianov, E. V., Olsnes, S., and Boquet, P. (1997) Membrane translocation of diphtheria toxin fragment A exploits early to late endosome trafficking machinery. *Mol. Microbiol.* **23,** 445–457.
65. Cover, T. L. and Blaser, M. J. (1995) *Helicobacter pylori*: a bacterial cause of gastritis, peptic ulcer disease and gastric cancer. *ASM News* **61,** 21–28.
66. Manetti, R., Massari, P., Burroni, D., Bernard, M., Marchini, A., Olivieri, R., Papini, E., Montecucco, C., Rappuoli, R., and Telford, J. L. (1995) *Helicobacter pylori* cytotoxin: importance of native conformation for induction of neutralizing antibodies. *Infect. Immun.* **63,** 4476–4480.
67. Labigne, A. and de Reuse, H. (1996) Determinants of *Helicobacter pylori* pathogenicity. *Infect. Agents Dis.* **5,** 191–202.
68. Schlievert, P. M. (1995) The role of superantigens in human disease. *Curr. Opin. Infect. Dis.* **8,** 170–174.
69. Stevens, D. L. (1995) Streptococcal toxic shock syndrome: spectrum of disease, pathogenesis and new concepts in therapy. *Emerg. Infect. Dis.* **1,** 69–78.
70. Müller-Alouf, H., Alouf, J. E., Gerlach, D., Ozegowski, J. H., Fitting, C., and Cavaillon, J.-M. (1994) Comparative study of cytokine release by human peripheral blood mononuclear cells stimulated with *Streptococcus pyogenes*

superantigen erythrogenic toxins, heat-killed streptococci and lipopolysaccharide. *Infect. Immun.* **62,** 4915–4921.

71. Müller-Alouf, H., Alouf, J. E., Gerlach, D., Ozegowski, J. H., Fitting, C., and Cavaillon, J.-M. (1996) Human pro- and anti-inflammatory cytokine patterns induced by *Streptococcus pyogenes* erythrogenic (pyrogenic) exotoxin A and C superantigens. *Infect. Immun.* **64,** 1450–1453.

72. Müller-Alouf, H., Gerlach, D., Desreumaux, P., Leportier, C., Alouf, J. E., and Capron, M. (1997) Streptococcal pyrogenic exotoxin A (SPE A) superantigen induced production of hematopoietic cytokines, IL-12 and IL-13 by human peripheral blood mononuclear cells. *Microb. Pathogen.* **23,** 265–272.

73. Coburn, J. and Gill, D. M. (1991) ADP-ribosylation of p21^{ras} and related proteins by *Pseudomonas aeruginosa* exoenzymes S. *Infect. Immun.* **59,** 4259–426.

74. Yahr, T. L., Vallis, L., Hancock, M. K., Barbieri, J. T., and Frank, D. W. (1998) Exo y, an adenylate cyclase secreted by the *Pseudomonas aeruginosa* type III system. *Proc. Natl. Acad. Sci. USA* **95,** 13,899–13,904.

75. Nakayama, K. (1997) Furin: a mammalian subtilisin Kex2p-like endoprotease involved in processing of a wide variety of precursor proteins. *Biochem. J.* **327,** 625–635.

76. Allured, V. S., Collier, R. J., Carroll, S. F., and McKay, D. B. (1986) Structure of exotoxin A of *Pseudomonas aeruginosa* at 3.0 Å resolution. *Proc. Natl. Acad. Sci. USA* **83,** 1320–1324.

77. Sixma, T. K., Pronk, S. E., Kalk, K. H., Wartna, E. S., van Zanten, B. A., Witholt, B., and Hol, W. G. (1991) Crystal structure of a cholera toxin-related heat-labile enterotoxin from *E. coli. Nature* **351,** 371–377.

78. Sixma, T. K., Kalk, K. H., van Zanten, B. A., Dauter, Z., Kingma, J., Witholt, B., and Hol, W. G. (1993) Refined structure of *Escherichia coli* heat-labile enterotoxin, a close relative of cholera toxin. *J. Mol. Biol.* **230,** 890–918.

79. Li, J., Carroll, J., and Ellar, D. J. (1991) Crystal structure of insecticidal delta-endotoxin from *Bacillus thuringiensis* at 2.5 Å resolution. *Nature* **353,** 815–821.

80. Stein, P. E., Boodhoo, A., Tyrrell, G. J., Brunton, J. L., and Read, J. R. (1992) Crystal structure of the cell-binding B oligomer of verotoxin-1 from *E. coli. Nature* **355,** 748–750.

81. Choe, S., Bennett, M. J., Fujii, G., Curmi, P. M., Kantardjieff, K. A., Collier, R. J., and Eisenberg, D. (1992) The crystal structure of diphtheria toxin. *Nature* **357,** 216–222.

82. Bennett, M. J., Choe, S., and Eisenberg, D. (1994) Refined structure of dimeric diphtheria toxin at 2.0 angstrom resolution. *Protein Sci.* **3,** 1444–1463.

83. Parker, M. W., Buckley, J. T., Postma, J. P., Tucker, A. D., Leonard, K., Pattus, F., and Tsernoglou, D. (1994) Structure of the *Aeromonas* toxin proarolysin in its water-soluble and membrane-channel states. *Nature* **367,** 292–295.

84. Stein, P. E., Boodhoo, A., Armstrong, G. D., Cockle, S. A., Klein, M. H., and Read, R. J. (1994) The crystal structure of pertussis toxin. *Structure* **2,** 45–57.

85. Stein, P. E., Boodhoo, A., Armstrong, G. D., Heerze, L. D., Cockle, S. A., Klein, M. H., and Read, R. J. (1994) Structure of a pertussis toxin-sugar complex as a model for receptor binding. *Nat. Struct. Biol.* **1,** 591–596.

86. Fraser, M. E., Chernaia, M. M., Kozlov, Y. V., and James, M. N. (1994) Crystal structure of the holotoxin from *Shigella dysenteriae* at 2.5 Å resolution. *Nat. Struct. Biol.* **1,** 59–64.

87. Zhang, R.-G., Scott, D. L., Westbrook, M. L., Nance, S., Spangler, B. D., Shipley, G. G., and Westbrook, E. M. (1995) The three-dimensional crystal structure of cholera toxin. *J. Mol. Biol.* **251,** 563–573.

88. Zhang, R.-G., Westbrook, M. L., Westbrook, E. M., Scott, D. L., Otwinowski, Z., Maulik, P. R., Reed, R. A., and Shipley, G. G. (1995) The 2.4 Å crystal structure of cholera toxin B subunit pentamer: choleragenoid. *J. Mol. Biol.* **251,** 550–562.

89. Song, L., Hobaugh, M. R., Shustak, C., Cheley, S'., Bayley, H., and Gouaux, J. E. (1996) Structure of staphylococcal alpha-hemolysin, a heptameric transmembrane pore. *Science* **274,** 1859–1866.

90. Vath, G. M., Earhart, C. A., Rago, J. V., Kim, M. H., Bohach, G. A., Schlievert, P. M., and Ohlendorf, D. (1997) The structure of the superantigen exfoliative toxin A suggests a novel regulation as a serine protease. *Biochemistry* **36,** 1559–1566.

91. Cavarelli, J., Prévost, G., Bourguet, W., Moulinier, L., Chevrier, B., Delagoutte, B., Bilwes, A., Mourey, L., Rifai, S., Piémont, Y., and Moras, D. (1997) The structure of *Staphylococcus aureus* epidermolytic toxin A, an atypic serine protease, at 1.7 Å resolution. *Structure* **5,** 813–824.

92. Petosa, C., Collier, R. J., Klimpel, K. R., Leppla, S. H., and Liddington, R. C. (1997) Crystal structure of the anthrax toxin protective antigen. *Nature* **385,** 833–838.

93. Umland, T. C., Wingert, L. C., Swaminathan, S., Furey, W. F., Schmidt, J. J., and Sax, M. (1997) Structure of the receptor-binding fragment Hc of tetanus neurotoxin. *Nat. Struct. Biol.* **4,** 788–792.

94. Rossjohn, J., Feil, C., McKinstry, W. J., Tweten, R.-K., and Parker, M. W. (1997) Structure of a cholesterol-binding thiol-activated cytolysin and a model of its membrane form. *Cell* **89,** 685–692.

95. Derewenda, Z. S. and Martin, T. W. (1998) Structure of the gangrene α-toxin: the beauty in beast. *Nat. Struct. Biol.* **5,** 659–662.

96. Lacy, D. B., Tepp, W., Cohen, A. C., Das Gupta, B. R., and Stevens, R. C. (1998) Crystal structure of botulinum neurotoxin type A and implications for toxicity. *Nat. Struct. Biol.* **5,** 898–902.

97. Swaminathan, S., Furey, W., Pletcher, J., and Sax, M. (1992) Crystal structure of staphylococcal enterotoxin B, a superantigen. *Nature* **359,** 801–806.

98. Jardetzky, T. S., Brown, J. H., Gorga, J. C., Stern, L. J., Urban, R. G., Chi, Y. I., Stauffacher, C., Strominger, J. L., and Wiley, D. C. (1994) Three-dimensional structure of a human class II histocompatibility molecule complexed with superantigen. *Nature* **368,** 711–718.

99. Hoffmann, M. L., Jablonski, L. M., Crum, K. K., Hackett, S. P., Chi, Y. I., Stauffacher, C. V., Stevens, D. L., and Bohach, G. A. (1994) Predictions of T-cell

receptor- and major histocompatibility complex-binding sites on staphylococcal enterotoxin C1. *Infect. Immun.* **62,** 3396–3407.

100. Papageorgiou, A. C., Acharya, K. R., Shapiro, R., Passalacqua, E. F., Brehm, R. D., and Tranter, H. S. (1995) Crystal structure of the superantigen enterotoxin C2 from *Staphylococcus aureus* reveals a zinc-binding site. *Structure* **3,** 769–779.

101. Schad, E. M., Zaitseva, I., Zaitsev, V. N., Dohlsten, M., Kalland, T., Schlievert, P. M., Ohlendorf, D. H., and Svensson, L. A. (1995) Crystal structure of the superantigen staphylococcal enterotoxin type A. *EMBO J.* **14,** 3292–3301.

102. Sundström, M., Hallen, D., Svensson, A., Schad, E., Dohlsten, M., and Abrahamsen, L. (1996) The co-crystal structure of staphylococcal enterotoxin type A with Zn^{2+} at 2.7 Å resolution. *J. Biol. Chem.* **271,** 32212–32216.

103. Sundström, M., Abrahmsen, L., Antonsson, P., Mehindate, K., Mourad, W., and Dohlsten, M. (1996) The crystal structure of staphylococcal enterotoxin type D reveals Zn^{2+} mediated homodimerization. *EMBO J.* **15,** 6832–6840.

104. Prasad, G. S., Earhart, C. A., Murray, D. L., Novick, R. P., Schlievert, P. M., and Ohlendorf, D. H. (1993) Structure of toxic shock syndrome toxin 1. *Biochemistry* **32,** 13,761–13,766.

105. Acharya, K. R., Passalacqua, E. F., Jones, E. Y., Harlos, K., Stuart, D. I., Brehm, R. D., and Tranter, H. S. (1994) Structural basis of superantigen action inferred from crystal structure of toxic-shock syndrome toxin–1. *Nature* **367,** 94–97.

106. Kim, J., Urban, R. G., Strominger, J. L., and Wiley, D. C. (1994) Toxic shock syndrome toxin–1 complexed with a class II major histocompatibility molecule HLA-DR1. *Science* **266,** 1870–1874.

107. Roussel, A., Anderson, B. F., Baker, H. M., Fraser, J. D., and Baker, E. N. (1997) Crystal structure of the streptococcal superantigen SPE-C: dimerization and zinc binding suggest a novel mode of interaction with MHC class II molecules. *Nature Struct. Biol.* **4,** 635–643.

107a. Papageargill, A., Collins, C. M., Gutman, D. M., Kline, J. B., O'Brien, D., Tranter, H., and Acharya, K. R. (1999) Structural basis for the recognition of superantigen streptococcal pyrogenic exotoxin A (Spe A1) by MHC class II molecules and T-cell receptors. *EMBO J.* **18,** 9–21.

107b. Sundberg, E. and Jardetzky, T. S. (1999) Structural basis ALA-DQ binding by the streptococcal superantigen SSA. *Nature Struct. Biol.* **6,** 123–129.

108. Munson, S. H., Tremaine, M. T., Betley, M. J., and Welch, R. A. (1998) Identification and characterization of staphylococcal enterotoxin types G and I from *Staphylococcus aureus*. *Infect. Immun.* **66,** 3337–3348.

109. Ren, K., Bannan, J. D., Pancholi, V., Cheung, A. L., Robbins, J. C., Fischetti, V. A., and Zabriskie, J. B. (1994) Characterization and biological properties of a new staphylococcal exotoxin. *J. Exp. Med.* **180,** 1675–1683.

110. Mollick, J. A., Miller, G. G., Musser, J. M., Cook, R . G., Grossman, D., and Rich, R. R. (1993) A novel superantigen isolated from pathogenic strains of *Streptococcus pyogenes* with aminoterminal homology to staphylococcal enterotoxins B and C. *J. Clin. Invest.* **92,** 710–719.

111. Norrby-Teglund, A., Norgren, M., Holm, S. E., Andersson, U., and Andersson, J. (1994) Similar cytokine induction profiles of a novel streptococcal exotoxin, MF, and pyrogenic exotoxins A and B. *Infect. Immun.* **62,** 3731–3738.

111a. Iwasaki, M., Igarashi, H., and Yutsudo. T. (1997) Mitogenic factor secreted by *Streptococcus pyogenes* is a heat-stable nuclease requiring His[22] for activity. *Microbiology* **143,** 2449–2455.

111b. Eriksson, A., Eriksson, B., Holm, S. E., and Norgren, M. (1999) Streptococcal DNase B is immunologically identical to superantigen Spe F but involves separate domains. *Clin. Diagn. Lab. Immunol.* **6,** 133–136.

111c. Nemoto, E., Rikiishi, H., Sugawara, S., Okamoto, S., Tamura, K., Maruyama, Y., and Kumagai, K. (1996) Isolation of a new superantigen with potent mitogenic activity to murine T cells from *Streptococcus pyogenes*. *FEMS Immunol. Med. Microbiol.* **15,** 81–91.

111d. Rikiishi, H., Okamoto, S., Sugawara, S., Tamura, K., Liu, Z. X., and Kumagai, K. (1997) Superantigenicity of helper T-cell mitogen (SPM-2) isolated from culture supernatants of *Streptococcus pyogenes*. *Immunology* **91,** 406–413.

111e. Kamezawa, Y., Nakahara, T., Nakano, S., Abe, Y., Nozaki-Renard, J., and Isono, T. (1997) Streptococcal mitogenic exotoxin Z, a novel acidic superantigenic toxin produced by a T1 strain of *Streptococcus pyogenes*. *Infect. Immunol.* **65,** 3828–3833.

112. Thanabalu, T., Berry, C., and Hindley, J. (1993) Cytotoxicity and ADP-ribosylating activity of the mosquitocidal toxin from *Bacillus sphaerieus* ssII-1: possible roles of the 27- and 70-kilodalton peptides. *J. Bacteriol.* **175,** 2314–2320.

2

Purification of *Clostridium botulinum* Type A Neurotoxin

Carl J. Malizio, Michael C. Goodnough, and Eric A. Johnson

1. Introduction

Botulinum neurotoxins produced by strains of the spore-bearing bacterium *Clostridium botulinum* have long been known to cause a distinctive paralytic disease in humans and animals *(1)*. In recent years, injection of crystalline botulinum toxin type A has been demonstrated to provide relief from certain involuntary muscle disorders, dystonic conditions, pain syndromes, and headaches *(2,3,4)*. The use of botulinum toxin for treatment of human disease and its usefulness in cell biology *(5)* has stimulated interest in the study of the toxin complexes and the neurotoxin component within the complexes. The use of botulinum toxin complex in medicine will benefit from the development of more potent, less antigenic, and longer-lasting toxin preparations. In this chapter we describe methods for production of high-quality botulinum type A toxin complex and neurotoxin. The procedures described in this chapter are not those used for preparation of botulinum toxin for medical use *(3)*.

C. botulinum produces seven serotypes of neurotoxin designated A–G. These neurotoxins exist in nature as toxin complexes, in which the neurotoxin is noncovalently bound to various nontoxic protein components and to ribonucleic acid *(1,3)*. The neurotoxins are considered the most potent poisons known for humans and certain animals *(3,6)*. The toxicity of type A botulinum neurotoxin has been estimated to be 0.2 ng/kg of body weight *(3,6)*, and as little as 0.1–1 µg may be lethal to humans *(3)*. Consequently, considerable care and safety precautions are necessary in working with botulinum neurotoxins.

Because the consequences of an accidental intoxication with botulinum neurotoxins are severe, safety must be a primary concern of scientists interested in the study of these toxins. The Centers for Disease Control and Prevention

From: *Methods in Molecular Biology, vol. 145: Bacterial Toxins: Methods and Protocols*
Edited by: O. Holst © Humana Press Inc., Totowa, NJ

(CDC) recommends Biosafety Level 3 primary containment and personnel precautions for facilities producing large (milligram) quantities of the toxins (7,8). All personnel who work in the laboratory should be immunized with pentavalent (A–E) toxoid available from the CDC, and antibody titers of immunized personnel should be confirmed.

In 1997, C. botulinum cultures and toxins were included in a group of select agents whose transfer is controlled by the CDC. To transfer these agents both the person sending and the person receiving them must be registered with the CDC (7). To ensure safety of personnel, a biosafety manual should be placed in the laboratory containing the proper emergency phone numbers and procedures for emergency response, spill control, and decontamination. All personnel should be trained in these procedures as well as in safe laboratory practices. When performing steps where aerosols may be created (such as in centrifugation) special precautions need to be taken. Toxins should be handled in sealed, unbreakable containers and manipulated in a Class II or III biological safety cabinet, and/or respiratory protection should be employed. The use of syringes and needles to perform bioassays using mice or for inoculation of rubber septum sealed tubes also requires caution and proper training.

The methods outlined here were developed or modified from previously described production and purification methods to limit the introduction of antigenic or contaminating material (3,9). Any steps that could be simplified, omitted, or improved from these earlier methods with equivalent or improved results were modified as appropriate.

Type A neurotoxin is produced in cultures as part of a protein complex (1,3,10). Under the growth conditions described, the neurotoxin molecule is noncovalently associated with 6–7 nontoxic proteins (**Fig. 1**). Some of these nontoxic proteins have hemagglutinating activity (3,10) and stabilize the neurotoxin molecule (3,9,10). The fully active neurotoxin is an approx 150-kDa dichain protein (10). The heavy (ca. 100 kDa) and light (ca. 50 kDa) chains are covalently bound by a disulfide bridge. C. botulinum Hall A strain produces proteolytic enzymes during culture. These proteases cleave the single-chain neurotoxin into the dichain form. The chains can be separated by reducing the sulfide bond as shown by sodium dodecyl sulfate-polyacrylamide gel electrophoresis (SDS-PAGE) (**Fig. 1**).

Toxin production is carried out in an undefined complex medium (3,11). After 4 d of incubation, acid precipitation is used to concentrate the toxin complex from the culture fluid. The precipitated toxin complex ("mud") is stable at these lower pHs (ca. 3.5) and can be stored in this crude form for several months to years. Toxin complex is solubilized from the mud in sodium phosphate buffer and precipitated with ammonium sulfate. The toxin complex is purified

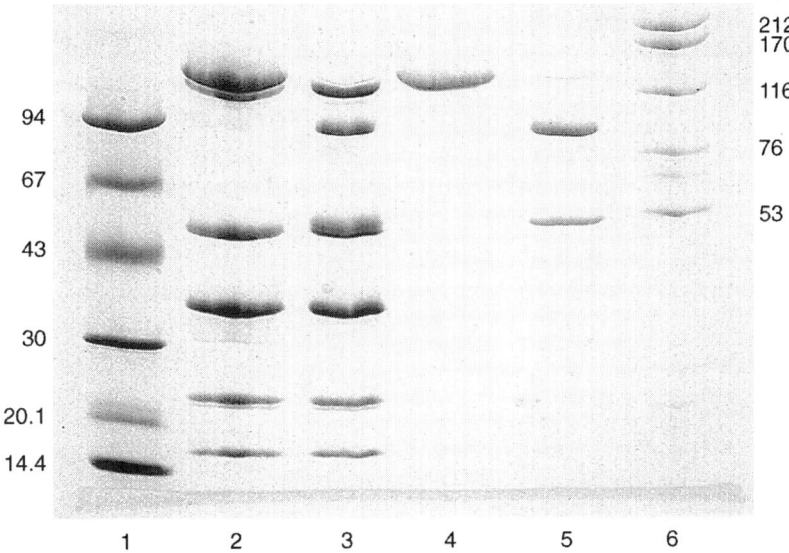

Fig. 1. Analytical SDS-polyacrylamide (10–15% gradient gel) of botulinum toxin complex and neurotoxin. Lanes: 1, Low molecular weight markers; 2, Crystalline complex; 3, Crystalline complex reduced; 4, Neurotoxin eluted from DEAE column; 5, Neurotoxin reduced; 6, Molecular weight markers (top to bottom): myosin, $M_r = 212,000$; α_2-macroglobulin, $M_r = 170,000$; β-galactosidase, $M_r = 116,000$; transferrin, $M_r = 76,000$; glutamic dehydrogenase, $M_r = 53,000$.

and the nucleic acids are removed by anion ion-exchange chromatography at an acidic pH. The toxin complex is then crystallized in 0.9 M ammonium sulfate *(3,12–14)*. These crystals are not true crystals, but consist of needle-shaped paracrystals that contain the neurotoxin and associated proteins *(3)*.

The neurotoxin is separated from the nontoxic complex proteins by chromatography at alkaline pH on an anionic exchange gel *(15–17)*. The use of a shallow sodium chloride gradient (0-0.3 M) increases the efficiency of this procedure. If present, trace contaminates can be removed by treatment with immobilized *p*-aminophenyl-β-D-thiogalactopyranoside *(18)* and/or chromatography on a cation-exchange column *(17)*.

The biological activity of botulinum toxin preparations can be assayed by the intravenous (IV) time to death method *(19,20)* or the intraperitoneal (IP) end point dilution method *(21)*. The IV method has a variance of ~15% when performed properly *(21)* and requires three to five mice per sample. The IP method results in a variance of as low as 5% if correct dilutions and sufficient animals are used *(22)*.

The best storage method for the neurotoxin is dependent on the end-use of the material. Neurotoxin standards with stabilizing protein added have been kept for years at 20–22°C in low-pH buffers with no appreciable loss of activity *(21)*. Ammonium sulfate precipitated neurotoxin can be stored at 4°C for several weeks. However, within a month, SDS-PAGE will show fragmentation and breakdown products of the neurotoxin stored in this manner. The neurotoxin can also be stored at –20°C with 50% glycerol as a cryoprotectant for years. Because crystals of the complex are very stable, purifying the neurotoxin from the complex as required could be the best option.

2. Materials
2.1. Toxin Production

1. Hall A strain: *Clostridium botulinum* (*see* **ref.** *[23]* for genetic characterization).
2. Inoculum medium: 500 mL of dH$_2$O, 2.0% casein hydrolysate, 1.0% yeast extract, 0.5% glucose (w/v), pH 7.2. Autoclave for 30 min at 121°C.
3. 50 g of glucose solution in 500 mL of dH$_2$O. Autoclaved for 30 min.
4. 10 L of toxin production medium in 13 L of carboy: 9.5 L of dH$_2$O, 2.0% casein hydrolysate, 1.0% yeast extract (w/v), pH 7.2. Autoclave for 90 min at 121°C.

2.2. Precipitation and Extraction

1. 3 *N* Sulfuric acid.
2. Sterile water.
3. 0.2 *M* Sodium phosphate buffer, pH 6.0.
4. 1.0 *N* Sodium hydroxide.
5. Ultrapure ammonium sulfate.

2.3. Purification of Type A Complex

1. 0.05 *M* Sodium citrate buffer, pH 5.5.
2. DEAE-Sephadex A-50 gel (80 g dry) swelled to a volume of 1–1.5 L and degassed.
3. Chromatography column (5 cm × 50 cm).
4. Dialysis tubing.
5. Fraction collector.
6. UV-spectrophotometer.
7. Ultrapure ammonium sulfate.

2.4. Crystallization of Type A Complex

1. 0.05 *M* Sodium phosphate buffer, pH 6.8.
2. Dialysis tubing.
3. UV-spectrophotometer.
4. 4 *M* Sterile ammonium sulfate solution.
5. 0.9 *M* Sterile ammonium sulfate solution.

2.5. Purification of Type A Neurotoxin

1. 0.02 M sodium phosphate buffer, pH 7.9.
2. Dialysis tubing.
3. Chromatography column (2.5 cm × 20 cm).
4. DEAE-Sephadex A–50 (19 g dry) swelled and degassed.
5. Gradient maker apparatus.
6. Fraction collector.
7. UV-spectrophotometer.

2.6. Additional Purification Steps

2.6.1. Treatment with Carbohydrate-Binding Affinity Gel

1. Washed and equilibrated p-aminophenyl-β-D-thiogalactopyranoside (pAPTG, Sigma).
2. 10-mL Chromatography column.
3. Fraction collector.
4. UV-spectrophotometer.

2.6.2. Chromatography on SP-Sephadex C-50

1. 25-mL Chromatography column.
2. Dialysis tubing.
3. 0.02 M Sodium phosphate buffer, pH 7.0.
4. SP-Sephadex C-50 gel swelled and degassed.
5. Gradient maker.
6. Fraction collector.
7. UV-spectrophotometer.

2.7. Bioassays

2.7.1. IV Assay

1. Female ICR mice, 18–22 g.
2. Mouse restraint ("mouse trap").
3. 30 mM Sodium phosphate buffer, pH 6.3, + 0.2% (w/v) gelatin.
4. IV Standard curve for crystalline type A toxin.
5. 1-mL Syringe with 25–26-gage needle.

2.7.2. IP Assay

1. Female ICR mice, 18–22 g.
2. Sterile 30 mM sodium phosphate buffer, pH 6.3, + 0.2% (w/v) gelatin.
3. Sterile serial dilution blanks.
4. 1-mL Syringe with 25–26-gage needle.

2.8. Precipitation and Storage of Toxin Preparations

2.8.1. Precipitation

Ultrapure ammonium sulfate.

2.8.2. Storage of Standard Toxin Solutions

Sterile 0.05 *M* sodium acetate buffer + 3% bovine serum albumin + 2% gelatin (w/v), pH 4.2.

2.8.3. –20° or –80°C Storage

1. Sterile glycerol.
2. Sterile vials.

3. Methods

3.1. Toxin Production

1. Inoculate 500 mL of inoculum medium with frozen stock culture of *C. botulinum* strain Hall A and incubate without shaking at 37°C until turbid (12–24 h).
2. Add 500 mL of cooled glucose solution to carboy.
3. Inoculate carboy with 500 mL of a 12–24-h inoculum culture.
4. Incubate production carboy for 4 d at 37°C (*see* **Note 1**). A schematic of toxin production is presented in **Fig. 2**.

3.2. Precipitation and Extraction

1. Adjust pH of the production medium to 3.4 by addition of 3 *N* sulfuric acid. Allow the precipitate to settle for 1–3 d (*see* **Note 2**). Remove the supernatant by siphoning and centrifugation (12,000*g* for 10 min at 20°C). Discard the supernatant to waste and decontaminate by autoclaving.
2. Wash the precipitated toxin with 1 L of sterile water and collect the washed precipitate by centrifugation (12,000*g* for 10 min at 20°C).
3. Resuspend pelleted toxin in 600 mL of 0.2 *M* sodium phosphate buffer, pH 6.0.
4. Adjust the pH of dissolved toxin to 6.0 with 1 *N* sodium hydroxide and gently stir for 1 h at 20°C.
5. Centrifuge (12,000*g* for 10 min at 20°C) the extracted toxin to clarify.
6. Save supernatant and repeat **steps 3–6** with the pellets using 400 mL of the buffer.
7. Pool clarified extracts and precipitate by bringing to 60% saturation (39 g/100 mL) with ammonium sulfate (*see* **Note 3**).

3.3. Purification of Type A Complex

1. Collect precipitated toxin by centrifugation (12,000*g* for 10 min at 8°C) and dissolve in 25 mL of 0.05 *M* sodium citrate buffer, pH 5.5.
2. Dialyze sample for 18 h at 4°C against 3× changes of 500 mL of the same buffer.
3. Pack and wash DEAE-Sephadex A-50 in the column with 1–2 column volumes of the citrate buffer.

Flow Chart of Type A Production

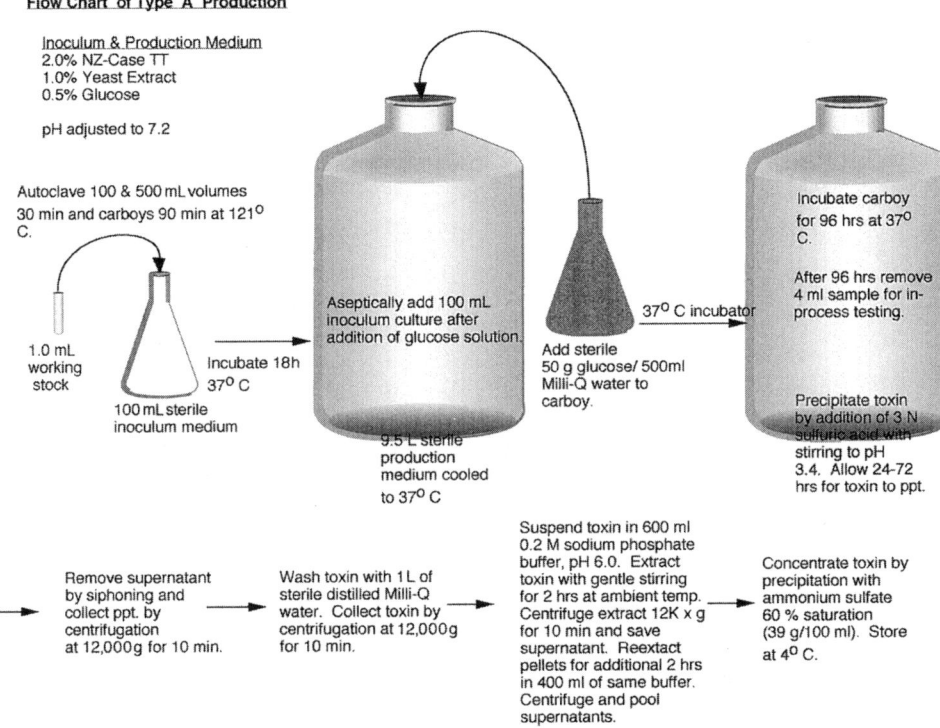

Inoculum & Production Medium
2.0% NZ-Case TT
1.0% Yeast Extract
0.5% Glucose

pH adjusted to 7.2

Autoclave 100 & 500 mL volumes
30 min and carboys 90 min at 121°
C.

1.0 mL working stock

100 mL sterile inoculum medium

Incubate 18h 37° C

Aseptically add 100 mL inoculum culture after addition of glucose solution.

9.5 L sterile production medium cooled to 37° C

Add sterile 50 g glucose/ 500ml Milli-Q water to carboy.

37° C incubator

Incubate carboy for 96 hrs at 37° C.

After 96 hrs remove 4 ml sample for in-process testing.

Precipitate toxin by addition of 3 N sulfuric acid with stirring to pH 3.4. Allow 24-72 hrs for toxin to ppt.

Remove supernatant by siphoning and collect ppt. by centrifugation at 12,000g for 10 min.

Wash toxin with 1 L of sterile distilled Milli-Q water. Collect toxin by centrifugation at 12,000g for 10 min.

Suspend toxin in 600 ml 0.2 M sodium phosphate buffer, pH 6.0. Extract toxin with gentle stirring for 2 hrs at ambient temp. Centrifuge extract 12K x g for 10 min and save supernatant. Reextact pellets for additional 2 hrs in 400 ml of same buffer. Centrifuge and pool supernatants.

Concentrate toxin by precipitation with ammonium sulfate 60 % saturation (39 g/100 ml). Store at 4° C.

Fig. 2. Schematic for toxin production.

4. Centrifuge (12,000g for 10 min at 20°C) dialyzed toxin solution to clarify.
5. Load sample onto the column at a flow rate at 35–45 mL/h (*see* **Note 4**). The toxin complex is eluted in the void volume in citrate buffer.
6. Start collecting fractions after 200 mL have passed through the column (*see* **Note 5**).
7. Measure absorbance of each fraction at 260 and 280 nm (A_{260} and A_{280}) with the UV-spectrophotometer.
8. Pool fractions from the first peak off the column with a A_{260}/A_{280} ratio of 0.6 or less (*see* **Note 6**).
9. Precipitate toxin pool by making 60% saturated with ammonium sulfate (*see* **Note 7**).

3.4. Crystallization of Type A Complex

1. Dissolve precipitated toxin at a concentration of ~10 mg/mL in 0.05 M sodium phosphate buffer, pH 6.8.
2. Dialyze sample thoroughly against 3× changes of 500 mL of the same buffer (*see* **Note 8**).

3. Centrifuge (12,000g for 10 min at 10°C) sample and dilute to a concentration of 6–8 mg/mL with 0.05 M sodium phosphate buffer, pH 6.8.
4. Slowly, with gentle stirring, add 4 M ammonium sulfate solution to a final concentration of 0.9 M (*see* **Note 9**).
5. Allow crystals to form for 1–3 wk at 4°C (*see* **Note 10**).

3.5. Purification of Botulinum Type A Neurotoxin

Recovery of neurotoxin from the complex is typically 10–13%. Because crystalline toxin is one of the more stable forms, only the required amount of neurotoxin (which is more labile than the crystals) is usually purified as needed.

1. Dissolve the required amount of crystals in 0.02 M sodium phosphate buffer, pH 7.9, for 1–2 h and dialyze for 18 h (10× with three buffer changes) to remove ammonium sulfate.
2. Pack and wash DEAE-Sephadex A-50 in a column sufficient in size to bind all the complex (*see* **Note 11**).
3. Centrifuge (12,000g for 10 min at 10°C) sample and load onto column. Wash column with at least 50 mL of starting buffer or until A_{280} is less than 0.01.
4. Set flow rate at 30 mL/h and begin collecting fractions.
5. Neurotoxin is eluted with a linear gradient of sodium chloride made of running buffer plus running buffer containing 0.3 M sodium chloride. Toxin elutes at ca. 0.15 M chloride ion. The volume of eluant is dependent on column size but is typically 4–5× the volume of the gel volume.
6. Read A_{280} of fractions and pool first peak eluted in the sodium chloride gradient (*see* **Note 12**).

3.6. Additional Purification Steps

The following methods can be used to eliminate any trace contaminants that may be present in the toxin pool that could interfere with sensitive protocols. Specific toxicity generally will not be improved, but trace contaminants are removed.

3.6.1. Treatment with Carbohydrate-Binding Affinity Gel

1. Mix pooled toxin with 1 mL of washed pAPTG gel and gently mix for 15 min at 20–22°C.
2. Load toxin–pAPTG slurry into the column and collect the toxin as it elutes using A_{280} to pool fractions. Neurotoxin does not bind to the gel matrix but nontoxic proteins including those with hemagglutinating activity do bind under these conditions.

3.6.2. Chromatography on SP-Sephadex C-50

1. Dialyze toxin against 0.02 M sodium phosphate, pH 7.0, at 4°C against several changes of buffer.

2. Pack and wash SP-Sephadex C-50 column with sodium phosphate buffer. Set flow rate at 30 mL/h.
3. Centrifuge toxin (12,000*g* for 10 min at 10°C) and load clarified solution onto column. Wash column with 2–3 column volumes of the running buffer. Elute neurotoxin with a 0-0.5 *M* sodium chloride gradient that is 4–5× the column volume.
4. Read A_{280} of fractions eluted with the sodium chloride gradient and pool fractions containing neurotoxin.

3.7. Bioassays

Extreme caution should be taken when performing these procedures. The IV method will give toxicity data in 30–70 min with concentrated toxin solutions. The IP method is completed in 3–4 d, but requires less technical expertise to perform accurately.

3.7.1. IV Mouse Assay

1. Dilute complex to 15 µg/mL and the neurotoxin to 4 µg/mL in 30 m*M* sodium phosphate buffer, pH 6.3, + 0.2% (w/v) gelatin (gel-phosphate buffer).
2. Inject 0.1 mL of the diluted sample into the lateral tail veins of three to five animals. Record the time the animals were injected and mark the animals (*see* **Note 13**).
3. Record the time to death and determine the average time to death (ATTD) for the mice (*see* **Note 14**).
4. Convert ATTD to IP LD_{50}/mL using the previously prepared standard curve (**Fig. 3**) (*see* **Note 15**).

3.7.2. IP Assay

1. Dilute toxin to ~10 LD_{50}/mL using gel-phosphate buffer (*see* **Note 16**).
2. Make several (approx five or six) serial twofold dilutions.
3. Inject 0.5 mL of diluted toxin IP into groups of 5–10 mice per dilution.
4. Record the number of deaths within 96 h (*see* **Note 17**).
5. The dilution that kills 50% of the animals is taken as the LD_{50}/0.5 mL. Multiply dilution ×2 to obtain LD_{50}/mL.

3.8. Storage of Botulinum Toxin Preparations

Purified neurotoxin is not as stable as the toxin complex. Precipitated neurotoxin will show degradation within a month when analyzed by SDS-PAGE. Diluted toxin solutions can be stored for years at 20°C by the addition of stabilizing protein excipients and adjustment of pH and ionic strength. Solutions of toxin in 50% glycerol stored at −20 to −70°C are stable for years. The preferred method of storage will depend on the project requirements.

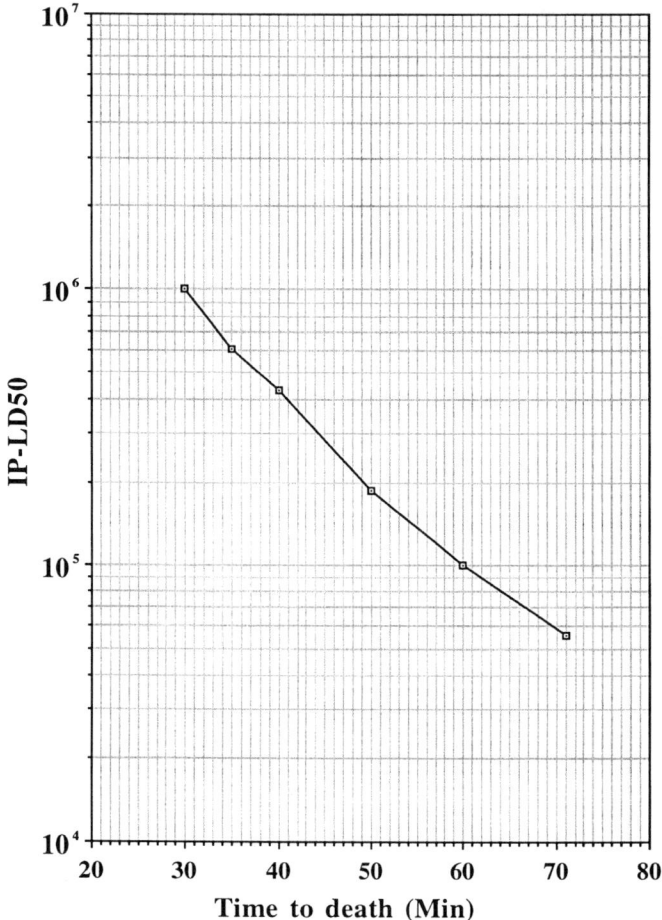

Fig. 3. IV time-to-death standard curve for botulinum type A crystalline toxin complex.

3.8.1. Precipitation

Make the toxin solution to 60% saturation with ammonium sulfate and store at 4°C.

3.8.2. Storage of Standard Toxin Solutions

1. Dilute filter-sterilized toxin (0.2 µm filter) to desired concentration in sterile 0.05 *M* sodium acetate buffer + 3% bovine serum albumin + 2% gelatin, pH 4.2.
2. Aliquot toxin solution to desired volumes and store at 4–25°C. Do not freeze these samples to avoid toxin inactivation.

3.8.3. –20 to –80°C Storage

1. Dilute toxin to twice the desired concentration in 0.05 M sodium phosphate buffer, pH 6.8.
2. Add an equal volume of sterile glycerol.
3. Aliquot toxin into convenient volumes and store at –20 to –80°C.

4. Notes

1. After 4 d, production culture fluid should have a pH of 5.6–5.9, a toxicity greater than $10^6 LD_{50}/mL$ and 1:1000 dilution should be neutralized by 1 IU of type A antitoxin (CDC).
2. NZ Amine A, NZ Amine B, and NZ Case TT all have lot-to-lot variation and need to be pretested to achieve adequate toxin production and precipitation qualities. In testing, toxin production should be $>10^6 LD_{50}/mL$. Culture supernatant should contain less than 10% of the toxicity 24–72 h after precipitation with acid.
3. To process larger culture volumes (up to 40 L) using the same volume of DEAE-Sephadex, treat extracts with RNase before dialysis. Dissolve toxin in 0.05 M sodium phosphate, pH 6.0, and add 0.05 mg/mL of RNase A (Sigma) and treat for 3 h at 37°C. Precipitate by bringing to 60% saturation with ammonium sulfate.
4. The volume of toxin solution loaded onto the column must be <10% of column volume.
5. Elution of protein from the column can be monitored at A_{280} with a UV detector. Collect fractions when protein (A_{280}) begins to elute.
6. The approximate toxin concentration in the sample eluting off the first DEAE column can be determined using an extinction coefficient of 1.65 for type A complex. The yield from 20 L of culture should be approx 350 mg of complex.
7. The crystallization procedure can be omitted and neurotoxin purification accomplished using toxin complex off of the first DEAE-Sephadex A-50 column if desired. Additional purification (*see* **Subheading 3.6.**) may be required.
8. Toxin precipitates if the concentration of ammonium sulfate is too high, so all residual ammonium sulfate must be removed prior to crystallization.
9. The last few milliliters of ammonium sulfate should be added very slowly. When the toxin solution becomes opalescent, the concentration of ammonium sulfate is usually sufficient to initiate crystallization of toxin complex.
10. Recovery of crystalline toxin is ~80% after 10 d. High quality crystalline toxin complex has a specific toxicity of approx $3.5 \times 10^7 LD_{50}/mg$. Crystals are stable in 0.9 M ammonium sulfate for several years.
11. DEAE-Sephadex A-50 will bind approx 0.9 mg/mL of the complex under these running conditions.
12. Protein concentration can be determined using an extinction coefficient of 1.63 for type A neurotoxin. Analysis of pure neurotoxin off of the second DEAE-Sephadex column by SDS-PAGE should show one band (M_r = 150 kDa) when unreduced (**Fig. 1**). Toxin should be >98% homogeneous and have a specific toxicity of $>10^8 LD_{50}/mg$.

13. If the needle is in the vein, the plunger of the syringe will slide easily. A drop of blood will appear immediately upon retraction of the needle. If the needle is not in the vein, the plunger will be difficult to depress. Repeat any injections where the vein was missed, using a new animal.
14. Animals injected with same sample should die within 5 min of each other if titration is valid.
15. IP LD_{50}/mL can be taken from the standard curve prepared with type A complex (**Fig. 2**), or calculated from the equation IP LD_{50}/mL $= 6.9 \times 10^6 - 3.4 \times 10^5(ATTD) + 5.7 \times 10^3(ATTD)^2 - 31.9 (ATTD)^3$ where ATTD = the average time to death.
16. The toxin complex is diluted to 4×10^{-6}/mg and neurotoxin 10^{-7}/mg.
17. Mice must die with clinical symptoms of botulism toxicity: contraction of abdominal muscles resulting in a wasp shape of the animal's waist, reduced mobility, labored breathing, convulsions, and eventual death.

References

1. Smith, L. D. S., and Sugiyama, H. (1988) *Botulism*, 2nd ed., Charles C. Thomas, Springfield, IL.
2. Schantz, E. J. and Johnson, E. A. (1997) Botulinum toxin: the story of its development for the treatment of human disease. *Perspect. Biol. Med.* **40**, 317–327.
3. Schantz, E. J. and Johnson, E. A. (1992) Properties and use of botulinum toxin and other microbial neurotoxins in medicine. *Microbiol. Rev.* **56**, 80–99.
4. Jankovic, J. and Hallett, M., eds. (1994) *Therapy with Botulism Toxin,* Marcel Dekker, New York, NY.
5. Montecucco, C., ed. (1995) *Clostridial Neurotoxins*, Springer-Verlag, Berlin.
6. Lamanna, C. (1959) The most poisonous poison, *Science* **130**, 763–772.
7. Centers for Disease Control and Prevention (1998) *Laboratory Registration and Select Agent Transfer Tracking System, 42 CFR 72*, U.S. Dept. of Health and Human Services, Washington, D. C.
8. Centers for Disease Control and Prevention/National Institutes of Health (1999) *Biosafety in Microbial and Biomedical Laboratories.* 4th ed. U.S. Dept. of Health and Human Services, pp. 94–95.
9. Schantz, E. and Johnson, E. A. (1993) Quality of botulinum toxin for human treatment, in *Botulinum and Tetanus Neurotoxins* (DasGupta, B. R., ed.), Plenum, New York, pp. 657–659.
10. Sugiyama, H. (1980) *Clostridium botulinum* neurotoxin. *Microbiol. Rev.* **44**, 419–448.
11. Schantz, E. and Scott, A. (1981) Use of crystalline type A botulinum toxin in medical research, in *Biomedical Aspects of Botulism* (Lewis, G. E., ed.), Academic Press, San Diego, CA, pp. 143–150.
12. Lamanna, C. McElroy, O. E., and Eklund, H. W. (1946) The purification and crystallization of *Clostridium botulinum* type A toxin. *Science* **103**, 613.
13. Duff, J., Wright, G., Klerer, J., Moore, D., and Bibler. R. (1957) Studies on immunity to toxins of *Clostridium botulinum*. I. A simplified procedure for isolation of type A toxin. *J. Bacteriol.* **73**, 42–47.

14. Sugiyama, H., Moberg, L., and Messer, S. (1976) Improved procedure for crystallization of *Clostridium botulinum* type A toxic complex. *Appl. Environ. Microbiol.* **33,** 963–966.

15. DasGupta, B. and Boroff, D. A. (1967) Chromatographic isolation of hemagglutinin-free neurotoxin from crystalline toxin of *Clostridium botulinum* type A. *Biochim. Biophys. Acta* **147,** 603.

16. Tse, C., Dolly, J., Hambleton, P., Wray, D., and Melling, J. (1982) Preparation and characterization of homogeneous neurotoxin type A from *Clostridium botulinum. Eur. J. Biochem.* **122,** 493–500.

17. DasGupta, B. and Sathyamoorthy, V. (1984) Purification and amino acid composition of type A botulinum neurotoxin. *Toxicon* **22,** 415–424.

18. Moberg, L. and Sugiyama, H. (1978) Affinity chromatography purification of type A botulinum neurotoxin from crystalline toxic complex. *Appl. Environ. Microbiol.* **35,** 878–880.

19. Boroff, D. and Fleck, U. (1966) Statistical analysis of a rapid in vivo method for the titration of the toxin of *Clostridium botulinum. J. Bacteriol.* **92,** 1580–1581.

20. Kondo, H., Shimizu, T., Kubonoya, M., Izumi, N., Takahashi, M., and Sakaguchi, G. (1984) Titration of botulinum toxins for lethal toxicity by intravenous injection into mice. *Jpn. J. Med. Sci. Biol.* **37,** 131–135.

21. Schantz, E. and Kautter, D. (1978) Standardized assay for *Clostridium botulinum* toxins. *J. Assoc. Off. Anal. Chem.* **61,** 96–99.

22. Pearce, L., Borodic, G., First, E., and MacCallum, R. (1994) Measurement of botulinum toxin activity: evaluation of the lethality assay. *Toxicol. Appl. Pharmacol.* **128,** 69–77.

23. Lin, W.-J. and Johnson, E. A. (1996) Genome analysis of *Clostridium botulinum* type A by pulsed-field gel electrophoresis. *Appl. Environ. Microbiol.* **61,** 4441–4447.

3

Shiga Toxins

David W. K. Acheson, Anne V. Kane, and Gerald T. Keusch

1. Introduction

The Shiga family of toxins is comprised of a group of genetically and functionally related molecules whose original family member was described 100 yr ago. Up until the early 1980s this group of toxins was little more than a scientific curiosity without a clear role in disease pathogenesis. However, since the discovery of these toxins in *E. coli* and other Enterobacteriaceae, and their association with identifiable clinical diseases such as hemorrhagic colitis and hemolytic uremic syndrome, their pathophysiological importance has become clear. In this chapter we describe briefly the clinical relevance and mechanisms of actions of the toxins and then focus on the utility of the available methods to purify and assay the toxins.

1.1. The Importance of Shiga Toxins in Disease Pathogenesis

The name *Shiga toxin* is derived from a toxic activity originally discovered in Shiga's bacillus, now termed *Shigella dysenteriae,* the prototypic *Shigella* species originally described by the Japanese microbiologist Kiyoshi Shiga in 1898 following an extensive epidemic of lethal dysentery in Japan *(1)*. Credit for the discovery of Shiga toxin is generally accorded to Conradi *(2)*, who described many of its properties in 1903. This activity was known as Shiga neurotoxin because when injected parenterally into mice or rabbits it resulted in limb paralysis followed by death of the animal. Although in 1978 *S. dysenteriae* was associated with the clinical condition known as hemolytic uremic syndrome (HUS), a triad of acute renal failure, thrombocytopenia, and hemolytic anemia *(3)*, it was not until the recognition of Shiga toxins in *E. coli* and their association as well with HUS in the early 1980s that the role of Shiga toxin from *S. dysenteriae* in HUS could be truly appreciated. Over the last 15

From: *Methods in Molecular Biology, vol. 145: Bacterial Toxins: Methods and Protocols*
Edited by: O. Holst © Humana Press Inc., Totowa, NJ

yr we have come to realize that the Shiga family of toxins are in fact a major cause of disease in many developed countries.

The discovery of the *E. coli* derived Shiga toxins followed the 1978 report by Konowalchuk et al. *(4)* that culture filtrates of several different strains of *E. coli* were cytotoxic for Vero cells. This cytolethal activity was heat labile but was not neutralized by antiserum to the classic cholera-like *E. coli* heat-labile enterotoxin. It soon became apparent that the cytotoxin was not due to a single protein, as careful investigation of one strain (*E. coli* O26:H30) revealed two variant toxin activities *(5)*. Following these observations, Wade et al. *(6)* in England noted the presence of cytotoxin-producing *E. coli* O26 strains in association with bloody diarrhea. This finding was supported by reports from Scotland et al. *(7)* in the United Kingdom and Wilson and Bettelheim *(8)* in New Zealand.

In 1983 Riley et al. *(9)* reported on the isolation of a rare serotype of *E. coli* (O157:H7) from outbreaks characterized by the development of hemorrhagic colitis. This strain of *E. coli* produced a cytotoxin that was shown to be neutralized by antisera to Shiga toxin. The outbreak strain was later shown to carry two bacteriophages each of which encoded a different type of Shiga toxin, thus confirming the earlier suggestions of multiple cytotoxins made by these strains. At around the same time Kamali et al. *(10)* demonstrated the association between infections due to Shiga toxin producing *E. coli* and the development of HUS. This heralded the onset of a new era in the Shiga toxin field. Currently more than 200 different types of *E. coli* have been shown to produce Shiga toxins, many of which have been associated with human disease *(11)*. Shiga toxin-producing *E. coli* (STEC) are, in fact, the most common cause of acute renal failure in the United States, and are estimated to result in at least 100 deaths annually in the United States alone.

The nomenclature for the Shiga toxin family has become confusing. In 1972, a toxin causing fluid secretion by rabbit small bowel was identified in *S. dysenteriae* type 1 and named Shigella enterotoxin. This toxin was subsequently proved to be identical to the originally described Shiga neurotoxin. Following the discovery that *E. coli* cytotoxins were active on Vero cells they were referred to as Verotoxins. This name is still used by many workers in the field who identify Verotoxin-producing *E. coli* as VTEC. However, when it became apparent in the early 1980s that these newly described *E. coli* toxins were very similar to Shiga toxin and were neutralized by antisera to Shiga toxin, other workers referred to them as Shiga-like toxins. By 1996, when the common mechanism of action and cellular binding site was proven, an international group of investigators decided to designate this group of biologically homogeneous toxins simply as Shiga toxins (Stx), irrespective of their bacte-

Table 1
The Shiga Toxin Family

Name	Gene	Protein	Comments
Shiga toxin	*stx*	Stx	From *S. dysenteriae*
Shiga toxin 1	*stx₁*	Stx1	
Shiga toxin 2	*stx₂*	Stx2	
Shiga toxin 2c	*stx₂c*	Stx2c	
Shiga toxin 2d	*stx₂d*	Stx2d	Mucus activatable
Shiga toxin 2e	*stx₂e*	Stx2e	Associated with porcine edema disease

rial origin, after the original description nearly 100 yr ago *(12)*. The gene designation (*stx*) for Shiga toxin from *S. dysenteriae* type 1 was already well established and the new nomenclature therefore maintained the *stx* gene designation for the *E. coli* derived toxins.

Currently there are five members of this family (**Table 1**). The toxins are divided into two main groups, based on antigenic differences. Shiga toxin from *S. dysenteriae* type 1 and Shiga toxin 1 form one group, and the Shiga toxin 2 family form the other group. As discussed below, Shiga toxin from *S. dysenteriae* and Stx1 from *E. coli* are virtually identical, whereas Shiga toxin 2 differs significantly and is made up of a number of subfamilies.

1.2. Diseases Associated with Shiga Toxins

The evidence implicating Shiga toxin in the pathogenesis of shigellosis is not conclusive, and is based on animal models and in vitro studies in cell culture–Koch's molecular postulates have not been satisfied. This is, in part, because none of the animal models used, including oral infection in primates, truly mimics human infection. Shiga toxin clearly causes fluid secretion when placed in the small bowel lumen of rabbits *(13)* and results in inflammatory enteritis in this model *(14)*. Although it is cytotoxic to human colonic epithelial cells *(15)* and thus can mimic colonic manifestations of clinical Shigellosis, the interpretation is complicated because *Shigella* are invasive and multiply within epithelial cells. Fontaine et al. *(16)* created a Shiga toxin deletion mutant strain of *S. dysenteriae* type 1, and compared its clinical effects in a Rhesus monkey model with those of a wild-type strain. The toxin-negative mutant resulted in clinical disease when fed to these animals, but both clinical and histological manifestations were less severe, particularly the lack of intestinal hemorrhage, compared with the wild-type. These data suggest but do not prove that the toxin plays a role in the hemorrhagic component of the dysenteric phase of shigellosis, and strongly argue that toxin is not a necessary factor for the initiation of clinical disease. The presence of Stx may therefore contribute to the extent of intestinal damage and hemorrhage in vivo in patients with shigellosis. There is strong epidemiological evidence to link only the Stx-producing species of *Shigella* species (*S. dysenteriae* type 1) with hemolytic uremic syndrome in many parts of the world and especially South Africa and Bangladesh, further strengthening the linkage between exposure to the toxin, which is present in the stool of these patients, and increased risk of HUS.

Evidence that the Shiga toxins from *E. coli* are involved in disease pathogenesis is much stronger. However, the evidence is predominantly epidemiological because it is unethical to challenge human subjects with STEC strains. There is a strong association of Stx1- and/or 2-producing STEC with both outbreaks and sporadic diarrheal disease. Typically the illness begins with watery diarrhea, abdominal pain, some nausea and vomiting, but little fever. This may or may not progress to bloody diarrhea, which in its most profound form causes an identifiable clinical syndrome, hemorrhagic colitis, which often progresses to the hemolytic uremic syndrome. Exposure to STEC has also been associated with the development of thrombotic thrombocytopenic purpura (TTP), a variant thrombotic microangiopathy believed to be related to the pathogenesis of HUS *(17)*. Of all the clinical associations with STEC, the development of hemolytic uremic syndrome is the most feared complication *(18)*. This association has subsequently been corroborated by many investigators *(19)*. However, it is not clear what predisposes an STEC-infected individual to develop HUS. In various large outbreaks of STEC and in dysentery due to *S. dysenteriae* type 1 it appears that between 5% and 10% of patients who become infected will go on to develop clinical and laboratory manifestations of HUS, and of these approx 5% will die during the acute phase of the disease.

Previously it was considered that STEC, especially O157:H7, were mainly transmitted via foodborne outbreaks. As surveillance for O157:H7 and other STEC has increased it has become clear that they cause considerable sporadic disease as well *(11)*. For example, a recent report from Virginia in the United States isolated 11 STEC from 270 diarrheal samples examined for bacterial enteric pathogens, but only six were O157:H7. Of the five non-O157:H7 isolates, one O111 and one O103 were cultured from patients with bloody diarrhea *(20)*. In this study STEC were also almost as common as *Salmonella* (n = 13) and exceeded both *Campylobacter* (n = 7) and *Shigella* (n = 4). Other, larger, studies have confirmed the prevalence of STEC in the United States to be 0.5–1% of stools submitted to clinical laboratories for testing. Of those samples from which any isolate is obtained, 30–50% are non-O157:H7 STEC *(20,21)*.

1.3. Shiga Toxin Pathogenesis and Mechanism of Action

1.3.1. Shiga Toxin-Producing Bacteria

Shiga toxin-producing bacteria fall into three major groups: (1) *S. dysenteriae* type 1, the cause of severe bacillary dysentery in many developing countries. (2) the major group of Shiga toxin-producing *E. coli* (STEC) of which there are currently more than 200 different types. (3) other members of the family Enterobacteriaceae which produce Stx and have been implicated in occasional cases of HUS (e.g., *Citrobacter freundii [22]* and *Enterobacter cloaceae [23]*).

1.3.2. STEC in the Gastrointestinal Tract

STEC are ingested by mouth, colonize portions of the lower intestine, and then produce Shiga toxins. Two recent reviews *(24,25)* discuss many aspects of the pathogenesis of STEC and their diagnosis, and the reader is referred to these articles for a more detailed discussion. However, there are several important issues that warrant emphasis. Toxins are produced in the intestinal lumen and act systemically, but the mechanisms by which Stx crosses the intestinal barrier are particularly poorly understood. Acheson et al. have used a simplified in vitro model of intestinal mucosa to address this question using cultured intestinal epithelial cells grown on permeable polycarbonate filters *(26)*. These monolayers develop functional tight junctions, with high transcellular electrical resistance. Biologically active Stx translocates across these epithelial cell barriers in an apical to basolateral direction without disrupting the tight junctions, apparently through the cells (transcellular) rather than between the cells (paracellular). This pathway appears to be energy dependent, saturable, and directional. Quantitative measurements of toxin transfer in this artificial system indicate that if the same events were to occur to the same extent in vivo, sufficient toxin molecules would cross the intestinal epithelial cell barrier to initiate the endothelial cell pathology associated with the microangiopathy seen in STEC-related disease.

1.3.3. Effects of Stx on Different Cell Types

Different cells have varying susceptibility to the members of the Stx family of toxins.

In the initial studies of the "enterotoxin" (fluid secretion) activity of Stx in ligated rabbit small bowel loops, epithelial cell damage and apoptosis of villus tip cells were noted *(13,14)*. Subsequent studies revealed that villus cell sodium absorption was diminished by Stx, with no alteration in substrate-coupled sodium absorption or active chloride secretion, suggesting that toxin acted on the absorptive villus cell and not on the secretory crypt cell *(27)*, an observation that relates to the villus cell brush border distribution of the toxin receptor *(28)*.

Obrig and co-workers *(29,30)* were among the first to recognize that HUS might be due to toxin effects on endothelial cells. They reported that preincubation of human umbilical vein endothelial cells (HUVEC) with lipopolysaccharide (LPS) or LPS-induced cytokines (interleukin-1 [IL-1]) or tumor necrosis factor (TNF) induced the toxin receptor Gb3, and converted the relatively Stx-resistant HUVEC cells into more responsive targets. These investigators later reported that human glomerular endothelial cells (GECs) constitutively produced Gb3 and were not further induced by preexposure to cytokines, suggesting a reason why these cells may be a preferred target for Stx *(31)*. The significance of this is now uncertain, as Monnens et al. *(32)* have reported conflicting data on their glomerular endothelial cell isolates. Similarly, Hutchinson et al. *(33)* found that cerebral endothelial cells were sensitive to Stx1 and that this sensitivity could be enhanced significantly with exposure to IL-1β and TNF-α.

Shiga toxin has also been shown to affect polymorphonuclear leukocytes (PMNs) and macrophages. Stx1 induces release of reactive intermediates from PMNs and causes a reduction in their phagocytic capacity *(34)*. Murine peritoneal macrophages, human peripheral blood monocytes, and human monocytic cell lines *(35,36)* are relatively refractory to the cytotoxic action of Stxs but respond by secreting TNF-α and IL-1 in a dose-dependent manner. Human peripheral blood monocytes have also been reported *(37)* to make TNF-α, IL-1β, IL-6, and IL-8 in response to Stx1. These cells may be the source of toxin-receptor-inducing cytokines in the course of infection with STEC or *S. dysenteriae* type 1.

1.4. Biochemical and Molecular Characterization of Shiga Toxins

1.4.1. Genetics of Shiga Toxins

Shiga toxins are either phage (Stx1 and 2) or chromosomally encoded (Stx, Stx2e) *(38–41)*. Stx-encoding phages are λ-like and the regulatory components relating to induction and phage gene control appear to be similar to those in λ. Clinical STEC isolates may contain one or more types of Stx, and many O157:H7 isolates harbor distinct Stx1 and 2 bacteriophages. The nucleotide sequences of the Stx genes have been published by various groups *(42–47)*. The different Stx operons have a similar structure and are composed of a single transcriptional unit which consists of one copy of the A-subunit gene followed by the B-subunit gene.

Shiga toxin from *S. dysenteriae* and Stx1 from *E. coli* are highly conserved. There are three nucleotide differences in three codons of the A-subunit of Stx1, resulting in only one amino acid change (threonine 45 for serine 45). The calculated masses for the processed A- and B-subunits are 32,225 and 7691 respectively. Signal peptides of 22 and 20 residues for the A- and B-subunits, respectively, are present.

Until recently, very little was known about the regulation of Shiga-family toxin production in vitro and essentially nothing about regulation in vivo. The production of both Stx from *S. dysenteriae* and Stx1 from STEC in vitro are regulated by *fur* via the concentration of iron in the growth medium. The iron regulation is mediated by the *fur* gene product, Fur, which acts as a transcriptional repressor *(47,48)*. The Stx2 operon does not have an upstream Fur binding site and is not iron regulated in vitro. Work from our own group *(49)* using a Stx2A–phoA fusion led to the conclusion that the location of the Stx2AB operon in the genome of a bacteriophage exerts an important influence on the expression of toxin. Two mechanisms by which the bacteriophage influenced toxin production were proposed. The first was through an increase in the number of toxin gene copies brought about by phage replication. The second proposed mechanism was the existence of a phage-encoded regulatory molecule whose activity was dependent on phage induction. Subsequent studies by Neely and Friedman *(50)* using the H19-B Stx1 encoding bacteriophage are consistent with this emerging picture of toxin regulation.

The gene for Stx2 was originally cloned from *E. coli* 933W and when compared with Stx1 was found to be organized in a similar way *(51)*. Comparison of the nucleotide sequence of the A- and B-subunits of Stx1 and 2 showed 57% and 60% similarity respectively, with 55% and 57% amino acid similarity *(45)*. Despite this degree of similarity, Stx1 and 2 are immunologically distinct, and neither is cross-neutralized by polyclonal antibody raised to the other toxin. Stx2c is similar to Stx 2 with identical A-subunits and 97% amino acid homology in the B-subunits. Stx2d has the same B-subunit as Stx2c although the A-subunit is slightly different (99% homology) *(52)*. The Stx2e gene from porcine strains when compared with Stx2 has 94% similarity of nucleotide sequence between the A-subunits and 79% nucleotide sequence similarity between the B-subunits *(53)*.

1.4.2. Protein Characteristics of Shiga Toxin

Purified Shiga toxins are heterodimeric proteins of approx 70 kDa in molecular mass that separate into two peptide subunits in sodium dodecyl sulfate (SDS) polyacrylamide gels under reducing conditions *(54)*. The larger A-subunit (31–32 kDa) is the enzymatically active subunit, mediating the inhibitory effect of the toxin on protein synthesis in cell-free systems *(55)*. The smaller pentameric B-subunit is responsible for the toxins' binding to cell surface receptors. The B-subunit binds to receptor-positive but not receptor-negative cells and competitively inhibits both the binding and cytotoxicity of holotoxin *(56,57)*. Crosslinking with heterobifunctional reagents results in a ladder of subunit–subunit configurations, with a maximum ratio of 1 A-5 B. X-ray crystallographic analysis of Shiga toxin and Stx1 B-subunits have confirmed this structure, in which 5 B-subunits form a pentameric ring that surrounds a helix at the C-terminus of the A-subunit *(58,59)*. Nicking the A-subunit with trypsin with reduction separates the larger A1 portion (approx 28 kDa) and a smaller A2 peptide (~4 kDa). The A1 fragment contains the enzymatically active portion of the toxin molecule; however, the A2 component is required to noncovalently associate the intact A-subunit with the B pentamer *(60)*. Whereas Stx B-subunits are not considered to be enzymatically active, serving primarily as the binding moiety for Stx holotoxin to intact cells, this may not be absolutely correct in view of work by Mangeny et al. *(61)* and others, who have reported that Stx 1 B-subunit alone is capable of inducing apoptosis in CD77-positive Burkitt's lymphoma cells. CD77 has been biochemically characterized as globotriaosylceramide (Gb3), the Stx1 and 2 receptor.

The Shiga toxin 1 and 2 subfamilies are separable by the fact that they are not crossneutralized by heterologous polyclonal antisera, although at least one monoclonal antibody has been described that is able to neutralize both Stx1 and Stx2 *(62)*. Stx1 proteins form a relatively homogeneous family and little significant variation has been reported within that family. In contrast, the Stx2 family is fairly diverse and contains a number of members with a variety of properties. The B-subunit is responsible for the majority of the significant functional differences within the Stx2 subfamily *(63,64)*. Different members of the Stx2 subfamily also have variable biological effects on tissue culture cells and may exhibit differential cell specificity *(52)*.

1.4.3. Mechanism of Action

Reisbig et al. *(55)* first reported that Shiga toxin could irreversibly inhibit protein synthesis by a highly specific action on the 60S mammalian ribosomal subunit. Brown et al. *(65)* and others *(66,67)* found that the A-subunit was activated by proteolysis and reduction, with the release of the A1 peptide from the A2-subunit. In these properties, Shiga toxin resembled the plant toxin ricin, and when the *stx1* gene was identified and sequenced a significant homology with ricin was noted. The following year Endo and Tsurugi *(68)* determined the enzymatic activity of ricin which was found to be an *N*-glycosidase that hydrolyzed adenine 4324 of the 28S ribosomal RNA of the 60S ribosomal subunit. Igarashi et al. *(69)* found that Stx1 from *E. coli* O157:H7 inactivated the 60S ribosomal subunits of rabbit reticulocytes and blocked elongation factor-1-dependent binding of aminoacyl-tRNA to ribosomes. Shortly thereafter, Endo and colleagues *(70)* reported that Shiga toxin had the identical enzymatic specificity as ricin. It is now known that all Stx proteins share this property;

hence this is one of the criteria used in the modern definition of the Shiga family of toxins.

A number of laboratories have cloned and expressed the Stx1 B-subunit *(71–73)*. In contrast, expression of Stx2 B-subunits has been much more difficult. Acheson et al. *(74)* cloned and expressed the Stx2 B-subunit in a variety of vectors using different promoters and, although it was possible to obtain modest levels of B-subunit expression, the multimeric forms of the B-subunit appeared to be unstable.

1.4.4. Toxin Receptors

The search for Shiga toxin receptors on mammalian cells began in 1977, when Keusch and Jacewicz *(75)* reported that toxin-sensitive cells in tissue culture removed toxin bioactivity from the medium, whereas toxin resistant cells did not. These studies also suggested that the receptor was carbohydrate in nature, and that the toxin was a sugar-binding protein or lectin. Jacewicz et al. *(76)* also extracted a toxin-binding constituent from toxin-sensitive HeLa cells and from rabbit jejunal microvillus membranes (MVMs). The MVM binding site was shown to be Gb3 by thin-layer chromatography methods *(76)* and later confirmed by high-performance liquid chromatography of derivatized glycolipids, which also demonstrated it to be the hydroxylated fatty acid variety of Gb3 *(77)*. Based on in vitro solid-phase binding studies using isolated glycolipids, Lindberg and colleagues *(78)* reported that toxin bound to the P blood group active glycolipid, Gb3. Gb3 consists of the trisaccharide α-Gal-$(1 \rightarrow 4)$-β-Gal-$(1 \rightarrow 4)$-Glc linked to ceramide. The functional receptor role of Gb3 has been shown in several ways including the correlation of Gb3 content of cells and sensitivity to Shiga toxins, and the parallel change in toxicity as Gb3 is reduced in cells treated with inhibitors of neutral glycolipids or increased by means of a liposomal delivery system. Most relevant to the intestinal effects of Shiga toxin is the work of Mobassaleh et al. *(57)*. They reported that newborn rabbits are not susceptible to the fluid secretory effects of Shiga toxin before 16 d of age, with a rapid rise thereafter to the sensitivity of adult rabbit intestine. This age-related sensitivity correlated with developmentally regulated Gb3 levels in rabbit intestinal MVMs, mediated by a sharp rise in the activity of the Gb3 biosynthetic galactosyltransferase enzymes and a concomitant decrease in the Gb3 degradative α-galactosidase enzyme activity.

One of the main differentiating features of Stx2 and Stx2e is their differential cytotoxic activities on HeLa and Vero cells. The latter is substantially more active on Vero cells. Stx2e binds preferentially to globotetrosylceramide (Gb4), another neutral glycolipid, which has a subterminal α-Gal-$(1 \rightarrow 4)$-Gal disaccharide and a terminal *N*-acetylgalactosamine residue. In addition, the cellular cytotoxic activity of Shiga toxins is modulated by the composition of the fatty acids of the receptor lipid ceramide moiety *(79)*. Fatty acid carbon chain length has been found to be particularly important, altering the intracellular uptake pathway of toxin and its subsequent biological activity.

After binding, toxin is internalized by receptor-mediated endocytosis at clathrin-coated pits. Brefeldin A, which interrupts the intracellular movement of vesicles at the Golgi stack, blocks Stx binding, uptake, and proteolysis *(80)*. This suggests that transport of toxin through the Golgi is necessary to reach the ribosomal target *(81,82)*. The requirement for transport of toxin beyond the Golgi to the cytoplasm has been confirmed by the use of ilimaquinone, a sea sponge metabolite that causes the breakdown of Golgi membranes and inhibits the retrograde transport of proteins to the endoplasmic reticulum *(83)*. This interesting substance also inhibits the action of Shiga toxins.

1.5. Shiga Toxin Purification

Initial purification of toxin was fraught with problems related to contamination with LPS. Definitive evidence that the "neurotoxin" was separable from LPS was not obtained until 34 yr after the original report of Shiga toxin, when chemical fractionation was employed *(84)*. Purification of Shiga toxin remained a problem until the early 1980s when small amounts of highly purified toxin were obtained by Olsnes and Eiklid *(85)* and O'Brien et al. *(66)* by ion-exchange column or antibody affinity chromatography, respectively. The first successful large-scale purification was reported by Donohue-Rolfe et al. in 1984 *(54)* using chromatofocusing as the principal technique. Their method resulted in a yield of nearly 1 mg of pure toxin from 3 L of culture. The purified toxin was isoelectric at pH 7.2, and consisted of two peptide bands of approx 7 and 32 kDa. Toxin production was enhanced by low iron concentration. Although O'Brien et al. *(66)* used chelex-treated medium to reduce iron content, Donohue-Rolfe et al. *(54)* found that the iron content of Syncase medium was sufficiently low to support maximum toxin production without the need for further manipulation of the media. In 1989, Donohue-Rolfe et al. *(62)* reported the use of a one-step affinity purification system that was suitable for the purification of all Shiga toxin family members. This involved the coupling of a glycoprotein toxin receptor analogue present in hydatid cyst fluid, the P1-blood group active glycoprotein (P1gp), to Sepharose 4B and using this as an affinity chromatography matrix. Toxin remained tightly bound to immobilized P1gp as 1 M salt or 4 M urea was used to remove nonspecifically bound proteins, but could be eluted with 4.5 M $MgCl_2$. Renatured toxin was fully biologically active, and as much as 10 mg of pure toxin were obtained from a 20-L fermentor culture. When the toxin expressing *E. coli* HB101 H19B (for Stx1) and *E. coli* C600 933W (for Stx2) lysogens are used the levels of toxin in the cultures can be increased significantly. The bacteriophage-inducing drug mitomycin C, when added to the culture, increases the levels of Stx2 expression up to 200 mg/20 L of fermentor culture. The P1gp purification method has been used to successfully purify Stx1, Stx2, 2c, 2d, and 2e. Others have published alternative purification systems for Shiga toxins that include the use of standard and high-performance liquid chromatography (HPLC) *(86)*, and receptor ligand based chromatography *(87)* that is based on the same principal as the P1gp system discussed previously.

2. Materials
2.1. Bacterial Strains
1. *E. coli* HB101 H19B.
2. *E. coli* C600W.

2.2. Media
1. Syncase broth: 5.0 g of Na_2HPO_4, 6.55 g of $K_2HPO_4 \cdot 3H_2O$, 10.0 g of casaminoacids, 1.18 g of NH_4Cl, 89 mg of Na_2SO_4, 42 mg of $MgCl_2$, 4 mg of $MnCl_2 \cdot H_2O$, 2.0 g of glucose, and 40 mg of tryptophan per liter.
2. Luria broth: 10 g of tryptone, 5 g of yeast extract, and 5 g of NaCl per liter.

2.3. Buffers, Stock Solutions, and Other Reagents
1. Hydatid cyst fluid (*see* **Note 1**).
2. Phenol.
3. Ethanol.
4. Cyanogen bromide activated Sepharose.
5. 10 m*M* Tris-HCl, pH 7.4.
6. 0.5 *N* NaCl in 10 m*M* Tris-HCl, pH 7.4.
7. 10 m*M* Phosphate-buffered saline (PBS) pH 7.5.
8. 4.5 *M* $MgCl_2$.
9. 20 m*M* NH_4HCO_3.
10. Mitomycin C.
11. Protease inhibitors: phenymethylsulfonyl fluoride (PMSF) and aprotinin.
12. Ammonium sulfate.
13. Blue Sepharose gel (Pharmacia/LKB).
14. 2% Formalin.
15. 0.13% Crystal violet, 5% ethanol, 2% formalin in PBS.
16. Leucine-free media.
17. [^{3}H]leucine.
18. 0.2 *M* KOH.
19. 10% Trichloroacetic acid.
20. 5% Trichloroacetic acid.
21. 1% Acetic acid.
22. Diethanolamine buffer: 97 mL of diethanolamine, 100 mg of $MgCl_2 \cdot 6H_2O$, and 200 mg of NaN_3 per liter, adjusted to pH 9.8 with HCl.

23. Nonfat powdered milk.
24. Coating buffer: 1.59 g of Na_2CO_3, 2.93 g of $NaHCO_3$, 200 mg NaN_3 per liter.
25. PBS–Tween: 8.0 g of NaCl, 0.2 g of KH_2PO_4, 0.2 g of KCl, 0.5 mL of Tween-20 per liter.
26. Goat anti-rabbit immunoglobulin G (IgG) alkaline phosphatase conjugate.
27. Sigma 104 phosphatase substrate.

2.4. Equipment

1. Shaking platform at 37°C.
2. French pressure cell press or sonicator.
3. Tangential flow membrane apparatus (0.45-μm filter, 10K mol mass cutoff filter) or centrifuge.
4. Liquid chromatography setup: peristaltic pump, fraction collector, UV monitor, and chart recorder.
5. Lyophilizer.
6. Enzyme-linked immunosorbent assay (ELISA) plate reader.
7. Millipore Multiscreen Assay system (plates, punch tip assembly, vacuum, and punch apparatus).
8. Beta counter.

3. Methods

The following descriptions of Shiga toxin 1 and 2 purification are the methods used in our laboratory to purify the two toxins. The strains we use are lysogens of the Stx1 phage (H19B) or Stx2 phage (933W). We have found that these methods are applicable to any Shiga toxin producing bacterial strains and are not restricted to the two strains discussed in the following subsections. The methods also work with Shiga toxin 2 variants (2c, 2d, and 2e).

3.1. Shiga Toxin 1 Purification

3.1.1. Preparation of P1 Glycoprotein Column

1. Crude hydatid cyst material (*see* **Note 1**) is dialyzed extensively (7–10 changes) against distilled water, then lyophilized.
2. Lyophilized material is redissolved in distilled water, then diluted to phenol/water (95:5, v/v).
3. The precipitate is collected, redissolved in water, then reprecipitated in 40% ethanol.
4. The partially purified P1 glycoprotein is bound to cyanogen bromide activated Sepharose according to the manufacturer's recommendations.

3.1.2. Stx1 Purification

1. *E. coli* strain HB101(H19B) is grown in syncase medium *(54)* in 20-L batches for 18 h (*see* **Note 2**). The bacteria are concentrated by passage over a 0.45-μm tangential flow membrane apparatus (Millipore) and a final pellet is collected by centrifugation (*see* **Note 3**).

2. Following a wash in 10 mM PBS, pH 7.5, the pellet is resuspended in 200 mL of the same buffer plus 2 mM PMSF and aprotinin (1:10,000). The bacteria are lysed by passage through an SLM/Aminco French pressure cell press, employing two passes at 24 Kpsi. Cell debris is removed by centrifugation (*see* **Note 4**).

3. Ammonium sulfate is added to the supernatant to 30% saturation and the preparation is held at least 4 h at 4°C. The supernatant is harvested by centrifugation, the ammonium sulfate concentration is increased to 70%, and the preparation is held for 18 h at 4°C. The precipitated proteins are harvested by centrifugation.

4. The pellet is resuspended in 100 mL of 10 mM Tris HCl pH 7.4, and dialyzed thoroughly against the same buffer, at least two changes of 4 L. The dialyzed proteins are then brought to a total volume of 300 mL by the addition of Tris buffer. The solution is loaded onto a 400-mL bed volume Blue Sepharose column (Pharmacia/LKB) at a rate of 60 mL/h followed by an 18-h wash with 10 mM Tris-HCl, pH 7.4, at the same rate (*see* **Note 5**). Crude toxin is eluted with 0.5 M NaCl in the same Tris buffer at a rate of 24 mL/h. The toxin appears in the first peak to elute off the column, which can be detected by monitoring the OD_{280} of eluted fractions or more simply by the appearance of a bright yellow pigment (*see* **Note 6**). The first 80–90 mL are combined and dialyzed against 10 mM PBS, pH 7.5. Following dialysis, the crude toxin is aliquoted into 10 mL lots and frozen at –70°C.

5. Toxin is loaded onto a P1-Sepharose column (*62*) in aliquots previously determined not to exceed the binding capacity of the column. Flow rate during loading is 5 mL/h. The column is washed with 10 mM PBS, pH 7.5, until the OD_{280} returns to baseline (this requires about two column volumes). Nonspecifically bound proteins are then eluted with 1 N NaCl in the PBS buffer, requiring again about two column volumes. Stx is eluted with 4.5 M MgCl$_2$ (*see* **Note 7**). Fractions comprising the eluted peak are pooled and dialyzed extensively (at least four changes of 4 L) against 20 mM ammonium bicarbonate.

6. Toxin concentration is determined by Lowry protein assay. Toxin is aliquoted into vials in 100-µg lots, frozen, and then lyophilized.

3.2. Shiga Toxin 2 Purification

1. *E. coli* strain C600W is grown in 10 L of Luria broth to an OD_{600} of 1.0, at which time the Stx phage is induced by the addition of mitomycin to a final concentration of 250 ng/mL.

2. Following induction, the culture is grown a further 14 h. Bacteria are then removed by passage over a 0.45-µm tangential flow membrane filter.

3. The filtrate, *that is* the culture supernatant, is concentrated by passage over a 10K molecular mass cutoff membrane in a parallel tangential flow membrane apparatus. Volume is reduced from 10 L to 500 mL (*see* **Note 8**).

4. Crude Stx2 material following the above procedure is dialyzed against 10 mM PBS, pH 7.5 and loaded onto the P1-Sepharose and eluted as for Stx1 described in **Subheading 3.1.**

3.3. Toxin Assays

3.3.1. Cytotoxicity Assays

Shiga toxins can be detected in a variety of ways. The use of tissue culture cytotoxicity assays is one of the classic methods and has the advantage of sensitivity. The disadvantage of cytotoxicity assays is that they require tissue culture, are relatively expensive because of the requirement for tissue culture plates and fetal calf serum in the medium, are time consuming, and require confirmation of specificity by neutralization with specific anti-Stx antibodies. Vero or HeLa cells are used most frequently for these assays. When cells are exposed to the toxin there are morphological changes, with detachment of the monolayer when the cells are killed. This can be assessed subjectively or by counting using microscopy or quantitated using vital cell stains such as crystal violet *(88)*. Measuring the inhibition of radiolabeled amino acid incorporation into protein is another approach, which is actually directly measuring the enzymatic activity of the toxins on protein synthesis, rather than cell detachment or death. This has the advantage of being faster in that the time of toxin exposure can be shorter (as little as 3 h in some cases), but the disadvantages of being somewhat cumbersome and costly, requiring use of radioactive isotopes and a need for radioactivity spectrometers *(89)*.

3.3.1.1. CRYSTAL VIOLET METHOD

1. Seed HeLa or Vero cells on 96-well tissue culture plates at a cell density of 10,000 cells per well.
2. Add purified toxin standards and samples to wells. Incubate at 37°C, 5% CO_2 for 18 h.
3. Wash monolayer 2× with PBS.
4. Add 100 μL of 2% formalin in PBS to each well. Hold for 1 min. Discard formalin.
5. Add 100 μL of a solution containing 0.13% crystal violet, 5% ethanol, and 2% formalin in PBS to each well. Hold for 20 min.
6. Wash 2× with tap water until blue color stops running off plate. Remove excess water by tapping inverted plates on a paper towel.
7. Add 100 mL of 95% ethanol per well.
8. Read OD 595 nm. Calculate OD as percent of control (media only).

3.3.1.2. LEUCINE INCORPORATION METHOD

1. Seed wells, add toxin standards and samples, and incubate 3–8 h as described in **Subheading 3.3.1.1.**
2. Wash wells 2× with 200 μL of PBS.
3. Add [^{3}H]leucine in leucine-free medium (1:200) at 100 μL/well. Incubate for 60 min at 37°C, 5% CO_2.
4. Wash wells 2× with PBS. Add 0.2 *M* KOH, 50 mL/well. Hold for 15 min at 20–22°C.
5. Add 10% trichloroacetic acid, 150 μL/well. Hold at 20–22°C for 30 min.

6. With a multichannel pipettor, transfer well contents to a Millipore multiscreen assay plate attached to the Millipore vacuum device (*see* **Note 9**).
7. Rinse wells 2× with 5% trichloroacetic acid, 200 μL/well, transferring well contents to the multiscreen plate with each rinse.
8. Rinse filters in multiscreen plate with 200 μL of 5% trichloroacetic acid.
9. Wash filters with 200 μL of 1% acetic acid.
10. Remove filter backing from multiscreen plate. Dry under heat lamp. Punch filters out into scintillation vials using Millipore multiscreen punch tip assembly and apparatus.
11. Add 4 mL of scintillation fluid per vial. Read in beta counter.

3.3.2. Immunoassays for Toxin (see **Note 10**)

A variety of enzyme immunoassays are also available for the detection and quantitation of Shiga toxins, although purified toxin is needed as standards for quantitative assays. Various capture systems have been described including the use of monoclonal or polyclonal antibodies, hydatid cyst material containing the P1 glycoprotein *(90)*, and the glycolipids Gb3 and Gb4 *(91–93)*.

3.3.2.1. ELISA METHOD: PREPARATION OF PLATES

1. Plates can be prepared in advance and stored at 4°C for months; we usually prepare 10 plates at a time. Our preferred brand is NUNC MaxiSorb Immunoplate.
2. Prepare coating buffer, at least 50 mL per plate. You will need 10 mL/plate (96 wells at 100 μL per well) for the capture agent and 40 mL/plate for the blocking agent.
3. Prepare a solution of capture agent (either 10 μg/mL of P1 glycoprotein or anti-Stx monoclonal antibody diluted, 1:1500–2000) in coating buffer. Add 100 μL to each well. Hold for 2–3 h at 20–22°C or at 4°C for 18 h.
4. Empty wells. Fill with 400 μL of 5% nonfat powdered milk in coating buffer. Hold for 2–3 h at 20–22°C. Rinse 5× with PBS–Tween. Blot on paper towel and air dry. Cover with parafilm. Store at 4°C.

3.3.2.2. ELISA METHOD: ASSAY

1. Prepare dilutions of standards and samples in PBS–Tween. For the standards, a stock Stx solution is diluted in twofold steps from 20 ng/mL to 0.16 ng/mL. Samples are generally prepared in a range of three 10-fold dilutions designed to span the expected concentration.
2. Samples and standards are plated in at least duplicate, 100 μL per well. Hold for 30 min at 20–22°C.
3. Rinse 5× with PBS–Tween, 200 μL/well. Prepare rabbit anti-Stx polyclonal antibody (1:5000–7500) in PBS–Tween. Add 100 μL/well. Hold for 30 min.
4. Rinse 5× with PBS–Tween, 200 μL/well. Prepare goat anti-rabbit IgG alkaline phosphatase conjugate (Sigma) in PBS–Tween (1:2000). Add 100 μL/well. Hold for 30 min.

5. Rinse 5× with PBS–Tween, 200 μL/well. Prepare Sigma 104 phosphatase substrate, 1 mg/mL in diethanolamine buffer, pH 9.8. Add 100 μL/well. Hold for 30 min.
6. Read on ELISA plate reader at A_{405}.

Alternative methods include the use of genetic techniques to detect Stx or STEC including polymerase chain reaction (PCR) or gene probe methodologies. These methods are discussed extensively in a recent review by Paton and Paton *(24)*. STEC colonies can be detected in mixed clinical samples using a colony immunoblot assay in which STECs are plated onto a pair of nitrocellulose filters placed on a culture plate. Overnight culture will result in colonies growing on the top filter that produce enough toxin to leach into the lower filter. The lower filter is then probed with anti-Stx antibodies. Any positive spots can then be matched up with the original colonies on the top filter to isolate the strain and confirm the findings on a pure culture *(94)*.

4. Notes

1. Hydatid cyst fluid is obtained from cysts in animals infected with *Echinoccocus granulosus.* This is common in Australia and parts of South America.
2. Toxin yield from a culture grown in syncase medium is at least fourfold higher than from a Luria broth culture at an equivalent cell density. If syncase medium is not available, simply use a larger volume of Luria broth for the starting material.
3. The tangential flow membrane apparatus is used as a rapid harvest technique for large-volume cultures. Centrifugation alone is a time consuming but perfectly acceptable alternative. The entire culture volume should be held at 4°C while centrifugation is in process.
4. Sonication is an alternative to lysis by French press.
5. Although we use a peristaltic pump to set buffer flow rates, the same parameters can be achieved by proper adjustment of reservoir and outlet heights in a gravity feed setup.
6. In the absence of a UV monitor, the initial peak can be detected by the appearance of a bright yellow pigment.
7. Toxin elutes with the $MgCl_2$ front. In the absence of a UV monitor, the front may be followed as it travels down the column by the change in appearance of the gel (opaque white to transparent grey) and by the increase in density of the collected fractions. Begin pooling with the fraction containing the first signs of high salt.
8. If an ultrafiltration apparatus is not available, crude toxin can be harvested from the supernatant by ammonium sulfate precipitation. The supernatant can be brought to 70% saturation and the precipitated proteins harvested by centrifugation.
9. If such a device is not available any filter that will "catch" trichloroacetic acid (TCA) precipitated protein can be used.
10. Currently there are two commercially available enzyme immunoassays available in kit form (Premier EHEC [Meridian Diagnostics, Cincinnati, O] and ProSpect

T Shiga toxin *E. coli* [Alexon-Trend, San Jose, CA]). Both are simple to use and detect Stx1 and 2. These enzyme immunoassays are designed for the detection of Stx in stool or food samples either directly or following culture amplification. However, both assays can be used to detect Shiga toxins 1 and 2 under many other conditions.

References

1. Shiga, K. (1898) Ueber den Dysenteriebacillus (*Bacillus dysenteriae*). *Zbl. Bakt. Parasit. Abt. 1 Orig.* **24,** 817–824.
2. Conradi, H. (1903) Über lösliche, durch aseptische Autolyse erhaltene Giftstoffe von Ruhr- und Typhus-Bazillen. *Dtsch. Med. Wochenschr.* **20,** 26–28.
3. Koster. F., Levin. J., Walker, L., Tung, K. S., Gilman, R. H., Rahaman, M. M., Majid, M. A., Islam, S., and Williams, R. C., Jr. (1978) Hemolytic-uremic syndrome after shigellosis. *N. Engl. J. Med.* **298,** 927–933.
4. Konowalchuk, J., Speirs, J. I., and Stavric, S. (1977) Vero response to a cytotoxin of *Escherichia coli. Infect. Immun.* **18,** 775–779.
5. Konowalchuk, J., Dickie, N., Stavric, S., and Speirs, J. I. (1978) Properties of an *Escherichia coli* cytotoxin. *Infect. Immun.* **20,** 575–577.
6. Wade, W. G., Thom, B. T., and Evens, N. (1979) Cytotoxic enteropathogenic *Escherichia coli. Lancet* **ii,** 1235–1236.
7. Scotland, S. M., Day, N. P., and Rowe, B. (1979) Production by strains of *Escherichia coli* of a cytotoxin (VT) affecting Vero cells. *Soc. Gen. Microbiol. Quart.* **6,** 156–157.
8. Wilson, M. W. and Bettelheim, K. A. (1980) Cytotoxic *Escherichia coli* serotypes. *Lancet* **ii,** 201.
9. Riley, L. W., Temis, R. S., Helgerson, S. D., McGee, H. B., Wells, J. G., Davis, B. R., Hebert, R. J., Olcott, E. S., Johnson, L. M., Hargrett, N. T., Blake, P. A., and Cohen, M. L. (1983) Hemorrhagic colitis associated with a rare *Escherichia coli* serotype. *N. Engl. J. Med.* **308,** 681–685.
10. Karmali, M. A., Steele, B. T., Petric, M., and Lim, C. (1983) Sporadic cases of hemolytic-ureamic syndrome associated with faecal cytotoxin and cytotoxin-producing *Escherichia coli* in stools. *Lancet* **ii,** 619–620.
11. Acheson, D. W. K. and Keusch, G. T. (1996) Which Shiga toxin-producing types of *E. coli* are important? *ASM News* **62,**302–306.
12. Calderwood, S. B., Acheson, D. W. K., Keusch, G. T., Barrett, T. J., Griffin, P. M., Swaminathan, B., Kaper, J. B., Levine, M. M., Kaplan, B. S., Karch, H., O'Brien, A. D., Obrig, T. G., Takeda, Y., Tarr, P. I., and Wachsmuth, I. K. (1996) Proposed new nomenclature for SLT (VT) family. *ASM News* **62,** 118–119.
13. Keusch, G. T., Grady, G. F., Mata, L. J., and McIver, J. (1972) The pathogenesis of *Shigella* diarrhea. I. Enterotoxin production by *Shigella dysenteriae* 1. *J. Clin. Invest.* **51,** 1212–1218.
14. Keusch, G. T., Grady, G. F., Takeuchi, A., and Sprinz, H. (1972) The pathogenesis of *Shigella* diarrhea II. Enterotoxin induced acute enteritis in the rabbit ileum. *J. Infect. Dis.* **126,** 92–95.

15. Moyer, M. P., Dixon, P. S., Rothman, S. W., and Brown, J. E. (1987) Cytotoxicity of Shiga toxin for primary cultures of human colonic and ileal epithelial cells. *Infect. Immun.* **55**, 1533–1535.

16. Fontaine, A., Arondel, J., and Sansonetti, P. J. (1988) Role of Shiga toxin in the pathogenesis of bacillary dysentery, studied by using a Tox-mutant of *Shigella dysenteriae* 1. *Infect. Immun.* **56**, 3099–3109.

17. Keusch, G. T. and Acheson, D. W. K. (1997) Thrombotic thrombocytopenic purpura associated with Shiga toxins. *Semin. Hematol.* **34**, 106–116.

18. Karmali, M. A., Petric, M., Lim, C., Fleming, P. C., Arbus, G. A., and Lior, H. (1985) The association between idiopathic hemolytic uremic syndrome and infection by verotoxin producing *Escherichia coli*. *J. Infect. Dis.* **151**, 775–782.

19. Karmali, M. A. (1989) Infection by verocytotoxin-producing *Escherichia coli*. *Microbiol. Rev.* **2**, 15–38.

20. Park, C. H., Gates, K. M., and Hixon, D. L. (1996) Isolation of Shiga-like toxin producing *Escherichia coli* (O157 and non-0157) in a community hospital. *Diagn. Microbiol. Infect. Dis.* **26**, 69–72.

21. Acheson, D. W. K., Frankson, K., and Willis, D. (1998) Multicenter prevalence study of Shiga toxin-producing *Escherichia coli*. Annual meeting of the American Society for Microbiology, Atlanta, GA, USA, C-205.

22. Tschäpe, H., Prager, R., Steckel, W., Fruth, A., Tietze, E., and Böhme, G. (1995) Verotoxinogenic *Citrobacter freundii* associated with severe gastroenteritis and cases of haemolytic uraemic syndrome in a nursery school. Green butter as the infection source. *Epidemiol. Infect.* **114**, 441–450.

23. Paton, A. W. and Paton, J. C. (1996) *Enterobacter cloaceae* producing a Shiga-like toxin II-related cytotoxin associated with a case of hemolytic-uremic syndrome. *J. Clin. Microbiol.* **34**, 463–465.

24. Paton, J. C. and Paton, A. W. (1998) Pathogenesis and diagnosis of Shiga toxin-producing *Escherichia coli* infections. *Clin. Microbiol. Rev.* **11**, 450–479.

25. Nataro, J. P. and Kaper, J. B. (1998) Diarrheagenic *Escherichia coli*. *Clin. Microbiol. Rev.* **11**, 142–201.

26. Acheson, D. W. K., Moore, R., DeBreuker, S., Lincicome, L., Jacewicz, M., Skutelsky, E., and Keusch, G. T. (1996) Translocation of Shiga-like toxins across polarized intestinal cells in tissue culture. *Infect. Immun.* **64**, 3294–3300.

27. Donowitz, M., Keusch, G. T., and Binder, H. J. (1975) Effect of Shigella enterotoxin on electrolyte transport in rabbit ileum. *Gastroenterology* **69**, 1230–1237.

28. Kandel, G., Donohue-Rolfe, A., Donowitz, M., and Keusch, G. T. (1989) Pathogenesis of *Shigella* diarrhea XVI. Selective targetting of Shiga toxin to villus cells of rabbit jejunum explains the effect of the toxin on intestinal electrolyte transport. *J. Clin. Invest.* **84**, 1509–1517.

29. Obrig, T. G., Del Vecchio, P. J., Brown, J. E., Moran, T. P., Rowland, B. M., Judge, T. K., and Rothman, S. W. (1988) Direct cytotoxic action of Shiga toxin on human vascular endothelial cells. *Infect. Immun.* **56**, 2372–2378.

30. Louise, C. B. and Obrig, T. G. (1991) Shiga toxin associated hemolytic uremic syndrome: combined cytotoxic effects of Shiga toxin, IL-1, and tumor necrosis factor alpha on human vascular endothelial cells *in vitro*. *Infect. Immun.* **59**, 4173–4179.

31. Obrig, T. G., Louise, C. B., Lingwood, C. A., Boyd, B., Barley-Maloney, L., and Daniel, T. O. (1993) Endothelial heterogeneity in Shiga toxin receptors and responses. *J. Biol. Chem.* **268,** 15,484–15,488.

32. Monnens, L., Savage, C. O., and Taylor, C. M. (1998) Pathophysiology of hemolytic-uremic syndrome, in Escherichia coli *O157:H7 and Other Shiga Toxin-Producing* E. coli *Strains* (Kaper, J. B. and O'Brien, A. D., eds.), American Society for Microbiology, Washington, DC, pp. 287–292.

33. Hutchinson, J. S., Stanimirovic, D., Shapiro, A., and Armstrong, G. D. (1998) Shiga toxin toxicity in human cerebral endothelial cells, in Escherichia coli *O157:H7 and Other Shiga Toxin-Producing* E. coli *Strains* (Kaper, J. B. and O'Brien, A. D., eds.), American Society for Microbiology, Washington, DC, pp. 323–328.

34. King, A. J., Sundaram, S., Cendoroglo, M., Acheson, D. W. K., and Keusch, G. T. (1999) Shiga toxin induces superoxide production in polymorphonuclear cells with subsequent impairment of phagocytosis and responsiveness to phorbol esters. *J. Infect. Dis.* **179,** 503–507.

35. Tesh, V. L., Ramegowda, B., and Samuel, J. E. (1994). Purified Shiga-like toxins induce expression of proinflammatory cytokines from murine peritoneal macrophages. *Infect. Immun.* **62,** 5085–5094.

36. Ramegowda, B. and Tesh, V. L. (1996) Differentiation-associated toxin receptor modulation, cytokine production, and sensitivity to Shiga-like toxins in human monocytes and monocytic cell lines. *Infect. Immun.* **64,** 1173–1180.

37. Van Setten, P. A., Monnens, L. A. H., Verstraten, G., van den Heuvel, L. P., and van Hinsberg, V. W. (1996) Effects of verocytotoxin–1 on nonadherent human monocytes: binding characteristics, protein synthesis and induction of cytokine release. *Blood* **88,** 174–183.

38. Newland, J. W., Strockbine, N. A., Miller, S. F., O'Brien, A. D., and Holmes, R. K. (1985) Cloning of Shiga-like toxin structural genes from a toxin converting phage of *Escherichia coli. Science* **230,** 179–181.

39. O'Brien, A. D., Marques, L. R., Kerry, C. F., Newland, J. W., and Holmes, R. K. (1989) Shiga-like toxin converting phage of enterohemorrhagic *Escherichia coli* strain 933. *Microb. Pathogen.* **6,** 381–390.

40. Weinstein, D. L., Jackson, M. P., Perera, L. P., Holmes, R. K., and O' Brien, A. D. (1989) *In vivo* formation of hybrid toxins comprising Shiga toxin and the Shiga-like toxins and a role of the B subunit in localization and cytotoxic activity. *Infect. Immun.* **57,** 3743–3750.

41. Smith, H. W. and Linggood, M. A. (1971) The transmissible nature of enterotoxin production in a human enteropathogenic strain of *Escherichia coli. J. Med. Microbiol.* **4,** 301–305.

42. Calderwood, S. B., Auclair, F., Donohue-Rolfe, A., Keusch, G. T., and Mekalanos, J. J. (1987) Nucleotide sequence of the Shiga-like toxin genes of *Escherichia coli. Proc. Natl. Acad. Sci.* USA **84,** 4364–4368.

43. De Grandis, S., Law, H., Brunton, J., Gyles, C., and Lingwood, C. A. (1989) Globotetrosylceramide is recognized by the pig edema disease toxin. *J. Biol. Chem.* **264,** 12,520–12,525.

44. Jackson, M. P., Neill, R. J., O'Brien, A. D., Holmes, R. K., and Newland, J. W. (1987) Nucleotide sequence analysis and comparison of the structural genes for Shiga-like toxin I and Shiga-like toxin II encoded by bacteriophages from *Escherichia coli* 933. *FEMS Microbiol. Lett.* **44,** 109–114.

45. Jackson, M. P., Newland, J. W., Holmes, R. K., and O'Brien, A. D. (1987) Nucleotide sequence analysis of the structural genes for Shiga-like toxin I encoded by bacteriophage 933J from *Escherichia coli. Microb. Pathogen.* **2,** 147–153.

46. Kozlov, Y. V., Kabishev, A. A., Lukyanov, E. V., and Bayev, A. A. (1988) The primary structure of the operons coding for *Shigella dysenteriae* toxin and temperate phage H30 Shiga-like toxin. *Gene* **67,** 213–221.

47. Strockbine, N. A., Jackson, M. P., Sung, L. M., Holmes, R. K., and O'Brien, A. D. (1988) Cloning and sequencing of the genes for Shiga toxin from *Shigella dysenteriae* type 1. *J. Bacteriol* **170,** 1116–1122.

48. Calderwood, S. B. and Mekalanos, J. J. (1987) Iron regulation of Shiga-like toxin expression in *Escherichia coli* is mediated by the *fur* locus. *J. Bacteriol.* **169,** 4759–4764.

49. Mühldorfer, I., Hacker, J., Keusch, G. T., Acheson, D. W. K., Tschäpe, H., Kane, A. V., Ritter, A., Olschlager, T., and Donohue-Rolfe, A. (1996) Regulation of the Shiga-like toxin II operon in *Escherichia coli. Infect. Immun.* **64,** 495–502.

50. Neely, M. N. and Friedman, D. I. (1998) Function and genetic analysis of regulatory regions of coliphage H–19B: location of shiga-like toxin and lysis genes suggest a role for phage functions in toxin release. *Mol. Microbiol.* **28,** 1255–1267.

51. Newland, J. W., Strockbine, N. A., and Neill, R. J. (1987) Cloning of genes for production of *Escherichia coli* Shiga-like toxin II. *Infect. Immun.* **55,** 2675–2680.

52. Melton-Celsa, A. R. and O'Brien, A. D. (1998) Structure, biology, and relative toxicity of Shiga toxin family members for cells and animals, in Escherichia coli *O157:H7 and Other Shiga Toxin-Producing* E. coli *Strains* (Kaper, J. B. and O'Brien, A. D., eds.), American Society for Microbiology, Washington, DC, pp. 121–128.

53. Weinstein, D. L., Jackson, M. P., Samuel, J. E., Holmes, R. K., and O'Brien, A. D. (1988) Cloning and sequencing of a Shiga-like toxin type II variant from an *Escherichia coli* strain responsible for edema disease of swine. *J. Bacteriol.* **170,** 4223–4230.

54. Donohue-Rolfe, A., Keusch, G. T., Edson, C., Thorley-Lawson, D., and Jacewicz, M. (1984) Pathogenesis of *Shigella* diarrhea. IX. Simplified high yield purification of Shigella toxin and characterization of subunit composition and function by the use of subunit specific monclonal and polyclonal antibodies. *J. Exp. Med.* **160,** 1767–1781.

55. Reisbig, R., Olsnes, S., and Eiklid, K. (1981) The cytotoxic activity of *Shigella* toxin. Evidence for catalytic inactivation of the 60S ribosomal subunit. *J. Biol. Chem.* **256,** 8739–8744.

56. Donohue-Rolfe, A., Jacewicz, M., and Keusch, G. T. (1989) Isolation and characterization of functional Shiga toxin subunits and renatured holotoxin. *Mol. Microbiol.* **3,** 1231–1236.

57. Mobassaleh, M., Donohue-Rolfe, A., Jacewicz, M., Grand, R. J., and Keusch, G. T. (1988) Pathogeneis of Shigella diarrhea: evidence for a developmentally regulated glycolipid receptor for Shigella toxin involved in the fluid secretory response of rabbit small intestine. *J. Infect. Dis.* **157**, 1023–1031.

58. Stein, P. E., Boodhoo, A., Tyrrell, G. J., Brunton, J. L., and Read, R. J. (1992) Crystal structure of the cell-binding B oligomer of verotoxin–1 from *E. coli. Nature* **355**, 748–750.

59. Fraser, M. E., Chernaia, M. M., Kozlov, Y. V., and James, M. N. (1994) Crystal structure of the holotoxin from *Shigella dysenteriae* at 2.5 Å resolution. *Nat. Struct. Biol.* **1**, 59–64.

60. Austin, P. R., Jablonski, P. E., Bohach, G. A., Dunker, A. K., and Hovde, C. J. (1994) Evidence that the A2 fragment of Shiga-like toxin type I is required for holotoxin integrity. *Infect. Immun.* **62**, 1768–1775.

61. Mangeney, M., Richard, Y., Coulard, D., Tursz, T., and Wiels, J. (1991) CD77: an antigen of germinal center B cells entering apoptosis. *Eur. J. Immunol.* **21**, 1131–1140.

62. Donohue-Rolfe, A., Acheson, D. W. K., Kane, A. V., and Keusch, G. T. (1989) Purification of Shiga toxin and Shiga-like toxin I and II by receptor analogue affinity chromatography with immobilized P1 glycoprotein and the production of cross-reactive monoclonal antibodies. *Infect. Immun.* **57**, 3888–3893.

63. Schmitt, C. K., McKee, M. L., and O'Brien, A. D. (1991) Two copies of Shiga-like toxin II-related genes common in enterohemorrhagic *Escherichia coli* strains are responsible for the antigenic heterogeneity of the O157:H- strain E32511. *Infect. Immun.* **59**, 1065–1073.

64. Lindgren, S. W., Samuel, J. E., Schmitt, C. K., and O'Brien, A. D. (1994) The specific activities of Shiga-like toxin type II (SLT-II) and SLT-II-related toxins of enterohemorrhagic *Escherichia coli* differ when measured by Vero cell cytotoxicity but not by mouse lethality. *Infect. Immun.* **62**, 623–631.

65. Brown, J. E., Ussery, M. A., Leppla, S. H., and Rothman, S. W. (1998) Inhibition of protein synthesis by Shiga toxin. Activation of the toxin and inhibition of peptide elongation. *FEBS. Lett.* **117**, 84–88.

66. O'Brien, A. D., LaVeck, G. D., Griffin, D. E., and Thompson, M. R. (1980) Characterization of *Shigella dysenteriae* 1 (Shiga) toxin purified by anti-Shiga toxin affinity chromatography. *Infect. Immun.* **30**, 170–179.

67. Olsnes, S., Reisbig, R., and Eiklid, K. (1981) Subunit structure of Shigella cytotoxin. *J. Biol. Chem.* **256**, 8732–8788.

68. Endo, Y. and Tsurugi, K. (1987) RNA *N*-glycosidase activity of ricin A-chain. Mechanism of actin of the toxic lectin ricin on eukaryotic ribosomes. *J. Biol. Chem.* **262**, 8128–8230.

69. Igarashi, K., Ogasswara, T., Ito, K., Yutsudo, T., and Takada, Y. (1987) Inhibition of elongation factor 1-dependent aminoacyl-tRNA binding to ribosomes by Shiga-like toxin I (VT1) from *Escherichia coli* O157:H7 and by Shiga toxin. *FEMS Microbiol. Lett.* **44**, 91–94.

70. Endo, Y., Tsurgi, K., Yutsudo, T., Takeda, Y, Igasawara, T., and Igarashi, E. (1988) Site of action of A Vero toxin (VT2) from *Escherichia* coli O157:H7 and of Shiga toxin on eukaryotic ribosomes. *Eur. J. Biochem.* **171,** 45–50.

71. Calderwood, S. B., Acheson, D. W. K., Goldberg, M. B., Boyko, S. A., and Donohue-Rolfe, A. (1990) A system for production and rapid purification of large amounts of Shiga toxin/Shiga-like toxin I B subunit. *Infect. Immun.* **58,** 2977–2982.

72. Acheson, D. W. K., Calderwood, S. B., Boyko, S. A., Lincicome, L. L., Kane, A. V., Donohue-Rolfe, A., and Keusch, G. T. (1993) A comparison of Shiga-like toxin I B subunit expression and localization in *Escherichia coli* and *Vibrio cholerae*, using *trc* or iron-regulated promoter systems. *Infect. Immun.* **61,** 1098–1104.

73. Ramotar, K., Boyd, B., Tyrrell, G., Gariepe, J., Lingwood, C., and Brunton, J. (1990) Characterization of Shiga-like toxin I B subunit purified from over-producing clones of the SLT-I B cistron. *Biochemistry* **272,** 805–811.

74. Acheson, D. W. K., DeBreucker, S. A., Jacewicz, M., Lincicome, L. L., Donhue-Rolfe, A., Kane, A. V., and Keusch, G. T. (1995) Expression and purification of Shiga-like toxin II B subunits. *Infect. Immun.* **63,** 301–308.

75. Keusch, G. T. and Jacewicz, M. (1977) Pathogenesis of *Shigella* diarrhea. VII. Evidence for a cell membrane toxin receptor involving β1-4 linked *N*-acetyl-D-glucosamine oligomers. *J. Exp. Med.* **146,** 535–546.

76. Jacewicz, M., Clausen, H., Nudelman, E., Donohue-Rolfe, A., and Keusch, G. T. (1986) Pathogenesis of *Shigella* diarrhea. XI. Isolation of a shigella toxin-binding glycolopid from rabbit jejunum and HeLa cells and its identification as globotriosylceramide. *J. Exp. Med.* **163,** 1391–1404.

77. Mobassaleh, M., Gross, S. K., McCluer, R. H., Donohue-Rolfe, A., and Keusch, G. T. (1989) Quantitation of the rabbit intestinal glycolipid receptor for Shiga toxin. Further evidence for the developmental regulation of globotriaosylceramide in microvillus membranes. *Gastroenterology* **97,** 384–391.

78. Lindberg, A. A., Brown, J. E., Stromberg, N., Westling-Ryd, M., Schultz, J. E., and Karlson, K. (1987) Identification of the carbohydrate receptors for Shiga toxin produced by *Shigella dysenteriae* type1. *J. Biol. Chem.* **262,** 1779–1785.

79. Kiarash, A., Boyd, G., and Lingwood, C. A. (1994) Glycosphingolipid receptor function is modified by fatty acid content. Verotoxin 1 and verotoxin 2c preferentially recognize different globotriaosyl ceramide fatty acid homologues. *J. Biol. Chem.* **269,** 1139–1146.

80. Garred, O., Dubinina, E., Holm, P. K., Olsnes, S., van Deurs, B., Kozlov, J. V., and Sandvik, K. (1995) Role of processing and intracellular transport for optimal toxicity of Shiga toxin and toxin mutants. *Exp. Cell Res.* **218,** 39–49.

81. Sandvig, K., Prydz, K., Ryd, M., and van Deurs, B. (1991) Endocytosis and intracellular transport of the glycolipid-binding ligand Shiga toxin in polarized MDCK cells. *J. Cell Biol.* **113,** 553–562.

82. Sandvig, K. and van Deurs, B. (1996) Endocytosis, intracellular transport, and cytotoxic action of Shiga toxin and ricin. *Physiol. Rev.* **76,** 949–966.

83. Nambiar, M. P. and Wu, H. C. (1995) Ilimaquinone inhibits the cytotoxicities of ricin, diphtheria toxin, and other protein toxins in Vero cells. *Exp. Cell Res.* **219,** 671–678.

84. Boivin, A. and Mesrobeanu, L. (1937) Recherches sur les toxines des bacilles dysentériques. Sur l'identité entre la toxine thermobile de Shiga et l'exotoxine presente dans les filtratés des cultures sur bouillon de la même bactérie. *C. Soc. Biol.* **126,** 323–325.

85. Olsnes, S. and Eiklid, K. (1980) Isolation and characterization of *Shigella shigae* cytotoxin. *J. Biol. Chem.* **255,** 284–289.

86. Noda, M., Yutsudo, Nakabayashi, N., Hirayama, T., and Takeda, Y. (1987) Purification and some properties of Shiga-like toxin from *Escherichia coli* O157:H7 that is immunologically identical to Shiga toxin. *Microb. Pathogen.* **2,** 339–349.

87. Ryd, M., Alfredsson, H., Blomberg, L,. Andersson, A., and Lindberg, A. A. (1989) Purification of Shiga toxin by alpha-D-galactose-(1-4)-beta-D-galactose-(1-4)-beta-D-glucose- receptor ligand-based chromatography. *FEBS Lett.* **258,** 320–322.

88. Gentry, M. K. and Dalrymple, J. M. (1980) Quantitative microtitre cytotoxicity assay for *Shigella* toxin. *J. Clin. Microbiol.* **12,** 361–366.

89. Keusch, G. T., Donohue-Rolfe, A., Jacewicz, M., and Kane, A. V. (1988) Shiga toxin: production and purification. *Methods Enzymol.* **165,** 152–162.

90. Acheson, D. W. K., Keusch, G. T., Lightowers, M., and Donohue-Rolfe, A. (1990) Enzyme linked imunosorbent assay for Shiga toxin and Shiga-like toxin II using P1 glycoprotein from hydatid cysts. *J. Infect. Dis.* **161,** 134–137.

91. Ashkenazi, S. and Cleary, T. G. (1989) Rapid method to detect Shiga toxin and Shiga-like toxin I based on binding to globotriaosylceramide (Gb3), their natural receptor. *J. Clin. Microbiol.* **27,** 1145–1150.

92. Ashkenazi, S. and Cleary, T. G. (1990) A method for detecting Shiga toxin and Shiga-like toxin I in pure and mixed culture. *J. Med. Microbiol.* **32,** 255–261.

93. Acheson, D. W. K., Jacewicz, M., Kane, A. V., Donohue-Rolfe, A., and Keusch, G. T. (1993) One step high yield purification of Shiga-like toxin II variants and quantitation using enzyme linked immunosorbent assays. *Microb. Pathogen.* **14,** 57–66.

94. Hull, A. E., Acheson, D. W. K., Echeveria., P., Donohue-Rolfe, A., and Keusch, G. T. (1993) Mitomycin C immunoblot colony assay for the detection of Shiga-like toxin producing *E. coli* in fecal samples: a comparison with DNA probes. *J. Clin. Microbiol.* **31,** 1167–1172.

4

Isolation and Detection of Microcystins and Nodularins, Cyanobacterial Peptide Hepatotoxins

Jussi Meriluoto, Linda Lawton, and Ken-ichi Harada

1. Introduction: *Toxin Producers, Structure, and Nomenclature*

Some species and strains within the freshwater cyanobacterial (blue-green algal) genera *Microcystis*, *Oscillatoria* (also known as *Planktothrix*), *Anabaena*, and *Nostoc* are known to produce cyclic heptapeptide liver toxins, microcystins. Closely related toxic cyclic pentapeptides, nodularins, have been isolated from the brackish water cyanobacterium *Nodularia spumigena*. In addition, some cyanobacteria produce neurotoxic compounds such as anatoxin-a, anatoxin-a(s), and saxitoxins. Recent reviews of the chemical and biological properties of microcystins, nodularins, and other cyanobacterial toxins are given in **refs.** *(1–4)*.

Most microcystins and nodularins are potent hepatotoxins and tumor promoters with an approximate LD_{50} value of 50–500 µg/kg (mouse, i.p.) *(3,5,6)*. The toxic mechanism of microcystins and nodularins is the inhibition of protein phosphatases 1 and 2A *(7)*. The general structure of microcystins (**Fig. 1**) is cyclo(-D-Ala-L-**aa'**-D-*erythro*-β-methylAsp(*iso*)-L-**aa''**-Adda-D-Glu(*iso*)-*N*-methyldehydroAla) where Adda stands for 3-amino-9-methoxy-2,6,8-trimethyl-10-phenyldeca-4(*E*),6(*E*)-dienoic acid *(1,8,9)*. The main structural variation of microcystins is seen in positions 2 (designated as **aa'**) and 4 (**aa''**) which are indicated by a two-letter suffix. For example, the most common microcystin containing leucine and arginine is called microcystin-LR. Variation other than in the residues 2 or 4 is usually described by a prefix with the residue number, *for example,* [D-Asp³]microcystin-LR (D-aspartic acid instead of D-methylaspartic acid). Typical variations include demethylations and changes in the stereochemistry and methylation degree of the Adda residue.

From: *Methods in Molecular Biology, vol. 145: Bacterial Toxins: Methods and Protocols*
Edited by: O. Holst © Humana Press Inc., Totowa, NJ

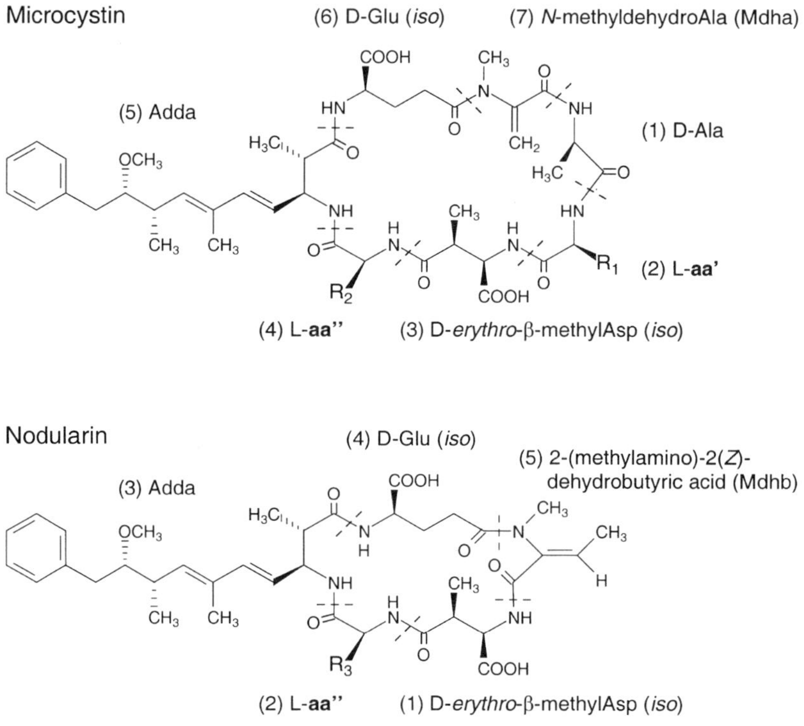

Fig. 1. General structures of microcystin and nodularin.

The total number of microcystins characterized thus far exceeds 60. The general structure of nodularins (**Fig. 1**) is cyclo(-D-*erythro*-β-methylAsp(*iso*)-L-**aa''**-Adda-D-Glu(*iso*)-2-(methylamino)-2(Z)-dehydrobutyric acid) although variability in parts other than **aa''** has also been observed *(1,9)*. The most common variant is nodularin-R with an arginine residue. Only a few microcystins and nodularins are available commercially (e.g., from Sigma [St. Louis, MO], Calbiochem [La Jolla, CA], or Alexis Corporation [San Diego, CA]). Purification procedures for some toxins are given in **Subheadings 2.1.–3.13**. These protocols can be modified fairly easily for other microcystins and nodularins.

The chapter introduces the reader to several chemical, biochemical, and biological methods for microcystin and nodularin detection. These different methods demonstrate an array of sensitivities ranging from microgram detection in bioassays through nanogram detection in high-performance liquid chromatography (HPLC) to picogram sensitivity in protein phosphatase inhibition assays and immunoassays. The chosen procedures are robust basic techniques that have been in use in several laboratories for many years. None of these methods are perfect in all aspects, nor can they be applied to all sample types such as water, cyanobacterial material, tissue samples, etc. Therefore we encourage that they are used in suitable combinations (e.g., a chemical technique followed by confirmation using a bioassay). References to some of the newest, not so thoroughly evaluated techniques are given in **Notes**.

2. Materials

All cyanobacterial material and purified toxins should be handled with great caution because of a risk of serious adverse acute and long-term effects. The major exposure routes include intraperitoneal, intravenous, and peroral exposure as well as exposure by inhalation. All solvents (including water) and salts should be of analytical grade except for the HPLC solvents, which should be of chromatographic grade. Buffer solutions for HPLC should be filtered through 0.45-μm filters. Chromatographic methods require access to chromatographic equipment but only special equipment is listed. The purified toxins should be kept refrigerated for short-term storage (weeks) and preferably at –20°C for long-term storage.

2.1. Samples for Methods A–M

Both fresh and preserved samples are suitable for microscopic identification of cyanobacteria. The recommended starting material for toxin isolation consists of thick bloom or culture material freeze-dried or freeze-thawed for complete toxin extraction. Samples of water containing cyanobacteria (200–500 mL) filtered on glass fiber filters (typically Whatman [Maidstone, UK] GF/C, diameter 25–70 mm) can be used for the assessment of intracellular toxin. Environmental or tap water samples for the analysis of extracellular toxin should be taken preferably in glass bottles to minimize the unwanted adsorption of toxin onto the container.

2.2. Method B

1. Extraction solvent: water–methanol–butanol (75:20:5, by vol).
2. C_{18} solid-phase extraction medium, e.g., Bond-Elut C_{18} (Varian, Harbor City, CA; typically 2–3 g self-packed in a 75-mL reservoir); methanol and water are used in cartridge preconditioning.
3. HPLC mobile phase: acetonitrile–0.0135 M (1.04 g/L) ammonium acetate (27:73, v/v).
4. HPLC stationary phase: Macherey–Nagel (Düren, Germany) Nucleosil-100 7C_{18}, 250 mm × 10 mm i.d., protected by a 0.5 μm in-line filter and a C_{18} guard column.
5. Thin-layer chromatography (TLC) mobile phase: ethyl acetate–isopropanol–water (9:6:5, by vol).
6. TLC stationary phase: Merck (Darmstadt, Germany) Kieselgel 60 F_{254}, 20 cm × 20 cm, 1 mm layer thickness, with concentrating zone.

2.3. Method C

1. Extraction solvent: Methanol. Water–methanol is used for the flash cartridge elution.
2. Flash chromatography system (Biotage, Charlottesville, PA) using Flash 40M KP-C_{18}-HS cartridges.
3. Preparative HPLC system, for example, Kiloprep 100 laboratory-scale HPLC with the 100 compression module (Biotage).
4. HPLC mobile phase: (A) ammonium acetate (0.1%, w/v) and (B) acetonitrile; proportions vary depending on microcystins being purified.
5. HPLC stationary phase: 150 mm × 75 mm i.d. cartridge packed with HS BDS C_{18} (Biotage, 12 μm particle size).

2.4. Method D

2.4.1. Isolation of Hydrophilic Microcystins

1. 5% Acetic acid for sample extraction. Water and methanol are needed for C_{18} silica gel column preconditioning, wash, and elution. Silica gel column mobile phases are chloroform–methanol–water (65:35:10, by vol, lower phase) and ethyl acetate–isopropanol–water (4:3:7, by vol, upper phase). Methanol is used for elution in HW-40 gel chromatography.
2. C_{18} Silica gel and silica gel for chromatography, for example, from Fuji Silysia Chemical, Tokyo, Japan. A suitable size of the C_{18} column is 10 g of column material per gram of freeze-dried cyanobacteria in the sample. The size of the slurry-packed silica gel column is 100–500 g of silica gel material per gram of fraction to be separated. (The size of column is determined by the required resolution between closely eluting microcystins).
3. Toyopearl HW-40 for gel chromatography (Tosoh, Tokyo, Japan), size 900 mm × 30 mm i.d. for samples of 30 mg or more microcystin (quantitated after the silica gel chromatography), 900 mm × 17 mm i.d. for smaller samples.

2.4.2. Analysis of Hydrophilic Microcystins

1. 5% Acetic acid for sample extraction. Water and methanol are needed for SPE cartridge preconditioning, wash, and elution.
2. C_{18} Silica gel cartridges from Baker (Phillipsburgh, NJ), Varian (Harbor City, CA), or Waters (Milford, MA). A suitable size of SPE cartridges is 14 mm × 8 mm i.d.
3. TLC phases: Kieselgel 60F$_{254}$ silica gel plates (Merck, Darmstadt, Germany). Two mobile phases, chloroform–methanol–water (65:35:10, by vol, lower phase) and ethyl acetate–isopropanol–water (4:3:7, by vol, upper phase) are used. Iodine and short-wavelength UV light source are required for plate visualization.
4. HPLC phases: column Cosmosil 5C$_{18}$-AR (150 mm × 4.6 mm i.d., Nacalai Tesque, Kyoto, Japan). Mobile phase alternatives: (1) methanol–0.05% trifluoroacetic acid (6:4, v/v); (2) methanol–0.05 M phosphate buffer, pH 3 (6:4, v/v); (3) methanol–0.05 M sodium sulfate (1:1, v/v); (4) acetonitrile–0.05% trifluoroacetic acid (1:1, v/v).

2.5. Method E

1. 10-mm Lightpath quartz cuvets.
2. Methanol free from UV-absorbing impurities.

2.6. Method F

1. 6 *M* Hydrochloric acid containing 0.1% (w/v) phenol.
2. Sigma P-0532 *o*-phthaldialdehyde reagent (prone to oxidation, but can be replenished with 2-mercaptoethanol).
3. Amino acid standards, for example, from Sigma.
4. Mobile phase: (A) tetrahydrofuran–methanol-0.03 *M* sodium acetate, pH 5.9 (1:19:80, by vol) and B) methanol-0.03 *M* sodium acetate, pH 5.9 (8:2, v/v).
5. Stationary phase: C_{18} analytical HPLC column, e.g., Phase Separations (Watford, UK) Spherisorb ODS2, 250 mm × 4.6 mm i.d.

2.7. Method G

1. Extraction solvent and HPLC mobile phase: acetonitrile–0.1 *M* (13.6 g/L) potassium dihydrogenphosphate, pH 6.8 (15:85, v/v).
2. Stationary phase: Regis (Morton Grove, IL) GFF-5-80 internal surface reversed-phase (ISRP) column, 250 mm × 4.6 mm i.d., protected by a 0.5 μm inline filter and a GFF guard column.

2.8. Method H

1. Extraction solvent: HPLC-grade methanol.
2. GF/C filters, diameter 47–70 mm, for example, from Whatman.
3. Residual chlorine removal: 1 g/100 mL of sodium thiosulfate, add 100 μL/L tap water.
4. Water modifier: 10% v/v trifluoroacetic acid, HPLC grade.
5. Isolute trifunctional, end-capped C_{18}(EC) SPE cartridges (1 g in a 3-mL syringe, International Sorbent Technology, Mid Glamorgan, UK). Preconditioned with methanol and water. Washed with 10%, 20%, and 30% methanol. Eluted with acidified methanol containing 0.01% trifluoroacetic acid.
6. Analytical gradient HPLC system with photodiode-array detector and column oven.
7. HPLC mobile phase: (A) 0.05% v/v trifluoroacetic acid and (B) acetonitrile containing 0.05% trifluoroacetic acid.
8. HPLC stationary phase: Symmetry C_{18} 250 mm × 4.6 mm i.d. (Waters, Milford, MA).

2.9. Method I

1. 4-Phenylbutyric acid (PB) or deuterated 3-methoxy-2-methyl-4-phenylbutyric acid (MMPB-d_3), used as internal standards, from Wako Pure Chemical Industries Ltd. (Osaka, Japan), methanol, GF/C filters.
2. Ozone generator, e.g., model 0-1-2 from Nihon Ozone (Tokyo, Japan).

3. Thermospray-liquid chromatography/mass spectrometry (TSP-LC/MS) is carried out under the following conditions: column, Cosmosil $5C_{18}AR$ (150 mm × 4.6 mm i.d., Nacalai Tesque, Kyoto, Japan), mobile phase methanol-0.5 M ammonium acetate (4:6, v/v); flow rate, 1.0 mL/min; mass spectrometer, Shimadzu (Kyoto, Japan); LC/MS-QP 1000; interface Shimadzu TSP-100; control temperature, 149–170°C; tip temperature, 190–212°C; block temperature, 249–270°C; vapor temperature 234–258°C; tip heater temperature, 281–298°C; SIM m/z 182 (PB), 226 (MMPB), and 229 (MMPB-d_3).

4. Electron ionization–gas chromatography/mass spectrometry (EI–GC/MS) is carried out under the following conditions: column, FFAP (15 m × 0.53 mm i.d., J&W Scientific, Folsom, CA); oven temperature, 200°C, carrier gas, He; flow rate, 10 mL/min; mass spectrometer, JMS-AX505W (JEOL, Tokyo, Japan); separator temperature, 250°C; electron energy, 70 eV; accelerating voltage, 3 kV; scan range, m/z 30–800. SIM experiments are carried out at 100 ms per mass unit of switching time. The ions at m/z 176 and 117 are selected for SIM.

2.10. Method J

Laboratory mice, for example, male Swiss Albino mice.

2.11. Method K

1. Dried brine shrimp (*Artemia salina*) eggs, often purchased from an aquarium shop but preferably bought from a scientific supplier.
2. Instant Ocean (Aquarium Systems, Sarrebourg, France) artificial seawater, 40.0 g/L in distilled water, pH adjusted to 6.5 with 0.1 M NaOH or HCl.
3. A Petri dish, which has been separated into two halves by a plastic divider with 1–2 mm diameter holes.
4. Clear flat-bottomed 96-well plates, size 400 µL/well.
5. 0.45-µm filters, for example, Pall Gelman Sciences (Ann Arbor, MI) LC PVDF.
6. KH_2PO_4, 0.1 g/mL, pH < 1.8.
7. $Cr_2K_2O_7$, 4 µg/mL in artificial seawater.

2.12. Method L

1. 90% Methanol in water.
2. Life Technologies (Gibco BRL, Gaithersburg, MD) Protein Phosphatase Assay System (product no. 13188016).
3. [γ-^{32}P]ATP (3000–6000 Ci/mmol, 5–10 mCi/mL; aqueous solution).
4. Crude cell extracts containing active protein phosphatases 1 and 2A, prepared, for example, according to (*10*).

2.13. Method M

1. Microcystin ELISA kit (enzyme-linked immunosorbent assay, monoclonal) from Wako Pure Chemical Industries, Osaka, Japan, or
2. EnviroGard microcystins kit (polyclonal ELISA) from Strategic Diagnostics, Newark, DE.

3. Methods

The chromatographic mobile and stationary phases are listed under **Subheading 2**. Additional illustrations and details are found in the original publications (listed in the method heading). The **Notes** section should be read in parallel with **Subheading 3**.

3.1. Method A: Microscopic Examination of Cyanobacteria

Use a microscope at a magnification of ×200–400 and a taxonomy book to identify the dominant cyanobacteria in the sample. Identification at the genera level is often adequate. At least the samples with notable amounts of known producers of cyanobacterial toxins (*see* **Subheading 1.** and **Note 1**) should be subjected to toxin tests.

3.2. Method B: Isolation of Microcystin-LR and -RR Variants and Nodularin-R, Small-Scale (11)

1. Extract freeze-dried cyanobacterial cells with water–methanol–butanol (75:20:5, by vol), 100 mL/g, in a bath sonicator [e.g., Branson (Danbury, CO) B2210] for 30 min. Centrifuge the extracts for 1 h preferably at 48,000g and 4°C. Collect supernatant.
2. Reextract the pellets and centrifuge as in **step 1**.
3. Rotary evaporate the volume of the combined supernatants to 50%. Some samples may benefit from GF/C filtration prior to the next step.
4. Concentrate on preconditioned (methanol followed by water) C_{18} cartridges. The flow is enhanced with a slight vacuum. Change to the next cartridge when the flow becomes considerably restricted.
5. Dry the cartridge by taking air through it (1 min) and elute with 30 mL of methanol per gram C_{18} material.
6. Rotary evaporate the methanol to dryness and dissolve the residue in the HPLC mobile phase (or methanol if TLC, **step 7**, is to be used) using 0.5–1.0 mL/g freeze-dried cyanobacterial starting material. Clarify the solution to be injected by centrifugation (10,000g, 10 min). Reextract the pellet twice with a smaller amount of mobile phase (for TLC–methanol) and centrifuge (10 min, 10,000g).
7. It is sometimes necessary to include a preparative silica gel TLC separation before the HPLC step (*see* **Note 2**). A useful solvent for the TLC is ethyl acetate–isopropanol–water (9:6:5, by vol) *(12)*. The TLC techniques listed under **Method D** can also be utilized. Visualize the plate under a 254 nm UV lamp and mark the separated bands with a pencil. Scrape off the bands, extract in water, reduce volume, test for biological activity typical for the toxins, filter through a 0.45-µm filter, and add acetonitrile to approx 27% of the volume to make the sample ready for the HPLC step.

8. Separate the extracts on C_{18} HPLC. Flow rate is 3.0 mL/min and injection volume ≤400 μL. Detection is performed at 238 nm or using a diode array detector scanning at 200–300 nm. A chromatogram of toxic extracts from *Anabaena* sp. can be seen in **Fig. 2**. Collect the fractions of interest and rotary evaporate at 40°C to a small volume (<20% of the original).

9. A repetition of **step 7** is often necessary, especially if a TLC step has not been employed, owing to slight discoloring from cyanobacterial pigments having unpredictable retention characteristics. An analytical HPLC system with a diode-array or mass spectrometric detector can provide information on toxin purity.

10. Desalting: Bind the purified toxins onto preconditioned C_{18} SPE cartridges, flush with water, and elute with methanol. Divide into suitable aliquots and remove the methanol using nitrogen and a heater block set at 40°C.

3.3. Method C: Isolation of Microcystins and Nodularins Using Flash Chromatography and HPLC, Preparative-Scale (13)

1. Collect thick bloom material or harvest cultured cyanobacterial cells (*see* **Note 3**) by tangential flow filtration and/or centrifugation (2000*g*, 30 min).

2. Add methanol (2 L per 30–50 g) to cyanobacterial cells (freeze-thawed or freeze-dried), stir regularly for 1 h, then centrifuge (1500*g* for 30 min) and retain supernatant.

3. Reextract pellet two more times, then combine the three methanolic extracts.

4. Dilute extract with water to give 20% methanol, and allow the sample to stand for approx 30 min before GF/C filtering.

5. Pass the diluted extract through preconditioned (wet with 1-L methanol followed by 1-L water wash) C_{18} flash cartridge.

6. Elute the flash cartridge with a methanol step gradient from 0% to 100% methanol in 10% increments. Each step is 1 L.

7. Analyze samples by analytical HPLC to determine microcystin content of each fraction (**Fig. 3**).

8. Combine relevant samples, dilute them to give 20% methanol, and load them onto a flash cartridge before eluting in 100% methanol (which is easier to rotary evaporate).

9. Purity of isolated microcystins can be enhanced by repeating the flash extraction and step elution.

10. High purity is achieved using laboratory-scale preparative C_{18} HPLC. To purify microcystin-LR use 22% B at a flow rate of 400 mL/min with a 35-mL injection loop and collect a number of samples across the peak.

11. Analyze the collected fractions by analytical HPLC, then pool those of suitable purity.

12. Collected samples of acceptable purity should be diluted with 1 vol of water, applied to a flash cartridge and washed with several column volumes of water to desalt the sample.

13. The bound microcystin should be eluted in methanol and dried.

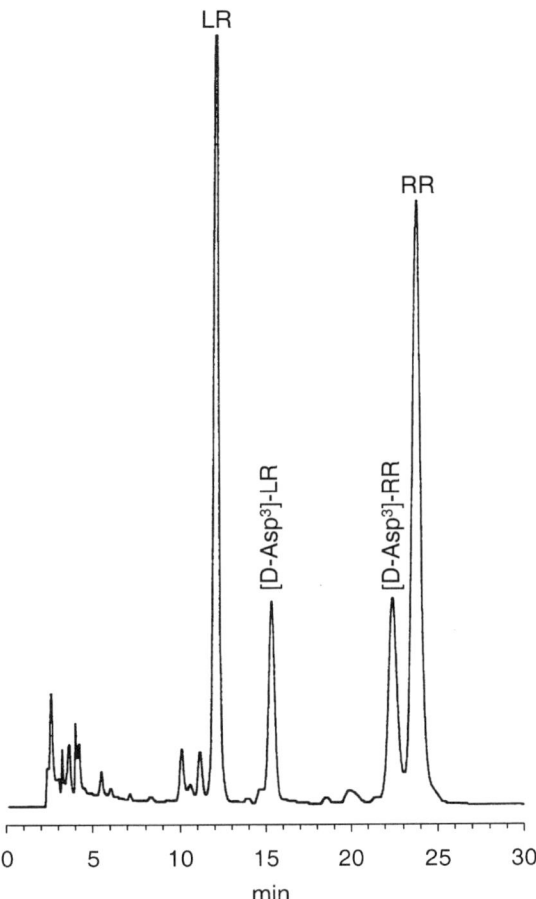

Fig. 2. Semipreparative C_{18} HPLC chromatogram of *Anabaena* sp. extracts containing microcystin-LR, [D-Asp³]microcystin-LR, [D-Asp³]microcystin-RR, and microcystin-RR monitored at 238 nm. Preliminary purification of the extracts was carried out on a preparative silica TLC plate (mobile phase ethyl acetate–isopropanol–water [9:6:5, by vol]) before the HPLC step, which was conducted as described in **Subheading 3.2.** The retention time of nodularin-R under similar conditions is about 9.5 min.

3.4. Method D: Isolation and Analysis of Hydrophilic Microcystins, Integrated System

3.4.1. Isolation (14)

1. Extract freeze-dried cells 3× with 5% aqueous acetic acid, 100 mL/g, for 30 min while stirring (*see* **Note 4**).
2. Centrifuge the combined extracts at 9300*g* for 1 h and collect the supernatant.

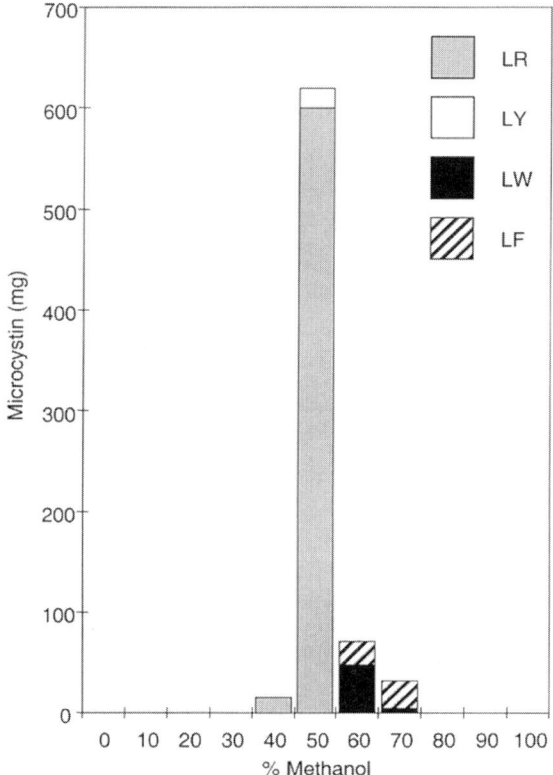

Fig. 3. Preparative C_{18} flash chromatography elution profile of the main micro-cystins extracted from a bloom of *Microcystis aeruginosa* (conditions as described in **Subheading 3.3.**).

3. Apply the supernatant directly to a preconditioned (1 vol of methanol followed by 2 vol of water) C_{18} silica gel column of necessary size using an appropriate pump at a flow rate of 3 mL/min.
4. Wash the column with water, 10 mL/g column material, followed by water-methanol (8:2, v/v), 10 mL/g column material.
5. Elute the toxins with water-methanol (1:9, v/v), 20 mL/g column material, evaporate to dryness, and dissolve in methanol.
6. The toxic residue should be analyzed by TLC as specified in **Subheadings 2.4.2.** and **3.4.2.** This gives an indication of the silica gel column size necessary for microcystin separation.
7. Chromatograph the toxin-containing fraction on silica gel with chloroform–methanol–water (65:35:10, by vol, lower phase). When the desired toxins are not purified by this step, chromatography should be repeated with another mobile phase, ethyl acetate–isopropanol–water (4:3:7, by vol, upper phase). Collect frac-

tions, evaporate them to dryness, and dissolve in a minimal amount of solvent (mobile phases or methanol).

8. Check the collected fractions by analytical HPLC and TLC as specified in **Subheadings 2.4.2. and 3.4.2.**
9. Chromatograph the collected toxic fractions on Toyopearl HW-40 gel using methanol as the mobile phase. Collect fractions and analyze them on an analytical system.
10. Evaporate the methanol using an evaporator.

3.4.2. Analysis (15)

1. Extract 50–100 mg of freeze-dried cells 3× with 5% aqueous acetic acid, 100 mL/g, for 30 min while stirring.
2. Centrifuge the combined extracts at 9300*g* for 1 h and collect the supernatant.
3. Apply the supernatant directly to a preconditioned (1 mL of methanol followed by 2 mL of water) C_{18} silica gel cartridge.
4. Wash the column with 5 mL of water followed by 5 mL of water–methanol (8:2, v/v).
5. Elute the toxins with 5 mL of water–methanol (1:9, v/v).
6. Evaporate the solvent under reduced pressure.
7. Dissolve the residue in 0.5 mL of methanol.
8. Run HPLC (**Fig. 4**) and/or TLC using the chromatographic phases described in **Subheadings 2.4.2.** HPLC flow rate is 1 mL/min, UV absorbance detection at 238 nm. The TLC plates are detected with iodine and short-wavelength UV light.

3.5. Method E: Spectrophotometric Determination of Concentration of Pure Microcystins

1. Dissolve purified microcystin in methanol to give an absorbance reading of 0.2–0.3 against methanol on a spectrophotometer at 238 nm (10 mm lightpath quartz cuvet).
2. Calculate the microcystin concentration using the following reported molar absorptivities: microcystin-LR, -RR, and -YR 39800, [D-Asp^3]microcystin-LR 31600, [dehydroAla^7]microcystin-LR 46800 *(16–18)* (*see* **Note 5**). An absorbance reading of 0.20 corresponds to 5.0 µg/mL of microcystin-LR in methanol.

3.6. Method F: Amino Acid Analysis of Purified Toxins (19)

1. Dissolve 0.1–1 µg of the purified toxin in 6 *M* HCl complemented with 0.1% (w/v) phenol (*see* **Note 6**). Transfer to a hydrolysis tube. Extrude air from the tube with nitrogen or argon gas and close the tube tightly.
2. Heat for 60 min at 150°C.
3. Evaporate the HCl solution with nitrogen or argon.
4. Dissolve the residue in 200 µL of water.
5. Derivatize a 100-µL aliquot of the liberated amino acids with an excess (e.g., 50 µL) of the *o*-phthaldialdehyde reagent.
6. Separate the derivatized amino acids immediately on reversed-phase HPLC and compare with derivatized amino acid standards. HPLC conditions: injection vol-

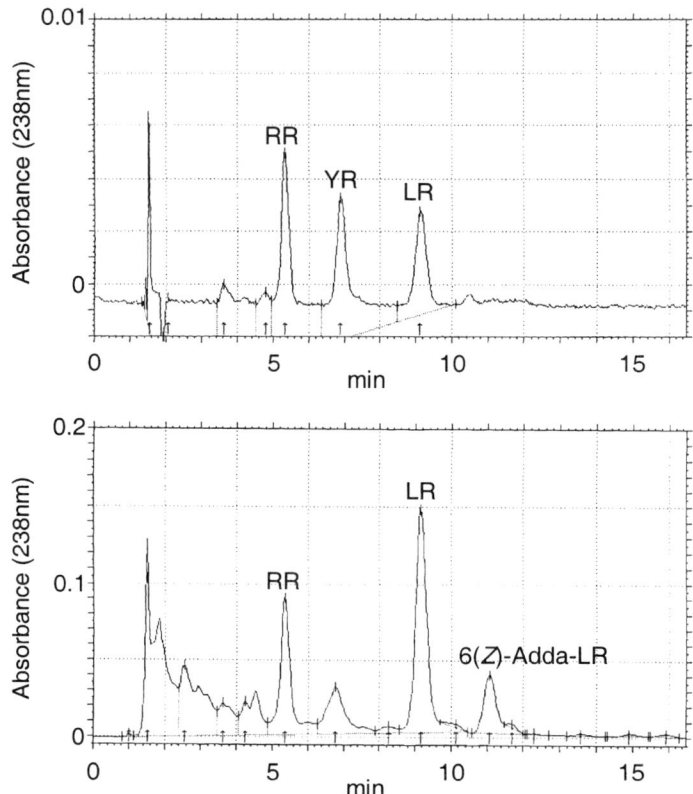

Fig. 4. C_{18} HPLC chromatograms of standard microcystins-RR, -YR, and -LR (upper panel) and of a natural sample from Lake Suwa in 1997 (lower panel). Conditions as described in **Subheading 3.4.2.**, mobile phase methanol-0.05 M phosphate buffer, pH 3 (58:42, v/v).

ume 20 µL, gradient elution from 0% B to 100% B in 40 min, hold at 100% B for 10 min, flow rate 0.5 mL/min, fluorimetric detection with excitation at 330 nm and emission at 418 nm.

3.7. Method G: Analysis of Microcystin-LA, -LR, -YR, -RR and Nodularin-R on Isocratic Internal Surface Reversed-Phase HPLC (20)

1. Weigh 4–10 mg of freeze-dried cyanobacteria in an 1.5-mL polypropylene microcentrifuge tube.
2. Add 100 µL of acetonitrile-0.1 M potassium dihydrogenphosphate, pH 6.8 (15:85, v/v, the HPLC mobile phase), per milligram of dry cyanobacteria, vortex-mix

vigorously, and extract partially sunk in an ultrasonic bath (Branson B2210) for 10 min. Vortex-mix once in the middle of the sonication.

3. Centrifuge for 10 min at 10,000*g*. Collect supernatant.
4. Resuspend and reextract the pellet as in **step 2**.
5. Combine the collected supernatants.
6. Analyze on internal surface reversed-phase HPLC (*see* **Note 7**). HPLC conditions: injection volume typically 20 µL, flow rate 1.0 mL/min, detection at 238 nm or 200–300 nm using a diode-array detector. Typical retention times: microcystin-LA, 3.4 min; -LR, 5.1 min; -YR, 5.8 min; and -RR, 7.4 min; nodularin-R, 4.3 min (**Fig. 5**).

3.8. Method H: Analysis of Microcystins and Nodularins in Cyanobacteria and Water on Reversed-Phase Gradient HPLC at Low pH (21)

1. Mix water sample immediately prior to filtering a 500-mL aliquot through a GF/C filter. Duplicate samples are recommended.
2. Filter samples: Extract freeze-thawed filters in 20 mL of methanol in a small glass beaker for 1 h, decant extract into a rotary evaporation flask, and then reextract filter a further 2×.
3. Combine the product of the three extractions, dry, then resuspend the residue in 2 × 250 µL methanol in a polypropylene microcentrifuge tube. Centrifuge (10,000*g*, 10 min). The sample is now ready for HPLC analysis. Go to **step 11**.
4. Water samples: If the water sample to be analyzed has been treated with chlorine (e.g., tap water; *see* **Note 8**) add 50 µL of sodium thiosulphate solution to the 500-mL sample, mix, and allow to stand for 5 min.
5. Add 5 mL of 10% trifluoroacetic acid solution.
6. Filter water through a GF/C filter, then add 5 mL of methanol prior to solid-phase extraction.
7. Precondition the specified C_{18} SPE cartridge with methanol, wash with water, then apply the 510-mL water sample using a vacuum manifold system.
8. Once all the water sample has passed through, wash the cartridge with 10 mL of 10%, 20%, and 30% (v/v) aqueous methanol.
9. Elute the cartridge with 3-mL of acidified methanol (0.01% trifluoroacetic acid) and dry the collected sample on a hot block (45°C) under a constant stream of nitrogen.
10. Resuspend samples in 200 µL of methanol prior to HPLC analysis.
11. HPLC separation is achieved over a linear gradient from 30% B to 35% B over the first 10 min followed by an increase to 70% B over the next 30 min. Flow rate is 1 mL/min and injection volume 25 µL. The HPLC column is maintained at 40°C and absorbance data are collected at 200–300 nm.
12. Identify microcystins or nodularins present in the sample by their characteristic absorbance spectra (**Fig. 6**). The elution order of some toxins is microcystin-RR, nodularin-R, microcystin-YR, -LR, -FR, -LA, -LY, -LW, and -LF.

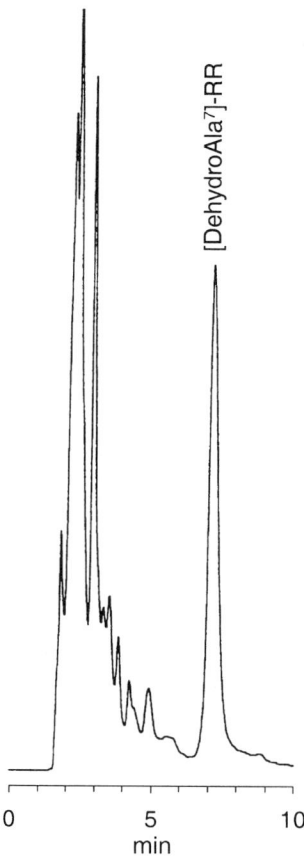

Fig. 5. ISRP HPLC chromatogram of *Oscillatoria agardhii* extracts containing [dehydro-Ala[7]]microcystin-RR monitored at 238 nm. Lake water (200 mL) was filtered on a 25-mm GF/C filter that was freeze-dried and then extracted twice with 1000 μL of mobile phase. The amount of intracellular toxin in the sample was 30 μg/L. HPLC conditions were as described in **Subheading 3.7.**

3.9. Method I: Adda Quantification (22)

1. Dissolve microcystins or samples containing microcystins (with a minimal water content to avoid freezing) in 2 mL of methanol and cool the resulting solution or suspension at −78°C under stirring.
2. Introduce a stream of ozone/oxygen into the solution or suspension at −78°C for 5–10 min. Usually, ozone is saturated within 5 min, showing purple color of the reaction solution. Ozonolysis of Adda yields 3-methoxy-2-methyl-4-phenylbutyric acid (MMPB).
3. Remove excess ozone/oxygen by a stream of nitrogen gas for 1.5–10 min. In the case of freeze-dried cells, filter the reaction solution through a GF/C filter.

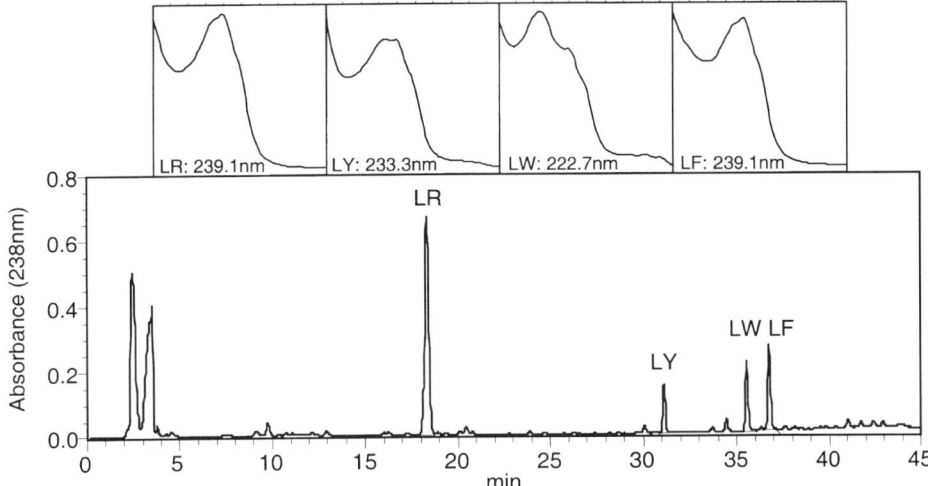

Fig. 6. C$_{18}$ HPLC chromatogram and UV absorbance spectra (200–300 nm) of microcystins present in a methanolic extract of *Microcystis aeruginosa* PCC7820. HPLC conditions were as described in **Subheading 3.8.** The local absorbance maxima around 238 nm are indicated in the spectra panels.

4. Subject an aliquot of the reaction solution directly to TSP-LC/MS analysis together with a solution of 4-phenylbutyric acid or deuterated 3-methoxy-2-methyl-4-phenylbutyric acid as an internal standard (*see* **Note 9**). EI-GC/MS is an alternative to TSP-LC/MS but in that context a more volatile derivative of MMPB such as methyl ester should be used instead of the underivatized molecule. TSP-LC/MS and EI-GC/MS are carried out using SIM mode according to the conditions listed in **Subheading 2.9.**
5. Estimate the amount of microcystin by comparing the ratio of the peak area of the sample and the internal standard with the corresponding ratio in the calibration curve. In the case of EI-GC/MS an absolute calibration method is used.

3.10. Method J: Mouse Bioassay (23)

1. Collect a dense portion of a cyanobacterial bloom (*see* **Note 10**).
2. Centrifuge at 9500*g* for 10 min.
3. Decant water (cells are concentrated at the bottom of the tube).
4. Freeze and thaw the pellet to rupture.
5. Centrifuge and collect supernatant.
6. Inject 1 mL i.p. into the mouse.
7. If hepatotoxin is present, death occurs within 3 h and liver weight is increased by 60% or more.
8. If neurotoxin is present, death occurs within 30 min (often faster) and no increase in liver weight is observed; animals show nervous signs.

3.11. Method K: Artemia salina Bioassay (24)

1. Add artificial seawater to the divided Petri dish and then a thick layer of dried brine shrimp (*Artemia salina*) eggs to the surface of the other half of the dish. Place a 60-W lamp above the covered dish (temperature 25–30°C). As larvae hatch they will swim to the lighter side of the dish. The number of the larvae should be appropriate, 10–40 per 50 µL, after about 30 h incubation. Use the larvae within 1 d after hatching.
2. Suspend freeze-dried cyanobacterial samples in artificial seawater (10 mg/mL) in small test tubes. Sonicate the suspensions for 10 min in an ultrasonic bath (e.g., Branson B2210), leave at 20–22°C for 18 h, and filter them next day through 0.45-µm filters.
3. Test the cyanobacterial suspensions at three different concentrations and in triplicate. Prepare the treatments in microtiter wells as follows: (a) 50 µL of *Artemia* suspension + 300 µL of cyanobacterial suspension (b) 50 µL of *Artemia* + 175 µL of cyanobacteria + 125 µL of artificial seawater, and (c) 50 µL of *Artemia* + 50 µL of cyanobacteria + 250 µL of artificial seawater. These concentrations correspond to ca. 8.5, 5.0, and 1.5 mg dry mass of cyanobacterial material per milliliter. Pure artificial seawater and 4 µg/mL of potassium dichromate are used in controls.
4. Incubate the samples under a 60-W lamp in 25–30°C for 24 h.
5. Count the number of dead or atypically moving (not constantly swimming) larvae under a preparation microscope.
6. Pipet 50 µL of the acidic phosphate buffer solution to the wells to kill all the larvae and count the total number of larvae.
7. Calculate the percentage of larvae killed by toxins in the cyanobacterial samples. The original authors assessed the tested cyanobacterial material as toxic if the concentration of 5.0 mg/mL affected more than 20% of the larvae and simultaneously 8.5 mg/mL affected more than 50% of the larvae (*see* **Note 11**). These values gave the best correlation with mice bioassay results. Spontaneous mortality in artificial seawater should be <3% and the potassium dichromate solution should kill approx 50% of the larvae.

3.12. Method L: Protein Phosphatase Inhibition Assay (7,10)

1. Sample preparation: Extract 20 mg of freeze-dried cyanobacteria in 1 mL of 90% methanol using 10 min bath sonication (e.g., Branson B2210) or 30 min standing with occasional shaking. Dilute the extracts in the assay buffer (preliminary suggestion: 1:1000, dilute further as required).
2. Proceed according to the kit instructions (*see* **Note 12**).

3.13. Method M: Enzyme-Linked Immunosorbent Assays (ELISA) (25,26)

Proceed according to the detailed manufacturer instructions supplied with the kits (*see* **Note 13**).

4. Notes

1. More than 50% of the cyanobacterial blooms have been found toxic in mouse bioassay, and the majority of the toxic ones are hepatotoxic. Field sampling of cyanobacteria should pay attention to the spatial, temporal, and vertical variations of cyanobacterial distribution and toxicity (for a further discussion, *see [27]*). Microcystins and nodularins are kept mainly inside the cyanobacteria until lysis.

2. Without the TLC step this protocol is a minimal procedure that we recommend to be used with samples having rather high toxin contents (several hundred micrograms of each toxin analog per gram of freeze-dried material or more). However, the procedure gives toxin pure enough for most experiments including nuclear magnetic resonance (NMR) work. Low toxin concentrations necessitate the use of TLC or other additional purification steps. Microcystin and nodularin peaks can be identified by their spectra (*see* **Note 5** and **Fig. 6**) and verified by testing the biological activities of the separated compounds. Monitor toxin recovery in this and other purification procedures on an analytical HPLC system. Hydrophilic microcystins and nodularins, for example, the arginine containing ones, are easily extracted from cyanobacteria by many aquatic solvents and methanol. 90% Methanol–10% water is a good universal solvent for the toxins. It is common to lyophilise purified toxins from water solutions but this procedure tends to lead to some loss of toxin (owing to the "fluffy" nature of the toxin).

3. Owing to the high quantities of microcystins that may be encountered in this procedure we recommend that most stringent safety precautions are adopted. After addition of water to methanolic extract it is important to leave the sample for 30 min before filtering, as this allows time for precipitation of compounds that will block the flash cartridge if not removed by filtration. The elution conditions for the preparative HPLC are optimized using an analytical scale column packed with identical stationary phase. Preparative chromatography of the toxins at low pH with trifluoroacetic acid in acetonitrile–water may lead to toxin destruction during the concentration of the collected peaks and to distorted peaks on some stationary phases.

4. The described preparative and analytical methods can be applied to nodularins without any modifications. Centrifugation in sample preparation is very important to prevent column clogging and difficult samples can also be filtrated through a GF/C filter after the centrifugation step. *Isolation*: In some cases, it is more effective to run silica gel chromatography using first ethyl acetate–isopropanol–water (4:3:7, by vol, upper phase) prior to using chloroform–methanol–water (65:35:10, by vol, lower phase). *Analysis*: If the mixture to be analyzed is composed of microcystins and other compounds that are not detected by UV absorbance at 238 nm, it is more effective to use the TLC with iodine vapor detection because this can detect such compounds in addition to microcystins. A tandem cleanup system employing both C_{18} and silica material has been developed for samples with trace amounts of microcystins but large amounts of impurities *(28)*.

5. The conjugated diene group in the Adda residue absorbing at 238 nm is the main chromophore of microcystins (and nodularins) *(2)*. The α,β-unsaturated carbonyl group of the *N*-methyldehydroalanine residue also absorbs strongly at the same wavelength, as evidenced by the lowered absorptivity of glutathione and cysteine conjugates of microcystins *(18)*. Microcystin variants containing tryptophan have an additional maximum at 222 nm *(21)*. *See* **Fig. 6** for typical UV spectra. An alternative to the spectrophotometric concentration measurement is quantitative amino acid analysis. Gravimetric measurement is useful for milligram amounts of toxins.

6. The primary purpose of this procedure is to identify which variable L-amino acids are present in the purified toxin. With suitable authentic amino acid standards this procedure can be useful also in determining the site of demethylation in a toxin (e.g., Asp instead of methylAsp as residue number 3). It should be noted that not all the amino acid residues in microcystins/nodularins are stable in the hydrolysis process (e.g., Adda and *N*-methyldehydroalanine). Advanced structural chemistry is outside the scope of this article and we direct the reader to the original work by Botes et al. *(8)*, Rinehart et al. *(9)*, Harada et al. *(29)*, and Namikoshi et al. *(30)*.

7. The later version of the ISRP column, GFFII, has been found unsuitable for the toxin application and therefore the original GFF column should be used. Owing to the short analysis time and the minimal sample preparation needed to maintain the ISRP column in a good working condition this method is especially useful in studies with a large number of samples, for example, in those concerning environmental effects on toxin production in laboratory cultures. Samples with low toxin concentration must be concentrated on solid-phase extraction cartridges before analysis. Isocratic elution makes mobile phase circulation possible. Guidelines for the interpretation of sample toxicity are given in **Table 1** *(31)*.

8. Practice the use of this method using samples spiked with microcystin to ensure adequate recoveries are obtained. It is important that where tap water is used as a practice matrix residual chlorine is removed prior to the addition of microcystin because residual free chlorine may adversely interact with the microcystin. Although the original published method recommended the use of an internal standard this is no longer considered necessary. Poor recovery is often observed with microcystin-LW and it is thought to be a problem with all tryptophan-containing microcystins owing to the tryptophan instability in acidic solutions.

9. It is now recommended to use MMPB-d_3 as the internal standard instead of PB. In addition to TSP-LC/MS, it is possible to use other types of LC/MS such as ESI (electrospray ionization) and Frit-FAB (fast atom bombardment). MMPB is also formed by the oxidation of microcystin with potassium permanganate and sodium metaperiodate *(32)*. Adda quantification has been used to assess microcystin levels in waterblooms of toxic cyanobacteria *(33)*.

10. A quantitative modification of the mouse bioassay (especially useful also for buoyant cells) is to freeze-dry the sample, make a suspension of 50 mg of the dry material per milliliter of 0.9% NaCl, and inject 1 mL of the suspension in a few

Table 1
Guidelines for the Interpretation of Sample Toxicity

Degree of toxicity	Mouse bioassay $(LD_{50}$ mg/kg)[a]	HPLC (mg/g)[b]	*Artemia* bioassay $(EC_{50}$ mg/ mL)[c]
Nontoxic	>1000	<0.01	>30
Low	500–1000	0.01–0.1	10–30
Medium	100–500	0.1–1.0	2–10
High	<100	>1.0	<2

[a]LD_{50} expressed in mg freeze-dried cyanobacteria per kilogram mouse body weight.
[b]Milligrams of microcystins/nodularins per gram of freeze-dried cyanobacteria.
[c]EC_{50} expressed in milligrams of freeze-dried cyanobacteria per milliter of test medium.

different dilutions intraperitoneally to mice. Quick-acting cyanobacterial neuro-toxins (anatoxin-a, anatoxin-a(s), saxitoxins) present in the sample can mask the toxic action of microcystins and nodularins. If the mice have neurotoxic symp-toms the sample has to be checked for hepatotoxicity using further tests. Neuro-toxins are commonly produced by the genera *Anabaena* and *Aphanizomenon* and sometimes by *Oscillatoria* (*Planktothrix*). Freshly isolated hepatocytes from rat have been suggested as an alternative to the mouse bioassay *(34)*. As cultured cell lines lose their multispecific bile acid transport system and therefore the abil-ity to take up microcystins and nodularins, freshly isolated hepatocytes have to be used in the cellular tests.

11. The EC_{50} of pure [dehydro-Ala[7]]microcystin-RR was reported to be 5.0 µg/mL *(24)*. Older larvae (24 h from hatching) are somewhat more sensitive to toxins than newly hatched larvae. The *Artemia* bioassay also detects anatoxin-a (although purified anatoxin-a hydrochloride does not appear to be toxic to *Artemia [24]*) and in addition some compounds not harmful to humans. How-ever, the response in the *Artemia* bioassay of extract fractions correlated well with the distribution of microcystin-LR *(35)*. Preliminary purification of cyanobacterial extracts on a C_{18} solid-phase extraction cartridge minimizes the number of false positives and also enhances the sensitivity of the *Artemia* assay. Possible differences in brine shrimp material of different origins remain uncharacterized but the differences do not appear to be a serious problem.

12. A new, revised edition of the book containing reference *(10)* is in preparation. Protein phosphatase inhibition assay has been successfully applied to the identi-fication of microcystin-containing fractions in HPLC *(36)*. There is also a colori-metric version of the assay for those laboratories that wish to avoid the use of radioisotopes *(37)*. The colorimetric method is less sensitive than the one using radiolabeled substrate and it requires the use of pure bacterially expressed pro-tein phosphatase 1.

13. The ELISA kits are meant primarily for quantitation of extracellular microcystin and nodularin in water—they are not fully characterized in the assessment of

intracellular toxin, tissue samples, etc. We have reason to believe that the kits will be useful for these categories of samples provided suitable extraction procedures are followed. The binding affinities of the different toxin analogs vary. Nontoxic microcystins with 6(Z)-Adda (a stereoisomer of Adda) are not recognized whereas nontoxic microcystins with esterified Glu residues give a positive reaction in the antibody tests *(26,37)*. Recently, a new ELISA kit for microcystins has been introduced by EnviroLogix (Portland, ME).

14. Future perspectives: The detection methods for microcystins and nodularins are developing rapidly. We anticipate seeing common use of mass spectrometry and other enhanced detection techniques such as fluorescence *(38)* and chemiluminescence detection *(39)* in HPLC. These techniques have not yet been extensively applied to natural samples. Matrix-assisted laser desorption/ ionization time-of-flight mass spectrometry (MALDI-TOF) has been used in the elucidation of secondary metabolites including microcystins in crude extracts of cyanobacteria *(40)*. We find this technique promising in the qualitative analysis of microcystins and nodularins. Another interesting technique to be applied to the analysis of the numerous cyanobacterial toxins is capillary electrochromatography, which possesses good selectivity characteristics for HPLC and high resolution characteristics for capillary techniques. Also the use of micellar electrokinetic capillary chromatography has been advocated *(41)*. Furthermore, we believe that new biotests based on (possibly genetically modified) plants, microbes, and invertebrates will be introduced in the near future. One challenging problem is the analysis of microcystins and nodularins in tissue material. Some issues to be solved include sample extraction, cleanup, and HPLC separation as well as the biological significance of the toxins bound in protein phosphatases. Good estimates of the total microcystin content in tissue have been obtained using Adda quantification (in a manner not fully identical to **Subheading 3.9.**) *(42)*.

References

1. Rinehart, K. L., Namikoshi, M., and Choi, B. W. (1994) Structure and biosynthesis of toxins from blue-green algae (cyanobacteria). *J. Appl. Phycol.* **6,** 159–176.
2. Harada, K.-i. (1996) Chemistry and detection of microcystins, in *Toxic Microcystis* (Watanabe, M. F., Harada, K.-i., Carmichael, W. W., and Fujiki, H., eds.), CRC, Boca Raton, FL, pp. 103–148.
3. Carmichael, W. W. (1997) The cyanotoxins. *Adv. Bot. Res.* **27,** 211–256.
4. Meriluoto, J. (1997) Chromatography of microcystins. *Anal. Chim. Acta* **352,** 277–298.
5. Falconer, I. R. (1991) Tumor promotion and liver injury caused by oral consumption of cyanobacteria. *Environ. Toxicol. Water Qual.* **6,** 177–184.
6. Nishiwaki-Matsushima, R., Ohta, T., Nishiwaki, S., Suganuma, M., Kohyama, K., Ishikawa, T., Carmichael, W. W., and Fujiki, H. (1992) Liver tumor promotion by the cyanobacterial cyclic peptide toxin microcystin-LR. *J. Cancer Res. Clin. Oncol.* **118,** 420–424.

7. MacKintosh, C., Beattie, K. A., Klumpp, S., Cohen, P., and Codd, G. A. (1990) Cyanobacterial microcystin-LR is a potent and specific inhibitor of protein phosphatases 1 and 2A from both mammals and higher plants. *FEBS Lett.* **264,** 187–192.
8. Botes, D. P., Tuinman, A. A., Wessels, P. L., Viljoen, C. C., Kruger, H., Williams, D. H., Santikarn, S., Smith, R. J., and Hammond, S. J. (1984) The structure of cyanoginosin-LA, a cyclic heptapeptide toxin from the cyanobacterium *Microcystis aeruginosa. J. Chem. Soc. Perkin Trans.* **1,** 2311–2318.
9. Rinehart, K. L., Harada, K.-i., Namikoshi, M., Chen, C., Harvis, C. A., Munro, M. H. G., Blunt, J. W., Mulligan, P. E., Beasley, V. R., Dahlem, A. M., and Carmichael, W. W. (1988) Nodularin, microcystin, and the configuration of Adda. *J. Am. Chem. Soc.* **110,** 8557–8558.
10. MacKintosh, C. (1993) Assay and purification of protein (serine/threonine) phosphatases, in *Protein Phosphorylation: A Practical Approach* (Hardie, D. G., ed.), IRL, Oxford, UK, pp. 197–230.
11. Meriluoto, J. A. O., Sandström, A., Eriksson, J. E., Remaud, G., Craig, A. G., and Chattopadhyaya, J. (1989) Structure and toxicity of a peptide hepatotoxin from the cyanobacterium *Oscillatoria agardhii. Toxicon* **27,** 1021–1034.
12. Ojanperä, I., Pelander, A., Vuori, E., Himberg, K., Waris, M., and Niinivaara, K. (1995) Detection of cyanobacterial hepatotoxins by TLC. *J. Planar Chromatogr.* **8,** 69–72.
13. Edwards, C., Lawton, L. A., Coyle, S. M., and Ross, P. (1996) Laboratory-scale purification of microcystins using flash chromatography and reversed-phase high-performance liquid chromatography. *J. Chromatogr. A* **734,** 163–173.
14. Harada, K.-i., Suzuki, M., Dahlem, A. M., Beasley, V. R., Carmichael, W. W., and Rinehart, K. L., Jr. (1988) Improved method for purification of toxic peptides produced by cyanobacteria. *Toxicon* **26,** 433–439.
15. Harada, K.-i., Matsuura, K., Suzuki, M., Oka, H., Watanabe, M. F., Oishi, S., Dahlem, A. M., Beasley, V. R., and Carmichael, W. W. (1988) Analysis and purification of toxic peptides from cyanobacteria by reversed-phase high-performance liquid chromatography. *J. Chromatogr.* **448,** 275–283.
16. Harada, K., Matsuura, K., Suzuki, M., Watanabe, M. F., Oishi, S., Dahlem, A. M., Beasley, V. R., and Carmichael, W. W. (1990) Isolation and characterization of the minor components associated with microcystins LR and RR in the cyanobacterium (blue-green algae). *Toxicon* **28,** 55–64.
17. Harada, K.-i., Ogawa, K., Matsuura, K., Nagai, H., Murata, H., Suzuki, M., Itezono, Y., Nakayama, N., Shirai, M., and Nakano, M. (1991) Isolation of two toxic heptapeptide microcystins from an axenic strain of *Microcystis aeruginosa,* K-139. *Toxicon* **29,** 479–489.
18. Kondo, F., Ikai, Y., Oka, H., Okumura, M., Ishikawa, N., Harada, K.-i., Matsuura, K., Murata, H., and Suzuki, M. (1992) Formation, characterization, and toxicity of the glutathione and cysteine conjugates of toxic heptapeptide microcystins. *Chem. Res. Toxicol.* **5,** 591–596.
19. Jones, B. N., Pääbo, S., and Stein, S. (1981) Amino acid analysis and enzymatic sequence determination of peptides by an improved *o*-phthaldialdehyde precolumn labeling procedure. *J. Liq. Chromatogr.* **4,** 565–586.

20. Meriluoto, J. A. O., Eriksson, J. E., Harada, K.-i., Dahlem, A. M., Sivonen, K., and Carmichael, W. W. (1990) Internal surface reversed-phase high-performance liquid chromatographic separation of the cyanobacterial peptide toxins microcystin-LA, -LR, -YR, -RR and nodularin. *J. Chromatogr.* **509,** 390–395.

21. Lawton, L. A., Edwards, C., and Codd, G. A. (1994) Extraction and high-performance liquid chromatographic method for the determination of microcystins in raw and treated waters. *Analyst* (Lond.) **119,** 1525–1530.

22. Harada, K.-i., Murata, H., Qiang, Z., Suzuki, M., and Kondo, F. (1996) Mass spectrometric screening method for microcystins in cyanobacteria. *Toxicon* **34,** 701–710.

23. Beasley, V. R., Dahlem, A. M., Cook, W. O., Valentine, W. M., Lovell, R. A., Hooser, S. B., Harada, K.-i., Suzuki, M., and Carmichael, W. W. (1989) Diagnostic and clinically important aspects of cyanobacterial (blue-green algae) toxicoses. *J. Vet. Diagn. Invest.* **1,** 359–365.

24. Kiviranta, J., Sivonen, K., Niemelä, S. I., and Huovinen, K. (1991) Detection of toxicity of cyanobacteria by *Artemia salina* bioassay. *Environ. Toxicol. Water Qual.* **6,** 423–436.

25. Chu, F. S., Huang, X., and Wei, R. D. (1990) Enzyme-linked immunosorbent assay for microcystins in blue-green algal blooms. *J. Assoc. Offic. Anal. Chem.* **73,** 451–456.

26. Nagata, S., Soutome, H., Tsutsumi, T., Hasegawa, A., Sekijima, M., Sugamata, M., Harada, K.-i., Suganuma, M., and Ueno, Y. (1995) Novel monoclonal antibodies against microcystin and their protective activity for hepatotoxicity. *Nat. Toxins* **3,** 78–86.

27. Meriluoto, J., Härmälä-Braskén, A.-S., Eriksson, J., Toivola, D., and Lindholm, T. (1996) Choosing analytical strategy for microcystins. *Phycologia* **35(6 Suppl.),** 125–132.

28. Tsuji, K., Naito, S., Kondo, F., Watanabe, M. F., Suzuki, S., Nakazawa, H., Suzuki, M., Shimada, T., and Harada, K.-i. (1994) A clean-up method for analysis of trace amounts of microcystins in lake water. *Toxicon* **32,** 1251–1259.

29. Harada, K.-i., Ogawa, K., Matsuura, K., Murata, H., Suzuki, M., Watanabe, M. F., Itezono, Y., and Nakayama, N. (1990) Structural determination of geometrical isomers of microcystins LR and RR from cyanobacteria by two-dimensional NMR spectroscopic techniques. *Chem. Res. Toxicol.* **3,** 473–481.

30. Namikoshi, M., Choi, B. W., Sakai, R., Sun, F., Rinehart, K. L., Carmichael, W. W., Evans, W. R., Cruz, P., Munro, M. H. G., and Blunt, J. W. (1994) New nodularins: a general method for structure assignment. *J. Org. Chem.* **59,** 2349–2357.

31. Lawton, L. A., Beattie, K. A., Hawser, S. P., Campbell, D. L., and Codd, G. A. (1994) Evaluation of assay methods for the determination of cyanobacterial hepatotoxicity, in *Detection Methods for Cyanobacterial Toxins (Proceedings of the First International Symposium on Detection Methods for Cyanobacterial [Blue-Green Algal] Toxins)* (Codd, G. A., Jefferies, T. M., Keevil, C. W., and Potter, E., eds.), The Royal Society of Chemistry, Cambridge, UK, pp. 111–116.

32. Sano, T., Nohara, K., Shiraishi, F., and Kaya, K. (1992) A method for micro-determination of total microcystin content in waterblooms of cyanobacteria (blue-green algae). *Int. J. Environ. Anal. Chem.* **49**, 163–170.

33. Tanaka, Y., Takenaka, S., Matsuo, H., Kitamori, S., and Tokiwa, H. (1993) Levels of microcystins in Japanese lakes. *Toxicol. Environ. Chem.* **39**, 21–27.

34. Heinze, R. (1996) A biotest for hepatotoxins using primary rat hepatocytes. *Phycologia* **35(6 Suppl.)**, 89–93.

35. Campbell, D. L., Lawton, L. A., Beattie, K. A., and Codd, G. A. (1994) Comparative assessment of the specificity of the brine shrimp and Microtox assays to hepatotoxic (microcystin-LR-containing) cyanobacteria. *Environ. Toxicol. Water Qual.* **9**, 71–77.

36. Boland, M. P., Smillie, M. A., Chen, D. Z. X., and Holmes, C. F. B. (1993) A unified bioscreen for the detection of diarrhetic shellfish toxins and microcystins in marine and freshwater environments. *Toxicon* **31**, 1393–1405.

37. An, J. and Carmichael, W. W. (1994) Use of a colorimetric protein phosphatase inhibition assay and enzyme linked immunosorbent assay for the study of microcystins and nodularins. *Toxicon* **32**, 1495–1507.

38. Harada, K.-i., Oshikata, M., Shimada, T., Nagata, A., Ishikawa, N., Suzuki, M., Kondo, F., Shimizu, M., and Yamada, S. (1997) High-performance liquid chromatographic separation of microcystins derivatized with a highly fluorescent dienophile. *Nat. Toxins* **5**, 201–207.

39. Murata, H., Shoji, H., Oshikata, M., Harada, K.-i., Suzuki, M., Kondo, F., and Goto, H. (1995) High-performance liquid chromatography with chemiluminescence detection of derivatized microcystins. *J. Chromatogr. A* **693**, 263–270.

40. Erhard, M., von Döhren, H., and Jungblut, P. (1997) Rapid typing and elucidation of new secondary metabolites of intact cyanobacteria using MALDI-TOF mass spectrometry. *Nat. Biotechnol.* **15**, 906–909.

41. Bouaïcha, N., Rivasseau, C., Hennion, M.-C., and Sandra, P. (1996) Detection of cyanobacterial toxins (microcystins) in cell extracts by micellar electrokinetic chromatography. *J. Chromatogr. B* **685**, 53–57.

42. Williams, D. E., Craig, M., Dawe, S. C., Kent, M. L., Holmes, C. F. B., and Andersen, R. J. (1997) Evidence for a covalently bound form of microcystin-LR in salmon liver and dungeness crab larvae. *Chem. Res. Toxicol.* **10**, 463–469.

5

Genetic Construction, Expression, and Characterization of Diphtheria Toxin-Based Growth Factor Fusion Proteins

Johanna C. vanderSpek and John R. Murphy

1. Introduction

The fusion protein toxins that have been described are generally composed of the catalytic and transmembrane domains of a bacterial toxin (e.g., diphtheria toxin [DT] or *Pseudomonas* exotoxin A) to which a polypeptide hormone, growth factor, or single-chain antibody (scFv) is genetically fused *(1–3)*. In these constructs, the native receptor binding domain of the toxin is genetically replaced with the targeting ligand. Our laboratory has focused almost exclusively on the construction, expression, and characterization of DT-based fusion toxins *(4–12)*. We have used these novel cytotoxic reagents as probes to study structure–function relationships of the DT catalytic and transmembrane domains, as well as to study the receptor binding components of the fusion proteins *(13–19)*. We are currently using these novel reagents to study the underlying mechanisms involved with delivery of the fusion protein toxin's catalytic domain across the membrane and into the cytosol of target eucaryotic cells. The catalytic domain of diphtheria toxin catalyzes the NAD$^+$-dependent ADP-ribosylation of the diphthamide residue in elongation factor 2, resulting in inhibition of protein synthesis *(20,21)*. As the delivery of a single molecule of the catalytic domain to the eucaryotic cell cytosol has been shown to result in the death of that cell, the fusion protein toxins represent a family of highly potent, receptor-specific, cytotoxic probes *(22)*.

Although a variety of promoter/operator genetic circuits have been used to direct the expression of the fusion protein toxins in recombinant *Escherichia coli*, in recent years we have used the T7 promoter-driven system developed by Studier and colleagues *(23–25)*. Following modification of the 5'-end of the

From: *Methods in Molecular Biology, vol. 145: Bacterial Toxins: Methods and Protocols*
Edited by: O. Holst © Humana Press Inc., Totowa, NJ

structural gene (i.e., the N-terminal end of the catalytic domain) by the introduction of a unique *NcoI* restriction endonuclease site, the genes encoding the fusion protein toxins are assembled in derivatives of the pET vector system (Novagen, Madison, WI). In this system, the expression of recombinant protein is under the control of T7 RNA polymerase, and following induction of gene expression, yields of fusion protein toxin may approach 20% of total cellular protein. The fusion toxins usually accumulate in the cytosol of recombinant *E. coli* as inactive insoluble inclusion bodies. As a result, initial partial purification may be achieved by a series of washes, followed by sonication in denaturing buffer to solubilize the protein, and dialysis in buffer to allow refolding into an active conformation.

The overall scheme used in the genetic construction of the gene encoding a fusion protein toxin, its subsequent expression, and refolding into an active conformation is as follows:

1. The catalytic and transmembrane domains of diphtheria toxin are encoded on a derivative of plasmid pET11d, pETJV127 (*see* **Fig. 1**). In this vector, the expression of fusion protein toxins are under the control of a T7 promoter, and the gene encoding a new receptor binding domain can be vectorially inserted into unique *SphI* and *HindIII* restriction endonuclease sites.

2. The gene encoding a new receptor binding domain is constructed such that an *SphI* restriction site is positioned at the 5' end to maintain correct translational reading frame between the diphtheria toxin transmembrane domain and the new surrogate receptor binding domain. A translational stop signal is introduced at the 3' end of the ligand coding sequence immediately upstream of a *HindIII* restriction site.

3. Once the gene encoding to the new receptor binding domain is ligated into pETJV127, the plasmid encoding the fusion protein toxin is transformed into *E. coli* host strains HMS174 and/or HMS174 (DE3). In some instances, a basal level of expression of the fusion protein toxin has been found to be toxic for *E. coli*. In those cases, the HMS174 strain is used as the bacterial host and chimeric fusion toxin gene expression is induced by infection with coliphage CE6. The CE6 derivative of coliphage λ carries the structural gene encoding T7 RNA polymerase.

4. Cultures of recombinant *E. coli* that carry the appropriate plasmid are grown to an OD_{600} of approx 0.8. In those instances in which the HMS174 strain is used for expression, induction of fusion protein toxin gene expression is induced by infection with λ CE6. In those instances in which fusion protein expression is performed in the HMS174 (DE3) host strain, recombinant protein expression is induced by the addition of isopropylthiogalactosidase (IPTG) to the growth medium. In HMS174 (DE3), the structural gene encoding T7 polymerase is under control of the *lacUV5* promoter which may be induced by the addition of the IPTG. While the HMS174 (DE3) host is more convenient to use, basal levels of expression from the gene of interest can be problematic; however, in both

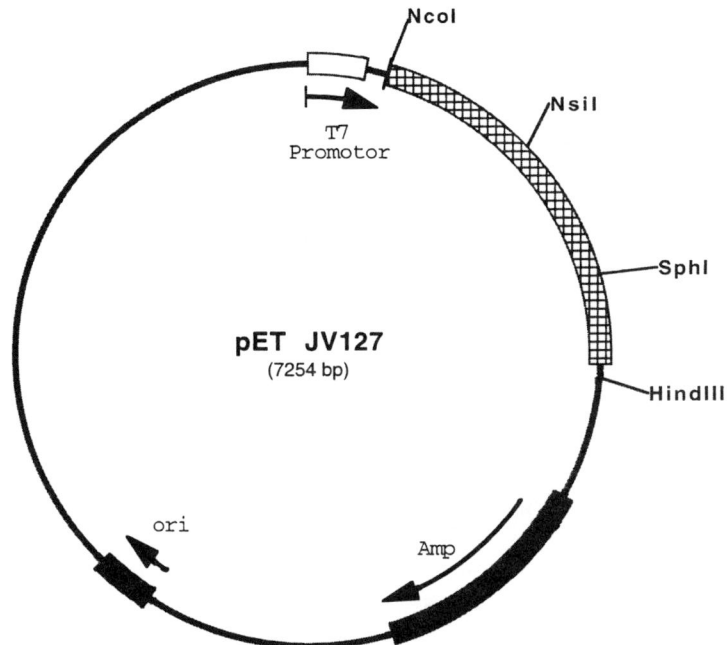

Fig. 1. Partial restriction endonuclease digestion map of plasmid map of pET–JV127. The structural gene encoding fusion protein toxin $DAB_{389}IL$-2 is shown (*hatched*). The *Nco*I restriction site includes the ATG translational initiation signal for the fusion protein toxin; the *Nsi*I restriction site defines the boundary between the catalytic and transmembrane domains of the fusion protein toxin; and the *Sph*I restriction site defines the fusion junction between the transmembrane domain and the human interleukin-2 receptor binding domain of the chimeric protein. This vector may be used as a cassette in the genetic construction of additional fusion protein toxins by the exchange of receptor binding domain encoding sequences using the unique *Sph*I and *Hin*dIII restriction sites.

instances, the fusion protein toxins generally accumulate in the cytosol of recombinant *E. coli* in inclusion body form on induction of gene expression.

5. Following induction of recombinant gene expression for 2–3 h at 37°C, the bacteria are harvested by centrifugation. The bacterial pellet is resuspended in STET buffer, treated with lysozyme, and the bacteria are then lysed by sonication. The inclusion bodies are harvested by centrifugation, and washed to remove weakly associated cytosolic proteins.

6. The purified inclusion body preparation is then solubilized by sonication in a guanidine–dithiothreitol (DTT) denaturing buffer. Following solubilization, the partially purified fusion protein toxin is diluted in refolding buffer and allowed to refold at 4°C for 18 h.

7. Following refolding, the fusion protein toxin is concentrated by diafiltration. If necessary, monomeric fusion protein toxin may be purified from aggregate forms by ion exchange chromatography on DEAE-Sepharose.
8. Once purified, the fusion protein toxin may be assayed for receptor specific cytotoxic activity using eucaryotic cells that display the targeted cell surface receptor.

2. Materials

The water used in the preparation of all reagents is purified by passage through a Milli-Q Water System (Millipore Corporation, Bedford, MA).

1. 1X TE: 10 mM Tris-HCl, pH 8.0, 1 mM EDTA, pH 8.0. Autoclave to sterilize.
2. LB–Amp: 1% Bacto-tryptone, 0.5% yeast extract, 1% NaCl, pH to 7.5 with NaOH. Autoclave to sterilize. Add ampicillin to 100 μg/mL just before use (1.0 mL of 100 mg/mL stock per liter of LB).
3. LB–Amp–maltose: Add maltose solution to LB–Amp for a final concentration of 0.2% just before use (10 mL of 20% maltose per liter of LB/Amp).
4. LB–maltose: Add maltose solution to LB for a final concentration of 0.2% just before use (10 mL of 20% maltose per liter of LB).
5. LB–maltose–MgSO$_4$: Add maltose solution to 0.2% and MgSO$_4$ to 10 mM just before use (10 mL 20% maltose and 10 mL 1 M MgSO$_4$ per liter LB).
6. Maltose solution: 20% Solution, filter sterilize, store at 20–22°C.
7. Glucose solution: 20% Solution, filter sterilize, store at 20–22°C.
8. MgSO$_4$: 1 M stock. Autoclave to sterilize, store at 20–22°C.
9. STET: 50 mM Tris-HCl, pH 8.0, 10 mM EDTA, 8% sucrose; and 5% Triton X-100. Store at 4°C.
10. Lysozyme: 10 mg/mL stock; store in 1.0-aliquots at –20°C.
11. Denaturing buffer: 7 M guanidine, 0.1 M Tris-HCl, pH 8.0, 10 mM EDTA, store at 4°C. Add dithiothreitol (DTT) to 6 mM just before use (30 μL of 1 M DTT stock to 5.0 mL).
12. Refolding buffer: 50 mM Tris-HCl, pH 8.0, 50 mM NaCl. Store at 4°C. Add reduced glutathione to 5 mM and oxidized glutathione to 1 mM just before use.
13. 10 mM phosphate Buffer: Prepare a 10 mM stock of KH$_2$PO$_4$ (monobasic) and a 10 mM stock of Na$_2$HPO$_4$ (dibasic). Add the KH$_2$PO$_4$ to the Na$_2$HPO$_4$ until a pH of 7.2 is obtained (this will require approximately 1 L of KH$_2$PO$_4$ per 1.7 L of Na$_2$HPO$_4$ stock). Store at 4°C.
14. KCl buffer: 0.8 M stock made up in the 10 mM phosphate buffer. Store at 4°C.
15. SM: 5.8 g of NaCl, 2 g of MgSO$_4$ · 7H$_2$O; 50 mL of 1 M Tris-HCl, pH 7.5, and 5 mL of 2% gelatin per liter. Autoclave to sterilize. Store at 20–22°C.
16. SM–glycerol: Add glycerol to SM to final a concentration of 8%.

3. Methods

3.1. Cloning

The molecular cloning of gene fragments is performed according to standard methods *(26)*. In brief, double-stranded plasmid DNA is prepared with a commercially available kit (e.g., QIAprep Spin Miniprep Kit, Qiagen, Santa

Clarita, CA) according the directions of the manufacturer. Oligonucleotides are synthesized on an Applied Biosystems PCR MATE synthesizer and puri-fied on Nensorb Prep cartridges (New England Nuclear, Boston, MA) as directed by the manufacturer.

Following the genetic construction of a new fusion protein toxin the *tox* gene is sequenced by the dideoxynucleotide chain termination method *(27)* as modified by Kraft et al. *(28)* using Sequenase (United States Biochemical, Cleveland, OH) to ensure that the correct translational reading frame is main-tained through the protein fusion junction.

3.2. Expression of Fusion Protein Toxin Genes from the T7 Promoter in Escherichia coli HMS174 Using λ CE6 Induction

1. Inoculate 7 mL of LB–Amp–maltose with a single colony of *E. coli* HMS174 that has been transformed with the appropriate recombinant plasmid, and incu-bate on a roller drum for 18 h at 37°C.
2. Add the 7 mL of the overnight culture to 500 mL LB–Amp–maltose in a 2-L Erlenmeyer flask, and incubate, with shaking, at 37°C until an $OD_{600} = 0.3$ is reached (this incubation usually takes 2–3 h).
3. Add glucose to 4 mg/mL (10 mL of 20% sterile stock solution).
4. Incubate the culture for an additional 2–3 h until the $OD_{600} = 0.8$. Remove 1 mL of the bacterial culture for analysis by sodium dodecyl sulfate-polyacrylamide gel electrophoresis (SDS-PAGE).
5. Add $MgSO_4$ to bacterial culture medium to a final concentration of 10 mM (5 mL of a 1 M solution), and then add 2×10^9 plaque-forming units of coliphage CE6 per milliliter of culture medium.
6. Incubate the culture at 37°C with shaking for an additional 2–3 h. Remove 1 mL for analysis by SDS-PAGE.
7. Harvest the bacterial culture by centrifugation at 5500g for 10 min (bacterial pellets may be stored at –20°C).
8. The 1-mL samples of the bacterial culture that were collected before and after induction of chimeric toxin gene expression are centrifuged and the bacterial pellets resuspended in loading buffer for SDS-PAGE analysis. Western blot analysis of expressed proteins may also be performed using anti-diphtheria tox-oid antibodies and/or antibody directed to the substitute receptor binding domain.

3.3. Expression of Fusion Protein Toxins from the T7 Promoter in Escherichia coli HMS174 (DE3) Using IPTG Induction

1. Inoculate 7 mL of LB–Amp with *E. coli* HMS174 (DE3) that has been trans-formed with the appropriate recombinant plasmid. Incubate, with shaking, at 37°C for 18 h.
2. Add the 7 mL of overnight growth to 500 mL of LB–Amp in a 2-L Erlenmeyer flask. Incubate, with shaking, at 37°C until an $OD_{600} = 0.8$ (this step usually takes

4–6 h). Remove 1 mL of the bacterial culture for SDS-PAGE and immunoblot analysis.

3. Add IPTG to the culture medium to a final concentration of 1 m*M*.
4. Incubate with shaking at 37°C for an additional 2–3 h, and then remove 1 mL of growth for gel analysis.
5. Pellet the cells by centrifugation at 5500*g* for 10 min (the bacterial pellet may be stored at –20°C).
6. The 1-mL samples of bacterial growth collected before and after induction of fusion protein toxin gene expression are centrifuged and the bacterial pellets re-suspended in loading buffer for SDS-PAGE and immunoblot analysis.

3.4. Preparation of Inclusion Bodies from Escherichia coli HMS174 and HMS174 (DE3) following Induction of Fusion Protein Toxin Expression

1. Resuspend the bacterial pellet in 20 mL of ice-cold STET buffer and transfer the mixture to a 30-mL polypropylene tube (Falcon 2059). Homogenize until the pellet is completely resuspended.
2. Add 500 µL of lysozyme stock solution and incubate the suspension on ice for 1 h.
3. Homogenize and then lyse the bacterial cells by sonication (e.g., Fisher 550 Sonic Dismembrator at 9% total output, 5 min total time, 1.5 s on, 1.0 s off.)
4. Centrifuge the suspension at 18,000*g* for 20 min at 4°C.
5. Carefully decant the supernatant fluid, and save an aliquot for SDS-PAGE and immunoblot analysis (*see* **Note 2**). Resuspend the crude inclusion body pellet fraction in 10 mL of ice-cold STET buffer.
6. Homogenize and then sonicate the pellet as described earlier.
7. Centrifuge the suspension at 10,000*g* for 20 min at 4°C.
8. Save the supernatant fraction and resuspend the pellet in 10 mL of ice-cold STET buffer.
9. Homogenize and sonicate as described previously.
10. Centrifuge the suspension at 10,000g for 20 min at 2°C. Save the supernatant fraction for SDS-PAGE and immunoblot analysis (the supernatant fractions may be stored at 4°C and the pellet at –70°C until gel analysis is performed).
11. Use a micropipet tip to remove a small sample of the pellet fraction (i.e., crude inclusion body preparation) and transfer into loading buffer for SDS-PAGE and immunoblot analysis. If the gel analysis indicates that the inclusion body prepa-ration is composed mainly of the fusion protein toxin, the inclusion bodies may then be denatured, and the proteins solubilized and refolded into an active con-formation. If the fusion protein toxin is found in the supernatant fraction(s), alter-native purification procedures must be used (*see* **Note 3**).

3.5. Refolding Partially Purified Fusion Protein Toxins into a Biologically Active Conformation

1. The inclusion body pellet is resuspended in 5 mL of denaturing buffer (*see* **Note 2**).

2. The resuspended pellet is sonicated as described previously to solubilize the recombinant protein.
3. The solubilized protein is then diluted to approx 50 mg/mL in refolding buffer and stirred gently at 4°C for 18 h.

3.6. Determination of Total Protein Concentration

1. Once refolded, the fusion protein toxins are concentrated to approx 50 mL using a 10,000 mol mass cutoff membrane filter (e.g., Omega series, Ultrasette tangential flow device from Filtron Technology Corporation, Northborough, MA).
2. The concentrated sample is then centrifuged at 10,000g, 30 min, 4°C to remove any insoluble material.
3. The fusion protein toxin concentration is determined using Coomassie Protein Assay Reagent (Pierce, Rockford, IL) and SDS-PAGE, native gel, and immunoblot analyses are performed. In general, the final concentration of fusion protein toxin partially purified from a 500-mL culture is in the range of 2–5 mg. Native gel electrophoresis is used to assess the relative concentration of monomeric fusion protein toxin in the preparation. If aggregate forms of the fusion protein are detected, DEAE-Sepharose ion exchange chromatography is performed to further purify the monomeric form.
4. Following purification, aliquots are stored at –70°C until use. (Because many of the fusion protein toxins are sensitive to repetitive cycles of freeze–thaw, aliquots are prepared such that they are thawed only once.)
5. If the concentration of the monomeric fusion protein toxin is too low for future study, the protein can be further concentrated by placing the sample in Spectra/ Por, 12–14,000 mol mass cutoff dialysis tubing (Fisher Scientific) and placing the bag in Carbowax brand polyethylene glycol (PEG) 8000 flakes (Union Carbide Corporation, Danbury, CT). Leave the dialysis bag buried in the PEG flakes until the desired volume is attained.
6. Remove the sample carefully from the dialysis bag, centrifuge the samples to remove any insoluble material, and save the supernatant.
7. Determine protein concentration and perform native gel analysis to assure aggregate forms were not created on further concentration.
8. Store aliquots of the partially purified fusion protein toxin at –70°C until use.

3.7. Purification of Fusion Protein Toxin
by DEAE-Sepharose Ion Exchange Chromatography *(See Note 3)*

1. Centrifuge and filter the 50 mL of concentrated protein to remove insoluble material.
2. Load the sample on a DEAE-Sepharose column (DEAE-Sepharose Fast Flow, Pharmacia Biotech, code no. 17-0709-01) that has been equilibrated in 10 mM phosphate buffer, pH 7.2. We use approx 30 mL of resin in a Bio-Rad chromatography column no. 737–2511. The top of the column is attached to the output of a Pharmacia LKB, P1, peristaltic pump. The bottom of the column is attached to the input of a Pharmacia LKB optical unit, UV-1. The sample is collected from the output of the optical unit. The optical unit is plugged into a Pharmacia

LKB control unit, UV-1, which is in turn connected to a Pharmacia chart strip recorder.

3. The column is washed with the 10 mM phosphate buffer until a baseline absorbance (A_{280nm}) is maintained.
4. The column is then developed with a 0–0.8 M KCl gradient is set up using a gravity gradient mixer with 50 mL, 10 mM phosphate buffer on the side that is attached to the pump and 50 mL of the KCl buffer on the other side.
5. The protein concentration of each fraction is determined and the samples analyzed by SDS-PAGE and native gel electrophoresis and immunoblot.
6. The fractions containing the purified fusion protein toxin are then divided into aliquots and stored at –70°C.

3.8. Preparation of Coliphage λ CE6

1. Inoculate one colony of *E. coli* LE392 into 5 mL of LB-maltose and incubate for 18 h, with shaking, at 37°C.
2. Add 2.5 mL of the overnight culture to 22.5 mL of LB-maltose and incubate at 37°C, with shaking, until an OD_{600} between 0.3 and 0.6 is reached.
3. To 300 μL of the log phase LE392 culture, add CE6 (Novagen). We generally set up four different CE6 concentrations to establish the correct ratio of phage to bacteria such that complete bacterial lysis will occur:
   Tube 1: 300 μL LE392 + 10^6 CE6
   Tube 2: 300 μL LE392 + 10^7 CE6
   Tube 3: 300 μL LE392 + 10^8 CE6
   Tube 4: 300 μL LE392 + 10^9 CE6
4. Incubate the tubes at 20–22°C for 20 min to allow the CE6 to adsorb to the bacteria.
5. Inoculate each tube of LE392 and CE6 into each of four flasks containing 500 mL of LB–maltose–$MgSO_4$. Incubate at 37°C with shaking. Bacterial lysis should occur after 12–16 h incubation.
6. After the incubation period, add 2 mL of chloroform to each flask and shake 15 min at 37°C. The chloroform will induce lysis in any cultures in which infection is completed but lysis has not yet occurred.
7. Combine the four flasks and centrifuge the culture at 4000g for 30 min to remove whole cells and debris.
8. Pour the supernatant fluid into clean flasks and add DNase I and RNase A to final concentrations of 1 μg/mL (Boehringer Mannheim). Agitate gently at 20–22°C for 30 min.
9. Add PEG 8000 to 10% and NaCl to 1 M. Stir gently at 20–22°C until the PEG and NaCl are in solution. The flasks are then incubated at 4°C for 18 h with gentle shaking to precipitate the phage.
10. The phage are harvested by centrifugation at 4000g, 4°C, for 30 min.
11. The supernatant fluid is decanted and the pellets are resuspend in SM buffer (5 mL of SM buffer for each 500 mL of starting material).
12. Transfer the CE6 phage preparation to a chloroform-resistant centrifuge tube and add an equal amount of chloroform for extraction. Vortex-mix well.

13. Centrifuge at 6000g for 5 min and remove the upper, aqueous phase, which contains the phage particles.
14. We generally extract CE6 preparations with chloroform 2 or 3 times until no PEG remains in the interface.
15. Titer the phage on lawns of *E. coli* LE392 according to standard methods. Determine the total number of plaque forming units (pfus) and adjust the concentration with SM–glycerol buffer to 10^{12}pfus/mL and store at –70°C, in 1-mL aliquots.

4. Notes

1. In the construction of a new fusion protein toxin, we routinely transform the recombinant plasmids into both HMS174 and HMS174 (DE3) and perform CE6 and IPTG induction of toxin gene expression. On occasion, we have found that expression of a given DT-based fusion toxin will inhibit bacterial growth in the IPTG system and the HMS174 (DE3) transformants will take longer (40 vs 16 h) to form visible colonies. In these instances, we only use the HMS174/CE6 expression system. We have found that leaky expression of the fusion protein toxin can result in selection of plasmids encoding mutant forms.
2. It is important to employ SDS-PAGE and native gel electrophoresis and immunoblot analysis to characterize the induction and purification of the desired protein. The 1-mL aliquots removed from the bacterial cultures before and after induction should be centrifuged and the bacterial pellet resuspended in 200 µL of denaturing, loading buffer. Because many of the fusion protein toxins are in the 60-kDa range, we routinely use 12% polyacrylamide under denaturing conditions and 7% polyacrylamide for native gel analysis to assess expression and refolding. If antibodies to either the diphtheria toxin component or the substitute ligand component are available, immunoblot analysis should be performed to further characterize the fusion protein toxin and possible proteolytic digestion products.
3. The purification of monomeric from aggregate forms of a fusion protein toxin can usually be accomplished by ion-exchange chromatography on DEAE-Sepharose. In those cases where separation proves difficult, native polyacrylamide gel electrophoresis on a Bio-Rad model 491 Prep Cell apparatus may be used an an alternative method *(15)*.

References

1. Brinkman, U., Reiter, Y., Jung, S-H., Lee, B., and Pastan, I. (1993) A recombinant immunotoxin containing a disulfide-stabilized Fv fragment (dsFv). *Proc. Natl. Acad. Sci. USA* **90,** 7538–7542.
2. Murphy, J. R. and vanderSpek, J. C. (1995) Targeting diphtheria toxin to growth factor receptors. *Semin. Cancer Biol.* **6,** 259–267.
3. Kreitman, R. J. and Pastan, I. (1995) Targeting Pseudomonas exotoxin to hematologic malignancies. *Semin. Cancer Biol.* **6,** 297–306.
4. Murphy, J. R., Bishai, W., Borowski, M., Miyanohara, A., Boyd, J., and Nagle, S. (1986) Genetic construction, expression and melanoma-selective cytotoxicity of

a diphtheria toxin-α-melanocyte stimulating hormone fusion protein. *Proc. Natl. Acad. Sci. USA* **83,** 8258–8262.

5. Williams, D., Parker, K., Bishai, W., Borowski, M., Genbauffe, F., Strom, T. B., and Murphy, J. R. (1987) Diphtheria-toxin receptor binding domain substitution with interleukin-2: genetic construction and properties of a diphtheria toxin-related interleukin-2 fusion protein. *Protein Engng.* **1,** 493–498.

6. Lakkis, F., Steele, A., Pacheco-Silva, A., Kelley, V. E., Strom, T. B., and Murphy, J. R. (1991) Interleukin-4 receptor targeted cytotoxicity: genetic construction and properties of diphtheria toxin-related interleukin-4 fusion toxins. *Eur. J. Immunol.* **21,** 2253–2258.

7. Aullo, P., Alcani, J., Popoff, M. R., Klatzman, D. R. Murphy, J. R., and Boquet, P. (1992) *In vitro* effects of a recombinant diphtheria-human CD4 fusion toxin on acute and chronically HIV-1 infected cells. *EMBO J.* **12,** 921–931.

8. Jean, L.-F. and Murphy, J. R. (1992) Diphtheria toxin receptor binding domain substitution with interleukin-6: genetic construction and interleukin-6 receptor specific action of a diphtheria toxin-related interleukin-6 fusion protein. *Protein Engng.* **4,** 989–994.

9. vanderSpek, J. C., Sutherland, J., Sampson, E., and Murphy, J. R. (1995) Genetic construction and characterization of the diphtheria toxin-related interleukin 15 fusion protein DAB_{389} sIL-15. *Protein Engng.* **8,** 1317–1321.

10. Fisher, C. E., Sutherland, J. A., Krause, J. E., Murphy, J. R., Leeman, S. E., and vanderSpek, J. C. (1996) Genetic construction and properties of a diphtheria toxin-related substance P fusion protein: *in vitro* destruction of cells bearing substance P receptors. *Proc. Natl. Acad. Sci. USA* **93,** 7341–7345.

11. vanderSpek, J. C., Sutherland, J. A., Zeng, H., Battey, J. F., Jensen, R. T., and Murphy, J. R. (1997) Inhibition of protein synthesis in small cell lung cancer cells induced by the diptheria toxin-related fusion protein DAB_{389} GRP. *Cancer Res.* **57,** 290–294.

12. Sweeney, E. B., Foss, F. M., Murphy, J. R., and vanderSpek, J. C. (1998) Interleukin 7 (IL-7) receptor-specific cell killing by DAB_{389} IL-7: a novel agent for the elimination of IL-7 receptor positive cells. *Bioconj. Chem.* **9,** 201–207.

13. Williams, D., Snider, C. E., Strom, T. B., and Murphy, J. R. (1990) Structure function analysis of IL-2 toxin (DAB_{486}IL-2): fragment B sequences required for the delivery of fragment A to the cytosol of target cells. *J. Biol. Chem.* **265,** 11,885–11,889.

14. Williams, D. P., Wen, Z., Watson, R. S., Boyd, J., Strom, T. B., and Murphy, J. R. (1990) Cellular processing of the interleukin-2 fusion toxin DAB_{486}-IL-2 and efficient delivery of diphtheria fragment A to the cytosol of target cells requires Arg[194]. *J. Biol. Chem.* **265,** 20673–20677.

15. vanderSpek, J. C., Mindell, J. A., Finkelstein, A., and Murphy, J. R. (1993) Structure/function analysis of the transmembrane domain DAB_{389}_IL-2, an interleukin-2 receptor-targeted fusion toxin. *J. Biol. Chem.* **268,** 12077–12082.

16. vanderSpek, J. C., Howland, K., Friedman, T., and Murphy, J. R. (1994) Maintenance of the hydrophobic face of the diphtheria toxin amphipathic transmembrane

helix 1 is essential for the efficient delivery of the catalytic domain to the cytosol of target cells. *Protein Engng.* **7,** 985–989.

17. vanderSpek, J. C., Cassidy, D., Genbauffe, F., Huynh, P. D., and Murphy, J. R. (1994) An intact transmembrane helix 9 is essential for the efficient delivery of the diphtheria toxin catalytic domain to the cytosol of target cells. *J. Biol. Chem.* **269,** 21,455–21,459.

18. vanderSpek, J. C., Sutherland, J. A., Ratnarathorn, M., Howland, K., Ciardelli, T. L., and Murphy, J. R. (1996) DAB$_{389}$ IL-2 receptor binding domain mutations: cytotoxic probes for studies of ligand/receptor interactions. *J. Biol. Chem.* **271,** 2145–12149.

19. Hu, H.-Y., Huynh, P. D., Murphy, J. R., and vanderSpek, J. C. (1998) The effects of helix breaking mutations in the diphtheria toxin transmembrane domain helix layers of the fusion toxin DAB$_{389}$ IL-2. *Protein Engng.* **11,** 101–107.

20. Robinson, E. A., Henriksen, O., and Maxwell, E. S. (1974) Elongation factor 2; amino acid sequence at the site of adenosine diphosphate ribosylation. *J. Biol. Chem.* **249,** 5088–5093.

21. Van Ness, B. G., Howard, J. B., and Bodley, J. W. (1980) ADP-ribosylation of elongation factor 2 by diphtheria toxin: isolation and properties of the novel ribosyl-amino acid and its hydrolysis products. *J. Biol. Chem.* **255,** 10,717–10,720.

22. Yamaizumi, K., Mekada, E., Uchida, T., and Okada, Y. (1978) One molecule of diphtheria toxin fragment A introduced into a cell can kill the cell. *Cell* **15,** 245–250.

23. Studier, F. W. and Moffatt, B. A. (1986) Use of bacteriophage T7 RNA polymerase to direct selective high-level expression of cloned genes. *J. Mol. Biol.* **189,** 113–130.

24. Rosenberg, A. H., Lade, B. N., Chui, D., Lin, S., Dunn, J. J., and Studier, F. W. (1987) Vectors for selective expression of cloned DNAs by T7 RNA polymerase. *Gene* **56,** 125–135.

25. Studier, F. W., Rosenberg, A. H., Dunn, J. J., and Dubendorff, J. W. (1990) Use of T7 RNA polymerase to direct expression of cloned genes. *Methods Enzymol.* **185,** 60–89.

26. Ausubel, F. M., Brent, R., Kingston, R. E., Moore, D. D., Seidman, J. G., Smith, J. A., and Struhl, K. (1988) *Current Protocols in Molecular Biology*, John Wiley & Sons, New York.

27. Sanger, F., Nicklen, S., and Coulsen, A. R. (1977) DNA sequencing with chain-terminating inhibitors. *Proc. Natl. Acad. Sci. USA* **74,** 5463–5467.

28. Kraft, R., Tardiff, J., Krauter, K. S., and Leinwand, L. A. (1988) Using mini-prep plasmid DNA for sequencing double stranded template with sequenase. *BioTechniques* **6,** 544–547.

6

Structure–Function Analysis
of Cysteine-Engineered Entomopathogenic Toxins

Jean-Louis Schwartz and Luke Masson

1. Introduction

The availability of bacterial toxin genes and, in many cases, their atomic structures permits, through properly designed structure–function studies, the precise mapping of the molecular determinants of their activity. The results of such studies are of particular importance for the elucidation of the mode of action of the toxins and the design of genetically engineered molecules displaying novel activity. For example, a large effort is currently underway to design, through mutagenesis and chimeric constructs, the next generation of *Bacillus thuringiensis* toxin-based pesticides with different specificities, improved toxicity, and the potential to delay pest resistance *(1)*.

Several bacterial toxins assume a globular structure in the water-soluble state. Among them, at least two representatives of *B. thuringiensis* toxins, Cry1Aa and Cry3Aa, as well as colicin A and diphtheria toxin, share structurally similar α-helical bundles in which one or more hydrophobic helices, long enough to span a membrane, are found embedded between the other α-helices *(2)*. This common structure suggests a similar mechanism of membrane penetration whereby these water-soluble toxins undergo a conformational change to permit two or more helices to partition into a lipid membrane and form a functional pore.

A powerful strategy to probe the role of conformational changes is to stabilize the protein by means of one or more strategically located disulfide bridges *(3–5)*. This approach, which makes use either of available cysteines or of cysteines introduced by mutagenesis, is particularly appealing when there are very few or no cysteine residues, such as in trypsin-activated *B. thuringiensis* Cry1A

From: *Methods in Molecular Biology, vol. 145: Bacterial Toxins: Methods and Protocols*
Edited by: O. Holst © Humana Press Inc., Totowa, NJ

toxins *(6)*. If the engineered disulfide bridges are sufficiently exposed to the solvent to be accessible to a reducing agent, it may be possible to break the S–S bond during a functional assay and therefore to investigate the relationship between increased flexibility and activity.

Important biophysical properties of channel-forming bacterial toxins can be derived from electrophysiological data *(7)*. The proteins of interest are reconstituted in an artificial phospholipid environment and their ion transport properties are measured under various conditions *(8)*. This provides unique information at the single-channel level on conductance (the capability to pass a given amount of charged particles per unit of time), selectivity (the ability of the channel to discriminate between different ion types), voltage dependence (the fact that the channel conductance may depend on applied voltage), and kinetic behavior (the mean open and closed times of the channel and its gating properties, i.e., how it alternates between its closed and conducing states). Furthermore, single-channel data analysis may also indicate the level of oligomerization of the molecules and whether they behave in a cooperative manner.

The judicious combination of structural data information, cysteine mutagenesis, and functional studies constitutes a powerful approach for the study of the role of flexibility in the mode of action of bacterial toxins. In this chapter, using a *B. thuringiensis* toxin as an example *(9)*, we describe the procedures used to assess *in situ* the effect of protein conformation changes on its channel activity in planar lipid bilayers. Cysteines are introduced by site-directed mutagenesis of the three-domain protein to create intramolecular disulfide bonds within domain I, the putative pore-forming region. To investigate the mechanism of toxin insertion and channel formation in planar lipid bilayers (PLBs), these disulfide bridges are strategically located to restrict the flexibility of helix movement within domain I. Single ion channel recording is performed before and after addition of the toxin in its oxidized state to the aqueous phase on one side of a lipid membrane formed using the painted bilayer technique *(10)*. The effect of a reducing agent on channel activity is then investigated.

2. Materials

2.1. Molecular Biology Equipment and Reagents

1. Clontech Transformer™ kit (cat. no. K1600-1), Clontech Laboratories, Palo Alto, CA.
2. CrylAa toxin from *B. thuringiensis* or any purified membrane pore-forming protein that lacks cysteines.
3. Sequencing reagents. We use an automated fluorescent sequencer from Applied Biosystems model 370A (Foster City, CA).

4. 15-mL Quick-Seal heat sealable tubes (Beckman Instr., Palo Alto, CA).
5. HPLC (Waters 650, Millipore, Milford, MA) equipped with a 10 cm × 5 mm Q-Sepharose anion exchange resin Fast Flow column (Pharmacia LKB, Uppsala, Sweden).
6. Mini-Protean II slab gel apparatus (Bio-Rad Laboratories, Mississauga, Ont, Canada).
7. French pressure cell (Spectronic Instruments, Rochester, NY).

2.2. Solutions and Buffers

1. Solubilization and activation solution: carbonate buffer: 40 mM Na$_2$CO$_3$, pH 10.0.
2. Trypsin (Boehringer-Mannheim, Laval, Qc, Canada): 1 mg/mL aliquots stored at –80°C. The activation protocol is readjusted for every new batch of trypsin (enzyme concentration and activation duration).
3. Protein assay dye reagent (cat. no. 500-0006), Bio-Rad, Mississauga, Ont, Canada.
4. Bovine serum albumin (BSA) stock solution for standard curve production (2.0 mg/mL in distilled water).
5. 10 mM potassium phosphate buffer, ph 6.5.
6. Solution for agar bridges: 200 mM KCl, 1 mM EDTA in 2% agar.
7. Filling solution for half-cell reservoirs: 200 mM KCl, 1 mM EDTA.
8. PLB KCl buffer: 150 mM KCl, 1 mM CaCl$_2$, 10 mM Tris-HCl, pH 9.5.
9. Stock solution for raising concentration of PLB KCl buffer: 3 M KCl.
10. Reducing agent stock solution: 500 mM β-mercaptoethanol (MEt) in water.

2.3. Lipids

1. Heart phosphatidylcholine (PC) and phosphatidylethanolamine (PE) in chloroform at 10 mg/mL from Avanti Polar Lipids (Alabaster, AL). Store at –20°C.
2. Cholesterol (Ch) in powder from Sigma (St. Louis, MO). Store at –20°C. Cholesterol stock solution (10 mg/mL in chloroform) stored at –20°C.
3. Open vials of lipids are always manipulated under nitrogen flow. Lipids are then used within 3 mo.

2.4. Planar Lipid Bilayer (PLB) Setup and Electronic Equipment

1. Air table to suppress mechanical vibrations.
2. Faraday enclosure to reduce electrical interference.
3. Low-power binocular microscope (×60 magnification) and fiberoptic illuminator to observe membrane thinning.
4. Bilayer experimental block in which a 1.8-mL *trans* chamber is milled. It holds the *cis* chamber made of a 1-mL Delrin cup. Three to five Delrin cups with a thinned wall 75 μm thick in which a circular hole 250 μm in diameter has been drilled. The bilayer block with Delrin cups is also available from Warner Instruments, Hamden, CT, (cat. no. BCH-13 or BCH-22).
5. Heat-sealed Pasteur pipet for painting the lipid membrane.
6. Injection system (hand pipettor mounted on a three-way precision micromanipulator and equipped with a polyethylene injection tube (1.5 mm outside diameter).

7. Magnetic fleas and stirring apparatus located under the PLB chambers.
8. Electrode system made of silver–silver chloride pellets, agar bridges, and small plastic reservoirs (heat-sealed pipettor tips).
9. Electronic recording, display and processing apparatus: bilayer/patch-clamp amplifier (BC-525C, Warner Instruments, Hamden, CT), dual-channel digital oscilloscope (OS-3020, GoldStar, Korea), eight-pole analog Bessel filter (Frequency Devices, Haverhill, MA), 125-kHz analog-to-digital converter (TL-125 Labmaster, Axon Instruments, Foster City, CA), and a personal computer.

3. Methods

3.1. Disulfide Bridge Mutants

The rationale for the design of bacterial toxin double-cysteine mutants and their functional assay using the PLB technique is explained in **Note 1**. Moreover, single-cysteine mutagenesis can also provide interesting mutants to be tested in PLBs for conformational changes taking place when the toxin integrates into the membrane, or for mapping the toxin's topology (*see* **Note 2**).

3.1.1. Mutagenesis

A protein with no cysteines is required for this work. If the protein of interest contains one or two cysteines, one could consider mutating these native cysteins to a different amino acid and ascertain any deleterious effect on the molecule. Knowledge of the atomic structure of the protein is also essential. Determine potential sites for cysteine conversion by meeting the following criteria:

a. Select amino acids within the regions targeted for bridging having a β-carbon distance within 3.6–5.4 Å.

b. Ascertain that the potential sulfur atoms are within 2 Å of each other.

In the case of the Cry1Aa toxin, the potential sites for amino acid conversion to cysteines are determined by examination of the Cry1Aa toxin atomic structure *(6)*.

1. Synthesize mutagenic oligonucleotides. A number of procedures and kits for in vitro oligonucleotide-directed mutagenesis are available. The Clontech "Transformer" kit is highly recommended, as it can be done directly on the double-stranded DNA plasmid that expresses the recombinant protein *(9)*.
2. In these experiments we create five bridged mutants in the cysteine-less Cry1Aa toxin with all mutants designed for restricted flexibility within the α-helical domain I (*see* **Fig. 1**): MP298 (Trp73Cys and Ile97Cys) links α_{2b} to α_3, MP159 (Arg99Cys and Ala144Cys) links the middle of α_3 to α_4, MP169 (Ile88Cys and Tyr153Cys) links the interhelical loop of $\alpha_2\alpha_3$ with that of $\alpha_4\alpha_5$, MP178 (Val162Cys and Ala207Cys) links the middle of α_5 to α_6, and MP206 (Ser176Cys and Ser252Cys) links the C-terminal ends of α_5 and α_7.
3. All Cry1Aa mutants were sequenced to avoid unwanted mutations.

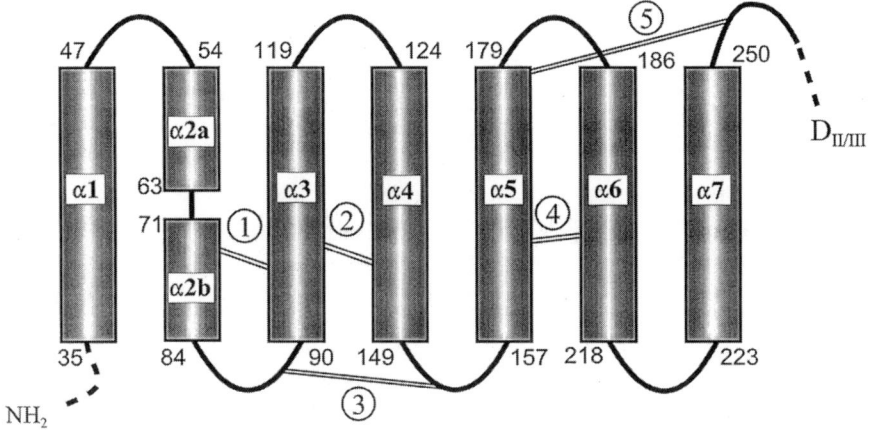

1. **MP298** Trp73Cys - Ile97Cys
2. **MP159** Arg99Cys - Ala144Cys
3. **MP169** Ile88Cys - Tyr153Cys
4. **MP178** Val162Cys - Ala207Cys
5. **MP206** Ser176Cys - Ser252Cys

Fig. 1. Schematic representation of domain I of *B. thuringiensis* bacterial toxin Cry1Aa. The atomic structure determined by X-ray crystallography *(6)* shows that it is made of seven α-helices arranged as a cylindrical bundle with helix α_5 in the central position. Next to the extremities of the helices are the residue numbers in the protein sequence. The *double lines* show the approximate positions of the engineered disulfide bonds. Numbers in *circles* refer to the table of mutants listed at the bottom of the figure. $D_{II/III}$ stands for domains II and III.

3.1.2. Protein Expression, Purification, and Activation

3.1.2.1. PROTOXIN PRODUCTION.

1. All mutant and native Cry1Aa protoxins are expressed as recombinant proteins in *Escherichia coli* HB101 (*supE*44 *hsdS*20 ($r_B^-m_B^-$) *recA*13 *ara*-14 *proA*2 *lacY*1 *galK*2 *rpsL*20 *xyl*-5 *mtl*-1) by growing for 2 d at 30°C in 1 L of double-strength yeast tryptone broth containing ampicillin at 100 μg/mL *(11)*.
2. *E. coli* cells are harvested by centrifugation (7000*g*, 10 min) and resuspended in 30 mL of 100 m*M* phosphate buffer, pH 6.5.
3. The cell suspension is lysed by two passages though a French pressure cell at 11,000 psi internal pressure.
4. The insoluble material is collected by centrifugation at 10,000*g* for 10 min, resuspended in 250 mL of water and recentrifuged at 10,000*g* for 10 min.
5. The previous wash step is repeated and the final pellet (protoxin) is stored in 10 mL of water at 4°C.

3.1.2.2. Protoxin Solubilization and Activation, and Toxin Purification

1. Mix 25 mg of protoxin in 1.5 mL of carbonate buffer containing 1 mL of trypsin in a 50-mL plastic test tube. Add ultrapure water to bring volume to 13 mL. Incubate at 34°C for 2 h. Check pH after 1 h and readjust to 10.0 with NaOH if necessary.
2. Transfer into 15-mL Quick-Seal tubes.
3. Centrifuge at 200,000g at 15°C for 90 min (L8-70M ultracentrifuge, Beckman Instr., Palo Alto, CA).
4. Collect supernatant in a 50-mL plastic test tube. Keep on ice until injection in HPLC apparatus.
5. After injection into the column (0.5 mL/min flow rate), the bound 65-kDa trypsin-resistant toxin is eluted using a gradient program of 50–500 mM aqueous NaCl. The fractions (1 mL) are collected and those corresponding to the activated toxin peak (280 nm peak at 350–400 mM NaCl) are pooled and dialyzed 3× against distilled water in a 2-L plastic beaker for at least 24 h until full precipitation of the toxin.
6. The content of the dialysis bag is transferred to a 1.5-mL Eppendorf tube for immediate use in PLB experiments. It may also be stored at 4°C for several weeks.
7. Toxin purity is tested by sodium dodecyl sulfate-polyacrylamide gel electrophoresis (SDS-PAGE). A single band is observed around 65 kDa.
8. It is verified that the bridged clones retain similar biochemical properties as the parental Cry1Aa protoxin with respect to expression, solubility at alkaline pH, and trypsin sensitivity.

3.2. Planar Lipid Bilayers

PLBs can be formed using the painted bilayer technique *(10)*, the folded bilayer technique *(12)* or the tip-dip technique *(13)*. We use the painted bilayer approach in which the membrane is formed by spreading from bulk solution (*see* **Note 6**). The folded bilayer approach is described in detail in Chapter 10.

3.2.1. Preparation of Toxin Stock Solution for PLB Experiments

1. A 250-µL aliquot is centrifuged in a table centrifuge (Heraeus Instr., Baxter Diagnostics, Montreal, Qc, Canada) for 2 min (13,000g) at 20–22°C. The pellet is reconstituted in 500 µL of PLB KCl buffer and vortex-mixed for 1 min.
2. Protein concentration is measured by the method of Bradford *(14)* using BSA as a standard. The final stock solution concentration is adjusted to 1 mg/mL of toxin.

3.2.2. Preparation of the Lipid Mixture for PLB Membranes

1. Mix 70 µL of PE, 20 µL of PC, and 10 µL of Ch in a 1-mL glass V-vial (Wheaton, Millville, NJ). Evaporate for 10 min under gentle nitrogen flow.
2. Reconstitute the lipids in 50 µL of decane in the same vial. Vortex-mix for 10 s. Store on ice in the dark.

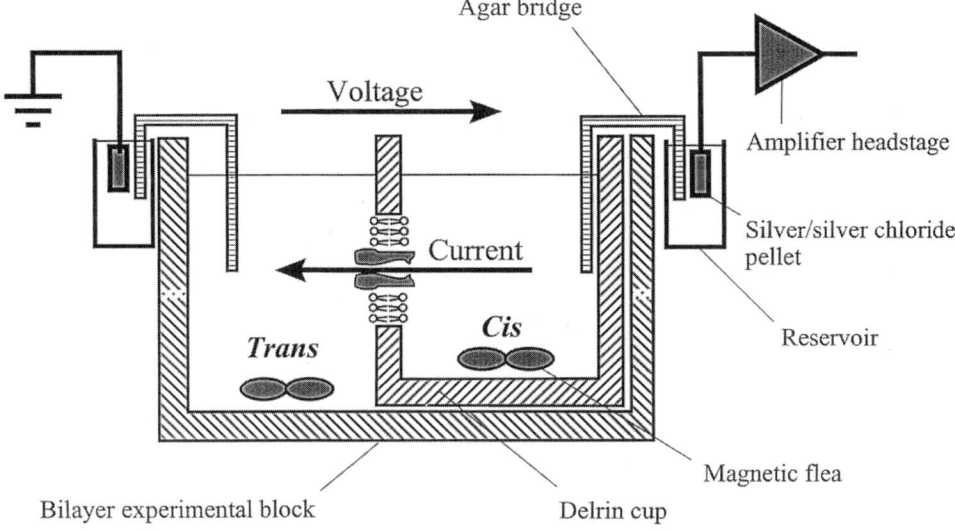

Fig. 2. Planar lipid bilayer experimental apparatus for painted bilayer membranes. The electronic equipment listed in **Subheading 2.4. (item 9)** is connected to the output of the amplifier headstage. A small battery-powered magnetic stirrer (*not shown*) is located under the bilayer experimental block to drive the magnetic fleas.

3.2.3. Pretreatment of Delrin Cup Apertures

Several cups are prepared in advance as follows:

1. Deposit a small drop (about 0.4 µL) of the lipid mixture on the cup aperture.
2. Evaporate under binocular microscope observation (×60 magnification) to ensure that a uniform ring of lipid has formed around the hole.
3. Dry cup under nitrogen flow for 10–20 min. Store in a clean container.

3.2.4. PLB Setup Preparation

The experimental apparatus used for studying channels in painted bilayer membranes is shown in **Fig. 2** (*see* Chapter 10 for description of the setup used with folded bilayer membranes).

1. Warm up electronic instruments for 30 min.
2. Fill electrode reservoirs with electrode reservoir solution. Place them in the bilayer experimental block.
3. Insert Delrin cup in bilayer experimental block.
4. Fill cup (the *cis* chamber) with 0.8 mL of PLB KCl buffer and the *trans* chamber with 1.6 mL of PLB KCl buffer.
5. Locate agar bridges between the electrode reservoirs and the chambers.

6. Immerse silver pellets in electrode reservoirs. Connect the *cis* chamber electrode to the recording amplifier headstage and the *trans* chamber electrode to ground.
7. Adjust junction potential using appropriate bilayer amplifier controls.
8. Paint membrane on cup aperture using heat-sealed tips of clean Pasteur pipets dipped in lipid-containing V-vial. Bilayer painting is done under binocular microscope observation of the aperture illuminated by a fiberoptic illuminator.
9. Monitor membrane formation visually through the binocular microscope. Thinning is complete when the membrane turns black.
10. Switch off the illuminator and close the lid of the Faraday enclosure tightly.
11. Check membrane quality through electrical capacitance determination using the capacitance measurement function of the bilayer amplifier. Membrane capacitance should be in the 150–200 pF range.

3.2.5. Experimental Protocol

It is assumed that, as a result of double-cysteine mutagenesis, disulfide bridges are successfully formed in the oxidized state (*see* **Note 3**). However, particularly when the atomic structure of the bacterial toxin under study is not available, it cannot be excluded that disulfide bridge formation does not occur (*see* **Note 4**).

All experiments are performed at 20–22°C.

1. Control period (20 min). Without added protein, membrane stability or contamination are monitored with ±50 mV applied square pulses.
2. Pretest period (20 min). The protein (1–20 µg/mL, i.e., 15–300 n*M*) is injected into the *cis* chamber in the close vicinity of the membrane. Stir for 30 s. Channel activity is continuously monitored by step changes in the current recorded with a –50 mV holding test voltage applied across the bilayer.
3. Add rapidly the reducing agent (5–10 m*M* MEt) to the *cis* chamber.
4. Test period (minimum 30 min). Apply a –50 mV holding voltage immediately. Following a 30-s observation period, stir for 30 s. Transmembrane current is continuously recorded.
5. If no activity is observed after 15 min, terminate experiment.
6. If channel activity is observed, use the standard procedures (as described in detail in Chapter 10) to characterize channel properties: apply a set of voltage steps for conductance measurement and vary the ionic gradient across the membrane for selectivity determination *(15)*.

3.2.6. Controls

Verify that:

1. The addition of 0.1–20 m*M* MEt alone to the *cis* chamber does not affect the electrical properties or the stability of the lipid bilayer and that it does not induce channel activity.

2. The channels formed by the parental toxin Cry1Aa *(6)* are unaffected by the reducing agent at doses ranging from 0.1 to 20 mM.

3.2.7. Data Analysis

Analysis is performed using pClamp and Axotape softwares (Axon Instruments, Foster City, CA). Examination of the channel properties of the mutant proteins may reveal that, even in the reduced state, they differ from the parental toxin (*see* **Note 5**).

4. Notes

1. Experimental rationale and interpretation of results: The rationale behind the design proposed in **Subheading 3.1.1.** is to test the hypothesis that activity of the mutants, that is, functional channel formation, is related to a conformational change taking place in domain I upon insertion in the membrane (refer to **Fig. 1** for mutation site locations). More precisely, it is proposed that membrane permeation requires the initial insertion of an α-helical hairpin. By locking this hairpin to a neighboring region with a disulfide bond, insertion will be prevented and no channel activity will be observed in the PLB. Upon reduction of the bond, the hairpin will be freed and allowed to insert in the membrane, which will be detected by the onset of channel activity. **Figure 3** shows the experimental data obtained with mutant MP159 in which a disulfide bridge has been engineered between helix α_3 and helix α_4, and mutant MP298 where helix α_{2b} and helix α_3 have been locked together. MP159 is silent in its oxidized state and becomes active upon reduction by 10 mM MEt, implying that either α_3 or α_4 (or both) are involved. On the other hand, MP298 is active both in the pretest and in the test periods, which suggests that neither α_{2b} nor α_3 movement is required for channel formation. Therefore the implication of α_4 involvement in addition to the PLB behavior of several other disulfide mutants constructed in Domain I of Cry1Aa led us to conclude that the insertion of the $\alpha_4\alpha_5$ hairpin is a necessary step for the formation of a functional channel in the bilayer membrane *(9)*.
2. Using cysteine mutagenesis to map the rearrangement or topology of a protein molecule within a membrane: It would be useful during the planning stage of the mutagenesis to consider the strategic location of single cysteines in addition to double-cysteine mutations. As there are several reagents commercially available that can covalently attach numerous side groups to an exposed sulfhydryl, complementary experiments to the bridged mutants could be possible. For example, a detectable group such as biotin can be attached to cysteines located in interhelical loop regions before or after membrane integration. One could then monitor for an effect on channel properties by adding avidin or streptavidin to either the *cis* or the *trans* side of the membrane to provide clues on helical hairpin translocations *(16)*. Alternatively, one could map the channel mouth or lumen by monitoring altered PLB activity *in situ* after the addition of charged alkylating methanethiosulfonate (MTS) reagents such as MTS-ES or MTS-EA. These

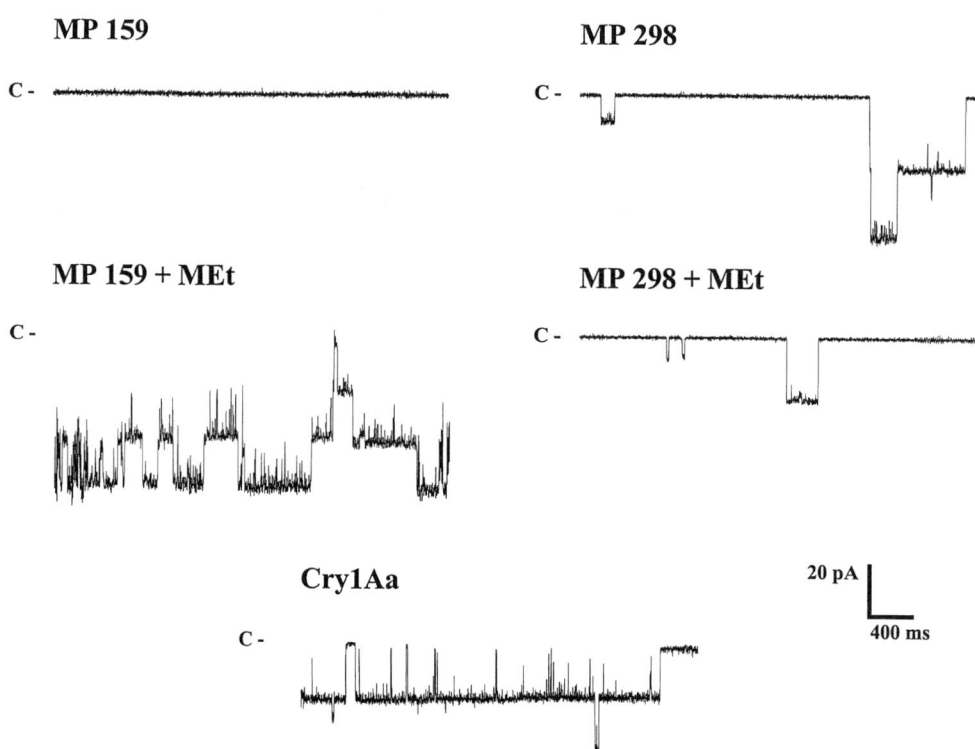

Fig. 3. Representative current recordings from mutant MP159 (*left*) and MP298 (*right*) before (*top traces*) and after addition (*bottom traces*) of 10 m*M* MEt. The *bottom trace* shows the channel current of the wild-type Cry1Aa toxin. Experiments were conducted under symmetrical conditions (150:150 m*M* KCl *cis:trans*). Applied voltage: –40 mV. The letter *c* on the left side of each trace indicates the closed state level.

reagents specifically add a negative or a positive charge to the free sulhydryl on cysteines and potentially alter the ion flow through the channel if the cysteine residue is exposed to the solvent in a sensitive region of the lumen *(17)*.

3. Disulfide bridge formation: The PLB approach provides an unequivocal, *a posteriori* demonstration of the presence of disulfide bridges in the protein under consideration if, in the pretest period, that is, before addition of MEt, no channel activity is recorded, whereas during the test period following the addition of the reducing agent, the protein displays normal electrophysiological activity. As the presence of a small amount of molecules lacking disulfide bonds cannot be excluded, it is recommended to conduct the PLB experiments using the lowest possible dose of mutant proteins to minimize the likelihood for these molecules to form channels in the pretest period. However, if this happens, a large increase in activity after addition of MEt would nevertheless demonstrate that most of the

mutant molecules were indeed stabilized by disulfide bonds before being exposed to the reducing agent.

4. Limitations of the method: The functional approach described in this chapter is a robust method for demonstrating the need for particular regional flexibility for channel activity as long as a clear change is observed in the mutant's activity in PLB before and after addition of a reducing agent. The ideal situation is when there is no activity with the oxidized mutant (in the pretest period) and a normal, wild-type activity after reduction of the protein by MEt in the test period (*see* **Fig. 3**, MP159). However, if normal activity is observed in the pretest period, failure of disulfide bond formation cannot be totally excluded. This is particularly true if the mutants were made in a protein for which the exact atomic structure is not available. Therefore the presence of disulfide bonds may have to be checked before conducting the PLB experiments. Several methods are available, including quantifying free thiols using Ellman's reagent (dithiobisnitrobenzoic acid) which forms a bright yellow aromatic thiol (nitrothiobenzoate) with free sulfhydryls (*18*) or secondary structure and thermal stability analysis by circular dichroism and ultraviolet absorbance spectroscopy (*19*).

5. Other considerations: Altered channel behavior in comparison to that of the wild-type protein may be observed with cysteine mutants. The introduction of cysteine residues in a pore-forming protein may conceivably affect the biophysical properties of the channel even if no disulfide bonds are formed or, alternatively, the engineered disulfide bridges may be located in a region of the protein that interacts with the functional region that makes the channel, thus modifying its activity. However, even with ion channel properties that deviate from those of the wild-type channel, if the region is important in membrane permeation, the oxidized form of the mutant should show little or no activity and the reduced mutant should form channels at a dose and speed similar to that of the wild-type protein.

6. PLB techniques: In the painted lipid bilayer technique (*10*), the membrane is formed on a pretreated aperture using a brush, a Teflon stick or a fire-polished glass rod dipped in the lipid mixture (lipids in decane). The painted bilayer setup, which is commercially available, is simpler than the one used for folded bilayers (*[12]*, Chapter 11 in this volume): there is no need for a delicate Teflon foil puncturing procedure and a rather complex chamber assembly, including the apparatus for liquid level adjustment as required for folded membrane formation. Also, the painted bilayer approach is obviously more economical than the tip-dip approach (*13*), which requires most of the sophisticated equipment of a full patch-clamp setup. However, thinning of painted membranes results in the accumulation of solvent in the annulus area of the Delrin cup aperture, and in some solvent remaining trapped in the bilayer itself. Therefore the membrane is thicker than folded membranes or biological membranes, and it may be sensitive to electro-compression. Furthermore, the painted bilayer approach cannot be used for asymmetrical membrane formation. On the other hand, folded bilayers made on Teflon foil apertures are not totally solvent free, as aperture pretreatment with pentane

or hexadecane is required to stabilize the bilayer. In the case of *B. thuringiensis* toxins, channels behave the same way in painted or in folded bilayers *(20)*.

Acknowledgments

The authors gratefully acknowledge the helpful advice of Dr. R. Brousseau, Dr. M. Cygler, and Dr. P. Grochulski from the Biotechnology Research Institute, National Research Council of Canada, of Dr. H. Kaplan from the University of Ottawa, and of Dr. R. Laprade and Dr. V. Vachon from the Groupe de recherche en transport membranaire, Université de Montréal. They wish to thank M. Juteau, A. Mazza, L. Potvin, and G. Préfontaine from the Biotechnology Research Institute, National Research Council of Canada, for expert technical assistance. This work was supported in part by Strategic Research Grant STR0167557 from the Natural Science and Research Engineering Council of Canada to R. Laprade and J. L. Schwartz.

References

1. Schnepf, E., Crickmore, N., Van Rie, J., Lereclus, D., Baum, J., Feitelson, J., Zeigler, D. R., and Dean, D. H. (1998) *Bacillus thuringiensis* and its pesticidal crystal proteins. *Microbiol. Mol. Biol. Rev.* **62,** 775–806.
2. Parker, M. W. and Pattus, F. (1993) Rendering a membrane protein soluble in water: a common packing motif in bacterial protein toxins. *Trends Biochem. Sci.* **18,** 391–395.
3. Duché, D., Parker, M. W., González-Mañas, J. M., Pattus, F., and Baty, D. (1994) Uncoupled steps of the colicin A pore formation demonstrated by disulfide bond engineering. *J. Biol. Chem.* **269,** 6332–6339.
4. Jacobson, B. L., He, J. J., Vermersch, P. S., Lemon, D. D., and Quiocho, F. A. (1991) Engineered interdomain disulfide in the periplasmic receptor for sulfate transport reduces flexibility. *J. Biol. Chem.* **266,** 5220–5225.
5. Pakula, A. A. and Simon, M. I. (1992) Determination of transmembrane protein structure by disulfide cross-linking: the *Escherichia coli* Tar receptor. *Proc. Natl. Acad. Sci. USA* **89,** 4144–4148.
6. Grochulski, P., Masson, L., Borisova, S., Pusztai-Carey, M., Schwartz, J. L., Brousseau, R., and Cygler, M. (1995) *Bacillus thuringiensis* CrylA(a) insecticidal toxin: crystal structure and channel formation. *J. Mol. Biol.* **254,** 1–18.
7. Menestrina, G. (1992) Ion transport and selectivity in model lipid membranes carrying incorporated cytolytic protein toxins. *J. Radioanal. Nucl. Chem.* **163,** 169–180.
8. Labarca, P. and Latorre, R. (1992) Insertion of ion channels into planar lipid bilayers by vesicle fusion, in *Methods in Enzymology—Ion Channels*, Vol. 207 (Rudy, B. and Iverson, L. E., eds.), Academic Press, San Diego, CA, pp. 447–463.
9. Schwartz, J. L., Juteau, M., Grochulski, P., Cygler, M., Préfontaine, G., Brousseau, R., and Masson, L. (1997) Restriction of intramolecular movements within the Cry1Aa toxin molecule of *Bacillus thuringiensis* through disulfide bond engineering. *FEBS Lett.* **410,** 397–402.

10. Mueller, P., Rudin, D., Tien, H. T., and Westcott, W. C. (1962) Reconstitution of excitable cell membrane structure *in vitro. Circulation* **26,** 1167–1171.

11. Masson, L., Préfontaine, G., Péloquin, L., Lau, P. C. K., and Brousseau, R. (1990) Comparative analysis of the individual protoxin components in P1 crystals of *Bacillus thuringiensis* subsp. *kurstaki* isolates NRD-12 and HD-1. *Biochem. J.* **269,** 507–512.

12. Montal, M. and Mueller, P. (1972) Formation of bimolecular membranes from lipid monolayers and a study of their electrical properties. *Proc. Natl. Acad. Sci. USA* **69,** 3561–3566.

13. Coronado, R. and Latorre, R. (1983) Phospholipid bilayers made from monolayers on patch-clamp pipettes. *Biophys. J.* **43,** 231–236.

14. Bradford, M. M. (1976) A rapid and sensitive method for the quantitation of microgram quantities of protein utilizing the principle of protein-dye binding. *Anal. Biochem.* **72,** 248–254.

15. Schwartz, J. L., Garneau, L., Masson, L., Brousseau, R., and Rousseau, E. (1993) Lepidopteran-specific crystal toxins from *Bacillus thuringiensis* form cation- and anion-selective channels in planar lipid bilayers. *J. Membr. Biol.* **132,** 53–62.

16. Slatin, S. L., Qiu, X.-Q., Jakes, K. S., and Finkelstein, A. (1994) Identification of a translocated protein segment in a voltage-dependent channel. *Nature* **371,** 158–161.

17. Mindell, J. A., Zhan, H., Huyn, P. D., Collier, R. J., and Finkelstein, A. (1994) Reaction of diphtheria toxin channels with sulfhydryl-specific reagents: observation of chemical reactions at the single molecule level. *Proc. Natl. Acad. Sci. USA* **91,** 5272–5276.

18. Creighton, T. E. (1989) Disulfide bonds between cysteine residues, in *Protein Structure: A Practical Approach* (Creighton, T. E., ed.), IRL Press at Oxford University Press, Oxford, UK, pp. 155–167.

19. Schmid, F. X. (1989) Spectral methods of characterizing protein conformation and conformational changes, in *Protein Structure: A Practical Approach* (Creighton, T. E., ed.), IRL Press at Oxford University Press, Oxford, UK, pp. 251–285.

20. Slatin, S. L., Abrams, C. K., and English, L. (1990) Delta-endotoxins form cation-selective channels in planar lipid bilayers. *Biochem. Biophys. Res. Commun.* **169,** 765–772.

7

Use of Fourier-Transformed Infrared Spectroscopy for Secondary Structure Determination of Staphylococcal Pore-Forming Toxins

Gianfranco Menestrina

1. Introduction

Pathogenic strains of *Staphylococcus aureus* produce a large number of exotoxins *(1)*. Among these a family of single-chain leukotoxins of 32–34 kDa has been identified. They include α-toxin (or α-hemolysin), which is probably one of the best studied toxins *(2)*, and a large group of bicomponent toxins formed by the leukocidins and the γ-hemolysins *(3)*. All these toxins share sequence homology *(4)* as well as functional similarity *(5)* which altogether suggest they probably have a significantly similar structural organization *(6)*.

These toxins belong to the so-called membrane-damaging/pore-forming supergroup. In fact, they produce well-defined pores in the cell membranes of attacked cells, increasing rather nonspecifically the cell permeability to ions and small molecules. The formation of such exogenous toxin channels in the cell membrane may lead to apoptosis and cell death if the permeability extends beyond the capacity of the cell to maintain its "milieu interieur" by active transport. The number of toxins known to belong to this group is large and still growing very quickly *(7)*.

A major feature of pore-forming toxins (PFTs) is represented by their ability to change from a water-soluble structure, suitable for diffusing from the bacterium through the body fluids to the target cell, to a membrane-embedded form, resembling intrinsic protein channels. Very often this step is accomplished via the assembling of a toxin aggregate.

Information on the secondary and three-dimensional structure of PFTs is paramount to understanding this mechanism. Accurate X-ray structures are beginning to become available, but only in a few cases at the present time *(8–11)*.

From: *Methods in Molecular Biology, vol. 145: Bacterial Toxins: Methods and Protocols*
Edited by: O. Holst © Humana Press Inc., Totowa, NJ

Fourier-transformed infrared spectroscopy (FTIR) spectroscopy provides a mean to estimate the secondary structure composition of proteins rather simply *(12,13)*. Although offering only a low resolution as compared to X-ray analysis (errors of FTIR determinations are typically ±5%) and providing no information on three-dimensional folding, it has the advantage of being applicable equally well to proteins in solution or adsorbed to the lipid phase. In addition, equipment for FTIR spectroscopy is less expensive and more readily available than X-ray crystallography. Therefore, this technique might be quite helpful to a number of investigators willing to investigate the secondary structure changes involved in the action of PFTs. This chapter, using *S. aureus* leukotoxins as an example, describes how to derive such information from simple experiments.

2. Materials
2.1. Reagents and Buffers

1. Tridistilled H_2O.
2. D_2O (minimum purity 99.7%).
3. Chloroform (of the purest available grade).
4. Acetone (of the purest available grade).
5. Ethanol (of the purest available grade).
6. Cleaning solution for optical cuvets (Hellmanex from Hellma, Forest Hills, NY).
7. EDTA, Sigma, St. Louis, MO.
8. Phosphatidylcholine (PC), Avanti Polar Lipids, Pelham, AL (99% pure).
9. Cholesterol (Cho), Fluka, Buchs, Switzerland (99% pure).
10. Lyophilized α-toxin, kindly supplied by Dr. Hungerer (Behring, Marburg, FRG) and used without further purification.
11. Leucotoxin components, a kind gift of Prof. G. Prevost (University of Strasbourg, France).
12. Buffer A: 10 mM *N*-2-Hydroxyethylpiperazine-*N'*-2-ethanesulfonic acid (HEPES), pH 7.0.
13. Buffer B: 100 mM NaCl, 30 mM Tris-HCl, 1 mM EDTA, pH 7.0.

2.2. Preparation of Lipid Vesicles

1. Rotary evaporator (Büchi, Uster, Switzerland).
2. Low vacuum source (e.g., water pump from Prolabo, Fontenay Sous Bois, France).
3. Clean high-vacuum source (e.g., Teflon diaphragm pump by Vacuubrand, Wertheim, Germany).
4. Vortex mixer (Prolabo).
5. Sonicator equipped with a titanium microtip sonotrode (e.g., Vibra Cell of Sonics & Materials, Danbury, CT, with steppered microtip).
6. Cylindrical 2-mL plastic tubes with conical bottom and screw cap (e.g., criovials of Nalge Nunc, Rochester, NY).
7. Liquid nitrogen.

8. Polycarbonate filters (Millipore, Bedford, MA).
9. Two-syringes extruder (LiposoFast from Avestin Inc., Ottawa, Canada).
10. Laser particle sizer (Z-sizer 3, Malvern, Southborough, MA).
11. Polysulfone ultrafilters of 300-kDa cutoff (NMWL, Millipore).

2.3. Hemolysis Assay

1. Freshly prepared rabbit red blood cells (RRBCs).
2. 96-Well microplates and absorbance microplate reader with stirrer (e.g., UVmax, Molecular Devices, Sunnyvale, CA).

2.4. FTIR Spectroscopy

1. Multirange high-resolution research spectrometer (e.g., FTS 185, Bio-Rad, Cambridge, MA, equipped with a KBr beamsplitter).
2. Detector: DTGS (deuterium triglycine sulfate) pyrometric bolometer with CsI window or linearized MCT (mercury cadmium telluride) photometric detector (both from Bio-Rad).
3. Molecular sieve gas purifier, able to remove water vapor and CO_2 from laboratory air (e.g., Ken 6 drier from Zander Filter Systems, Norcross, GA).
4. ATR attachment for attenuated total reflectance sampling (e.g., 10-reflections 45° ATR by Graseby Specac, Fairfield, CT, equipped with Ge crystal).
5. Four ports demountable liquid cell holder for ATR crystal (Specac).
6. High-efficiency rotating polarizer (e.g., the wire-grid polarizer in KRS-5 from Specac).
7. Specialized FTIR spectroscopy software (e.g., Bio-Rad Win-Ir, derived from the general purpose spectroscopic package GRAMS of Galactic Industries, Salem, NH).

3. Methods

3.1. Preparation of Protein and Lipid Samples for FTIR Investigation

3.1.1. Protein

1. Extensively dialyze the toxins, in 1-mL aliquots, against three changes of 200 mL of buffer A, for 12 h each time at 4°C.
2. Adjust the final protein concentration to 1 mg/mL.

3.1.2. Lipid Vesicles

3.1.2.1. PREPARATION OF MULTILAMELLAR LIPOSOMES

1. Suspend PC and Cho separately in chloroform.
2. Mix PC and Cho in a 1:1 molar ratio.
3. Dry onto the walls of a glass vial using a rotary evaporator connected to a low-vacuum source (water pump).
4. Remove any trace of residual organic solvent from the film by an overnight exposure to high vacuum in a desiccator connected to a Teflon diaphragm pump.

5. Resuspend the dried lipid film in buffer A using a vortex mixer (*see* **Note 1**).
6. Adjust the final lipid concentration to 6 mg/mL.

3.1.2.2. PREPARATION OF SMALL UNILAMELLAR LIPID VESICLES (SUV)

1. Collect 1 mL of the multilamellar liposomes solution described in **Subheading 3.1.2.1.** in a 2-mL crio-vial plastic tube with a conical bottom.
2. Sonicate for 45–60 min with a microtip sonotrode using a typical output power of 25 W and a 50% duty cycle (*see* **Note 2**).
3. Centrifuge for 10 min at 4000*g* in a tabletop centrifuge to eliminate titanium particles released by the sonotrode.
4. Collect the supernatant.

3.1.2.3. PREPARATION OF LARGE UNILAMELLAR VESICLES (LUV)

1. Collect 1 mL of the multilamellar liposomes solution as described in **step 1** of **Subheading 3.1.2.2.**
2. Freeze the sample by immersing in liquid nitrogen and thaw at 20–22°C.
3. Repeat **step 2** 5 or 6 times (*see* **Note 3**).
4. Extrude several times through two stacked polycarbonate filters with holes of average diameter 100 nm, using a two-syringes extruder (*see* **Note 4**).

3.1.2.4. ASSAY VESICLE SIZE BY PHOTON CORRELATION SPECTROSCOPY

Use a laser dynamic light scattering particle sizer *(14)* to measure vesicle size and dispersity (*see* **Note 5**).

3.1.3. Preparation of Membrane Bound Toxin

1. Prepare aliquots of 200 µL of SUV or LUV at a 3-mM lipid concentration, and mix with either 15 µM α-toxin or 7.5 µM of each member of the bicomponent toxins in buffer A (final molar lipid/toxin 200:1).
2. Incubate at 37°C for 1 h.
3. Remove unbound toxin by spinning through polysulfone ultrafilters of 300 kDa cutoff at 2000*g* for 20 min.
4. Collect the filtrates and test their hemolytic activity on RRBCs (to evaluate the presence of free toxin).
5. Resuspend the retentates in 200 µL of buffer A.
6. Repeat the washing **steps 3–5**, until there is no hemolytic activity in the filtrates.
7. Collect the retentates containing vesicles and bound toxin only, and bring to a final volume of 200 µL with buffer A.

3.1.4. Determination of the Hemolytic Activity

1. Collect fresh rabbit venous blood adding 6 mM EDTA.
2. Separate rabbit red blood cells (RRBC) washing thrice (10 min centrifugation at 700*g*, 20–22°C) in buffer B.
3. Resuspend the pellet at a final concentration of 0.26% (v/v) in buffer B.

4. Take a 96-well microplate with flat optical bottom and fill each row with twofold serial dilutions of the toxins in a final volume of 100 µL of buffer B.
5. Add 100 mL of RRBC from **step 3**.
6. Place in a microplate reader and estimate the turbidity by measuring the optical absorption at 650 nm (the initial value of A_{650} should be approx 0.1).
7. Follow the time course of hemolysis for 45 min, stirring and reading the microplate approx every 8–10 s.
8. Calculate the extent of hemolysis as follows:

$$\% \text{ hemolysis} = 100 \, (A_i - A_f)/(A_i - A_w) \tag{1}$$

where A_i and A_f are the absorbances at the beginning and at the end of the reaction, and A_w is the value obtained lysing the RRBCs with pure water *(15)*.
9. Plot these values vs toxin dose and extrapolate C_{50}, the concentration producing 50% hemolysis.

3.2. Secondary Structure of Protein Toxins by FTIR Spectroscopy

3.2.1. Interferogram Recollection Conditions and Fourier Transform

1. Collect FTIR spectra in the region between 4000 and 1000 cm^{-1}, at a resolution of 0.5 cm^{-1} using a multirange high-resolution spectrometer.
2. Use a DTGS detector for normal samples or a linearized MCT detector (providing 5× higher response rate and 5× higher sensitivity) for low transmitting samples (*see* **Note 6**).
3. To avoid environmental interferences, constantly purge the optical bench and the sample compartment of the instrument each with a flux of around 10 L/min of air deprived of water vapor and CO_2.
4. Apply Fourier transform to the interferogram and convert the result into an absorption spectrum by dividing by an appropriate background spectrum (see **Subheading 3.2.2.5.**).
5. Co-add successive acquisitions and monitor in real time.
6. Proceed with the accumulation until a good signal/noise ratio is achieved (typically >500:1 around 1650 cm^{-1}) (*see* **Note 7.**)
7. Digitize and store the spectra for off-line analysis with a specialized FTIR software package.

3.2.2. Sampling Technique

Best results are obtained when the spectra are collected in the configuration of attenuated total reflectance (ATR). In this method the sample is held in contact with a long, thin, infrared (IR)-transparent crystal, called the internal reflection element (IRE). The beam enters the IRE at one end and passes through its length in a zig-zag pattern, internally reflecting off the two opposite surfaces several times before leaving the IRE at the other end. At each reflection a portion of the radiation, the evanescent wave, crosses the interface and penetrates the sample for a short distance whereby absorbance occurs. For most purposes, an appropriate IRE is a 10-reflections Ge crystal with parallelogram shape and 45° cut (*see* **Note 8**). Proceed as follows:

1. Take 40 μL of a 1 mg/mL protein solution (*see* **Note 9**) dialyzed against buffer A as for **Subheading 3.1.1.** and deposit it on one side of the Ge crystal (*see* **Notes 10** and **11**).
2. Using a thin cylindrical Teflon bar gently spread the sample on the IRE surface under a smooth stream of nitrogen until it dries in a thin layer (*see* **Note 12**).
3. Assemble the crystal in a demountable liquid cell using a 0.5-mm Teflon gasket.
4. Position the liquid cell in the ATR attachment and house it in the sample compartment.
5. Acquire a protein vibrational spectrum as described in **Subheading 3.2.1.** using a previously stored background spectrum collected with the same ATR crystal and no sample.
6. The spectrum shows the typical amide I and II bands of the protein, in the region $1700–1600$ cm^{-1} and $1600–1500$ cm^{-1}, respectively, plus a number of other bands attributable also to the organic buffer and water.
7. Through one of its ports flush the liquid cell with nitrogen saturated with D_2O for approx 45 min (*see* **Note 13**).
8. Continuously collect spectra during the deuteration process, to verify that the exchange is taking place, until a steady state is attained.
9. Collect the final spectrum in deuterated conditions; the bands of interest are now termed amide I' and II'.
10. Disassemble the liquid cell and clean the IRE for a later use (*see* **Note 14**).

3.2.3. Obtaining the Toxin Secondary Structure from Analysis of the Amide I' Band

3.2.3.1. Preliminary Processing of the Whole Spectrum

1. If present, remove the sharp bands of residual environmental H_2O vapor by subtracting a previously acquired absorbance spectrum of a crystal flushed with water vapor. Use an appropriate multiplying coefficient to obtain a smooth baseline between 2000 and 1700 cm^{-1}.
2. Subtract similarly the absorbance spectrum of HEPES alone (deposited on the IRE), choosing the coefficient that minimizes the residual at the position of the typical band at 1200 cm^{-1}.
3. Finally, subtract a linear baseline between 1720 and 1500 cm^{-1}, two positions at which the protein themselves do not absorb appreciably.

3.2.3.2. Evaluation of the Amide I' Band, Exemplified for α-Toxin in Fig. 1

1. Delimit the region between 1700 and 1600 cm^{-1}, comprising the amide I' band.

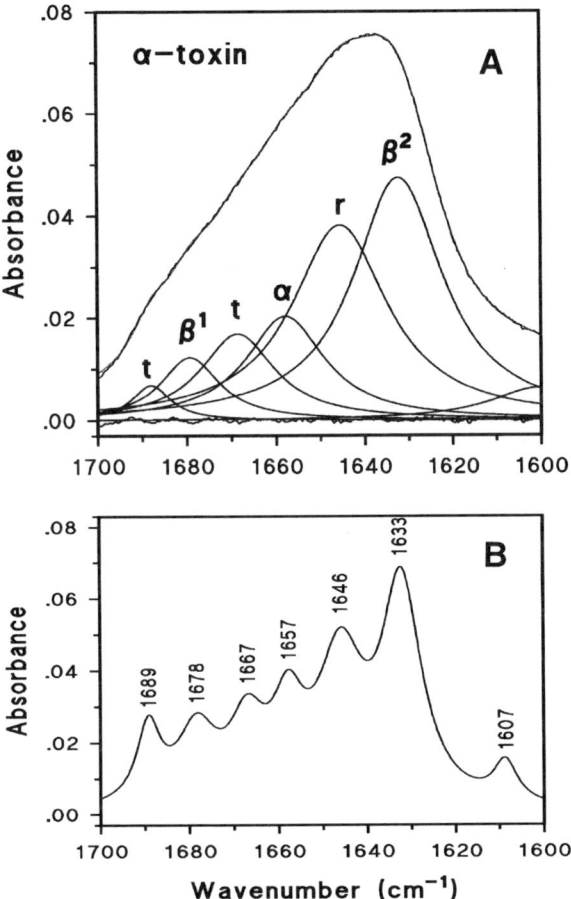

Fig. 1. IR-ATR spectrum in the amide I' region of deuterated *S. aureus* α-toxin. **(A)** The spectrum (*solid line*), seven Lorentzian best-fit components (*thin solid lines*) with their sum (*dotted line*), and the residual (*bottom trace*) are shown. Typically the curves have FWHH (full width at half height) of 20 ± 8 cm^{-1}. The assignments of the Lorentzian components are indicated next to each band as follows: β-turn (t); β-sheet (β^1 and β^2); α-helix (α); random coil (r). The minor band around 1600 cm^{-1} is attributed to side chains *(16)*. The evaluated secondary structures are reported in **Table 1**. **(B)** Deconvoluted spectrum showing the presence and location of the seven components used to fit the original data. Deconvolution parameters were: resolution enhancement 1.5 and Bessel smoothing, with factor 92%.

2. Use the appropriate program routine perform a Fourier self-deconvolution of this region, choosing a resolution enhancement factor between 1.0 and 2.0 and a suitable smoothing factor (*see* **Fig. 1B**).

Table 1
FTIR Determination of the Secondary Structure
of *S. aureus* α-Toxin Calculated as Shown in Fig. 1

Toxin	β^1	β^2	t	α	r	$\beta_{tot}{}^a$
α-Toxin	6	36	13	14	31	55
α-Toxin + SUV	6	43	16	16	19	65
α-Toxin + DOC[b]	β-Sheet = 56		9	6	29	65

[a]β_{tot} = β-Structure total = $\beta^1 + \beta^2 + t$.
[b]Determined by X-ray crystallography of the DOC-induced oligomer *(8)*.

3. Using a curve-fit routine, determine the number, position, amplitude, and width of the Lorentzian components appearing in the deconvoluted spectrum (normally from seven to nine components).
4. Take the set of Lorentzian curves previously determined as initial values to perform a nonlinear least-squares fit to the original spectrum, using again the curve-fit routine. Choose the Levenberg–Marquardt method and leave all the parameters free to vary (*see* **Fig. 1A**).
5. Calculate the residual.

3.2.3.3. EVALUATION OF THE PROTEIN SECONDARY STRUCTURE

1. Assign each Lorentzian component to a particular secondary structure on the basis of its center frequency according to the following criteria: bands in the regions 1696–1680 cm^{-1} and 1670–1660 cm^{-1}: β-turn (t); band at 1675 ± 4 cm^{-1}: antiparallel β-sheet (β^1); band at 1655 ± 4 cm^{-1}: α-helix (α); band at 1645 ± 4 cm^{-1}: random coil (*r*); bands in the region 1640–1620 cm^{-1}: parallel plus antiparallel β-sheet (β^2) (*see* **Note 15**).
2. Ignore, if present, the minor bands between 1614 cm^{-1} and 1600 cm^{-1} which should be attributed to the contribution of side chains (*16*).
3. Calculate the relative content of each secondary structure element by first summing the areas of all the individual components assigned to that structure and then dividing by the area of the sum of all components between 1696 cm^{-1} and 1614 cm^{-1}. As an example, the secondary structure of α-toxin is reported in **Table 1**.

3.2.4. Vibrational Spectra of Lipid Layers

To obtain an absorption spectrum of the lipid matrix, proceed as follows:

1. Take 60 µL of SUV or LUV prepared as explained in **Subheadings 3.1.2.2.** and **3.1.2.3.** and apply to the surface of the IRE as described in **Subheading 3.2.2.1.** (*see* **Note 16**).
2. Gently dry the sample and obtain the spectrum of the hydrated form repeating **steps 2–5** of **Subheading 3.2.2.** (*see* **Note 17**).
3. The spectrum shows some typical features: symmetric and antisymmetric CH$_2$ stretching bands (centered at 2851 cm^{-1} and 2925 cm^{-1}, respectively); antisym-

metric CH_3 stretching bands (centered at 2958 cm^{-1}); carboxyl groups band (centred at 1738 cm^{-1}); and a number of other bands attributable also to the organic buffer and water (*see* **Note 18**).

4. Continue repeating **steps 7–9** of **Subheading 3.2.2.** to collect also the spectrum of the deuterated form.

3.2.5. Vibrational Spectra of Mixed Lipid Protein Preparations

To obtain an absorption spectrum of the protein inserted into the lipid matrix, proceed as follows:

1. Take 90 μL of SUV or LUV prepared as explained in **Subheading 3.1.3.** and apply to the surface of the IRE as described in **Subheading 3.2.2.1.** (*see* **Note 19**).
2. Gently dry the sample and obtain the spectrum of the hydrated form, repeating **steps 2–5** of **Subheading 3.2.2.**
3. The spectrum shows all the lipid and protein features previously discussed: the CH_2 and CH_3 stretching bands (in the region from 2960 cm^{-1} to 2850 cm^{-1}); the lipid carboxyl groups band (1738 cm^{-1}); the amide I and II bands, and those of the organic buffer and water.
4. Repeat **steps 7–9** of **Subheading 3.2.2.** to collect the spectrum of the deuterated form.
5. Before analyzing the protein spectrum remove the contribution of the lipid in the amide I' region by subtracting the spectrum of the lipid alone multiplied by a factor that minimizes the residual signal of the phospholipid carboxyl groups at 1738 cm^{-1}.
6. Repeat all the steps of **Subheadings 3.2.3.2.** and **3.2.3.3.** and calculate the secondary structure of the membrane-inserted protein. As an example, the secondary structure of α-toxin embedded in a lipid membrane is given in **Table 1**. It contains more β-structure than the soluble form and is very similar in composition to the deoxycholate (DOC)-induced oligomer as it was determined by X-ray crystallography *(8)*. Results with the bicomponent *S. aureus* leukotoxins are quite similar *(5)*.

3.3. Determination of the Average Molecular Orientation of Lipids and Toxins

When multilayers of lipid, or lipid plus bound toxin, are deposited on the IRE, they maintain some orientation with respect to the plane of the IRE surface and with respect to each other. Such average orientation can be determined by polarization experiments and may provide information on the structure adopted by the toxin in the membrane.

3.3.1. Collecting Polarized ATR-IR Spectra of Lipid or Lipid-Bound Toxin

1. Perform all steps necessary to obtain a deuterated FTIR spectrum of layers of either lipid alone or lipid with protein (i.e., **step 4** of **Subheadings 3.2.4.** and **3.2.5.**, respectively).

2. Interpose on the beam line, between the liquid cell in the ATR holder and the detector, a rotating polarizer.

3. Manually position the polarizer with orientation either parallel (0°) or perpendicular (90°) to the plane of the internal reflections and acquire an absorption spectrum in both positions.

4. In the case of lipid/toxin samples there is a cross contribution (albeit small) of the lipid and the protein components in their respective characteristic regions. Henceforth, before analysis the spectra have to be corrected to isolate the contribution of either the lipid or the protein alone. If one is interested in the lipid spectrum, the contribution of the protein in the region 2800 cm^{-1} to 3000 cm^{-1} is eliminated by subtracting the spectrum of the protein alone multiplied by a factor that minimizes the remaining signal in the amide I' region. To obtain the protein spectrum the contribution of the lipid in the amide I' region is eliminated as described in **step 5** of **Subheading 3.2.5.**

5. Repeat **step 4** for the parallel (0°) and the perpendicular (90°) spectrum.

3.3.2. Orientation of Lipid Molecules by Quantitative Analysis of the Polarized ATR-IR Spectra

1. Consider the lipid spectrum in the region 2800 cm^{-1} to 3000 cm^{-1} both at 0° and 90°. and best fit the sum of mixed Gaussian-Lorentzian components (as in **step 4** of **Subheading 3.2.3.2.**).

2. Integrate the bands centred at 2851 cm^{-1} and 1925 cm^{-1} (corresponding to the symmetric and antisymmetric CH$_2$ stretching, respectively).

3. Calculate the dichroic ratio R for the two vibrations, defined as $R = A_{0°}/A_{90°}$, where $A_{0°}$ and $A_{90°}$ are the integrated absorbtion bands in the parallel and perpendicular configuration, respectively.

4. From the dichroic ratio derive the order parameter for the lipid chains, S_L, using *(17–19)*:

$$S = \frac{E_x^2 - RE_y^2 + E_z^2}{\frac{1}{2}(3\cos^2\theta - 1)(E_x^2 - RE_y^2 - 2E_z^2)} \tag{2}$$

where θ is the angle between the longest axis of the molecule under consideration and the transition moment of the investigated vibration. In the case of the lipid chains θ is 90° for the symmetric and antisymmetric CH$_2$ stretching and 0° for the symmetric CH$_3$ stretching. E_x, E_y, and E_z are the components of the electric field of the evanescent wave in the three directions (the z-axis being perpendicular to the plane of the crystal) (*see* **Note 20**). For practical purposes **Fig. 2A** can be used to calculate S from R, at any θ, in the experimental conditions described here, that is, incidence angle $\alpha = 45°$, refractive index of the Ge crystal $n_1 = 4$, and of the deposited layer $n_2 = 1.43$.

5. From the order parameter S calculate the average tilt angle γ of the molecular axis with respect to the z-axis (*i.e.*, the perpendicular to the plane of the membrane) according to *(19)*:

$$S = \tfrac{1}{2}(3 \cos^2 y - 1) \tag{3}$$

Figure 2B can be used to calculate the angle γ from S.

As typical values we obtained: R = 1.30 and 1.33 for the bands at 2851 cm^{-1} and 1925 cm^{-1} respectively, providing S_L = 0.52 and 0.47 and finally γ = 34.5° and 36.5° which is a value characteristic for phospholipids in the liquid crystalline state *(20)* induced in this case by the presence of cholesterol. Upon insertion of the toxin the lipid layers result only slightly less oriented (**Table 2**).

3.3.3. Relative Orientation of Proteins in Lipid Layers by Quantitative Analysis of the Polarized ATR-IR Spectra

1. Consider the amide I' band and integrate between 1600 and 1700 cm^{-1} both at 0° and 90°.
2. Best fit it with the sum of Lorentzian components (as described in **Subheading 3.2.3.2.**).
3. Calculate the dichroic ratio R for the whole amide I' band and for the β-strands, using the sum of the Lorentzian components at 1678 ± 4 cm^{-1} and 1631 ± 3 cm^{-1} as described in **step 3** of **Subheading 3.3.2.**
4. Calculate $S_{\text{amide I'}}$, for the whole amide I' band, with $\theta = 0°$ *(21,22)*; and S_β, for the β-strands, with $\theta = 70°$ *(22)* as described in **step 4** of **Subheading 3.3.2.** (use **Fig. 2A**).
5. Calculate the average tilt angle γ of the molecular axis with respect to the z-axis for the whole amide I' band and for the β-strands, as described in **step 5** of **Subheading 3.3.2.** (use **Fig. 2B**), *see* **Note 21**.
6. Alternatively, calculate the mean angle σ of the β-structure axis with respect to the direction of the lipid chains from *(19)*:

$$S' = \frac{S_\beta}{\tfrac{1}{2}(3 \cos^2 \gamma_L - 1)} = \tfrac{1}{2}(3 \cos^2 \sigma - 1) \tag{4}$$

where γ_L is the angle formed by the lipid chains with the z-axis (i.e., 37°). For practical purposes use **Fig. 2C** to calculate S' from S, at any γ_L, and again **Fig. 2B** to calculate the angle σ from S'. In the case of α-toxin we derive 22° ± 12°, which indicates a substantial orientation of the β-strands perpendicular to the plane of the membrane as deduced also by the three-dimensional structure of the oligomer *(8)*.

3.3.4. Determination of Lipid to Toxin Ratio and Partition Coefficient

1. Calculate the lipid to toxin ratio (r) from the perpendicular spectrum $A_{90°}$ by the following algorithm *(21,22)*:

$$r = L / T_b = \frac{(n_{res} - 1)}{9.6} \frac{1 - S_{\text{amide I'}}}{1 + S_L/2} \frac{\int_{2800}^{2980} A_{90°}(v_L)dv}{\int_{1600}^{1690} A_{90°}(v_{\text{amide I'}}) \, dv} \tag{5}$$

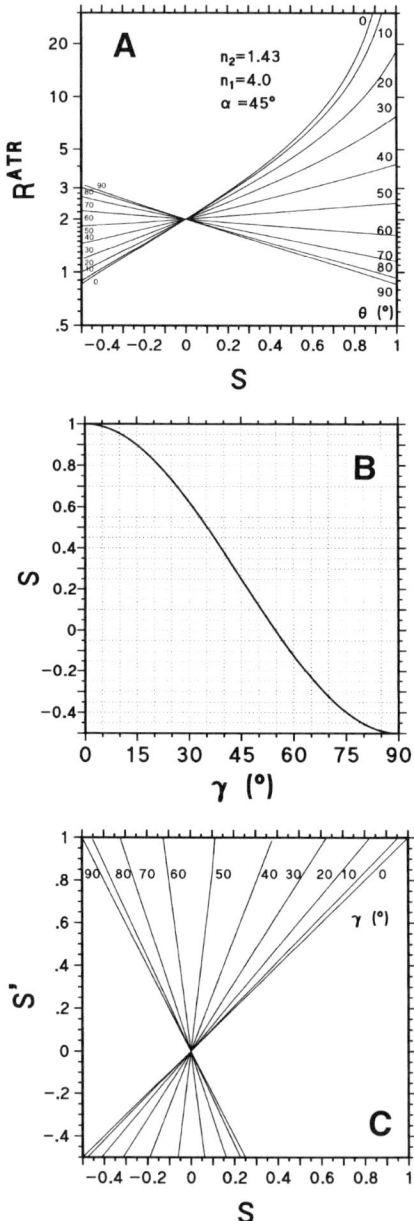

Fig. 2. Graphical determination of the average molecular orientation from polarized ATR-IR spectra. **(A)** Derivation of the order parameter S from the dichroic ratio R^{ATR} according to Eq. (2) for different values of the angle θ between the longest axis of the molecule and the transition moment of the investigated vibration. **(B)** Derivation of the average tilt angle γ of the molecular axis with respect to the z-axis (i.e., the

where L is the total concentration of lipid present and T_b is the concentration of bound toxin; n_{res} is the total number of residues in the toxin, S_L and $S_{amide\ I'}$ have been defined above and the integrals refer to the corrected spectra obtained as in **steps 4** and **5** of **Subheading 3.3.1.** Under the experimental conditions described in **Subheading 3.1.3.** we obtained $L/T_b = 310$.

2. Evaluate the partition coefficient (K) of the protein from the water to the lipid phase as follows:

$$K = \frac{T_b}{(T_0 - T_b) \cdot L} = \frac{1}{T_0 \cdot r - L} \tag{6}$$

where T_0 is the total concentration of toxin applied. Using r determined earlier we obtain $K \cong 600\ M^{-1}$ for α-toxin.

4. Notes

1. For a faster action introduce a few small glass beads.
2. To avoid excessive sample heating the tube must be cooled by immersing in chilly water during sonication.
3. This step increases the interlamellar space and hydration *(23,24)*.
4. Owing to the construction of the extruder, consisting of two syringes connected head-to-head through the filter, it is important to perform an odd number of passages; in fact, in this way the material will end up in the syringe opposite to the one in which it was loaded, thus ensuring that no unfiltered material is present. Normally, approx 19 passages are recommended to produce monodisperse vesicles *(23)*.
5. We routinely estimated an average diameter of 40 ± 10 nm and 90 ± 10 nm, respectively for SUVs and LUVs prepared as described previously.
6. The DTGS detector operates at 20–22°C whereas the MCT detector requires cooling with liquid nitrogen.
7. A variable number of acquisitions is required for this, depending on the sampling technique and the concentration of the specimen. It should be remembered that the signal-to-noise ratio improves with the square root of the number of measurements co-added. Typically, 64 runs should be enough.
8. For a shorter penetration depth the refractive index of the IRE should be as different as possible from the refractive index of the deposited film. Because the refractive index of biological samples is approx 1.45, Ge is an optimal material from this point of view, because it has a refractive index of 4 at 2000 cm⁻¹ (higher than any other IR transparent material). In these conditions the penetration depth at 1650 cm⁻¹ is approx 0.4 µm *(25)* which ensures that all the field decays within the thickness of the deposited layer (*see also* **Note 17**). Ge is also insoluble and transparent in the range 550–5000 cm⁻¹. One problem is that it is very fragile and

perpendicular to the plane of the membrane) from the order parameter S according to Eq. (3). (**C**) Calculation of the form factor S' with respect to the direction of the lipid chains from the form factor S and γ_L, the angle formed by the lipid chains with the z-axis, according to Eq. (4).

Table 2
Orientation and Dichroic Ratio of Some IR Bands Observed with Depositions of Lipid Alone and Lipid Plus α-Toxin

Wavenumber (cm^{-1})	Vibration	Direction[a]	Lipid alone			α-Toxin + lipid		
			Dicroic ratio	Form factor	Angle[b]	Dicroic ratio	Form factor	Angle[b]
2925	as CH$_2$ Stretching[c]	90°	1.33 ± 0.02	0.47	$36.5° \pm 1°$	1.34 ± 0.03	0.46	$37° \pm 1°$
2850	s CH$_2$ Stretching[c]	90°	1.30 ± 0.02	0.52	$34.5° \pm 1°$	1.41 ± 0.02	0.43	$38.5° \pm 1°$
1675 + 1630	Amide I' β-sheet	70°	—	—	—	1.63 ± 0.11	0.35 ± 0.11	$42° \pm 5°$

[a]Direction of the variation of the dipole moment associated to the vibration with respect to the main molecular axis (aliphatic chains or β-strand).
[b]Average angle between the direction of the molecular axis and the perpendicular to the crystal plane (which is the same as the membrane plane).
[c]as, Antisymmetric; s, symmetric.

should be handled with great care. Other possible choices are ZnSe or KRS-5 which, however, have refractive index 2.43 and 2.38, respectively.

9. The amount of protein required is typically 50–100 µg, with a lower limit at 5 µg. If the layer is too thin, the relative contribution from the molecules directly adsorbed to the Ge surface becomes too important and may lead to artefacts. For practical purposes a concentration of approx 25 µg/cm^2 can be recommended.

10. The amount of salt present in the buffer should be kept to a minimum because it might crystallize during the drying process, disturbing the formation of a good protein film or of a regularly stacked lipid multilayer. In addition, the investigated sample should not be excessively dispersed in other materials because this decreases the available signal. In practice, the salt should not exceed the sample (lipid or protein or both) by weight.

11. Molecules containing COOH groups, for example, EDTA, EGTA, trifluoroacetic acid (TFA), or urea, should not be present in the buffer, as they absorb in, or near, the amide I region (1700–1600 cm^{-1}). In practice, a low concentration of a buffer such as HEPES would be suitable in most cases.

12. The protein hydration water is retained during this procedure *(26)* as indicated by the presence in the IR spectrum of a broad band centered around 3300 cm^{-1} representing the stretching of the OH group of water *(12)*.

13. This procedure completely exchanges hydration water with D$_2$O, as indicated by the progressive disappearance of the band centred around 3300 cm^{-1} replaced by a new one around 2500 cm^{-1} for the OD stretching. Replacing D for H will shift from around 1655 cm^{-1} to around 1645 cm^{-1} the disordered components of the protein, that normally form hydrogen bonds with free water molecules *(12,26)*. A typical flux would be 30–50 mL/min of N$_2$ which was bubbled twice through D$_2$O. Although a complete exchange of D$_2$O for H$_2$O would take several hours the shift of the disordered components, especially if they are not membrane inserted, occurs very rapidly (typically within 20–30 min).

14. Cleaning the IRE is very important. We wash our Ge crystals with a basic soap and hot water, and rinse with abundant tap water followed by distilled water. They are then soaked for 2–3 h in 2% Hellmanex cleaning solution. This step is necessary to remove an adsorbed protein layer that is resistant to washing and otherwise could accumulate with time. Finally the crystals are thoroughly rinsed again in distilled water and organic solvents, dried, and stored for use.

15. This standard interpretation is based on *ab initio* calculations of the amide I' frequency of vibration for the different structures as well as on the experimental refinement of the technique by comparison with proteins of known crystallographic three-dimensional structure *(12,27,28)*. The values reported are for the *S. aureus* leukotoxin family and may vary slightly from one protein family to the other.

16. Control lipid vesicles should receive the same treatment as protein/lipid samples, for example, the same number of incubations, filtration, etc.

17. These conditions correspond to depositing approx 100 double layers with a total thickness of about 0.5–1 µm depending on the amount of hydration water retained.

18. Normally there is no contribution from the lipid molecules in the absorption spectrum at the amide I (or I') region except for the case of lipids containing the ceramide moiety, for example, sphingomyelin and gangliosides *(29)*.

19. For the same reasons exposed in **Note 10,** the protein/lipid ratio in the mixed toxin–vesicles samples should be larger than 0.1 by weight.

20. The terms E_x^2, E_y^2, and E_z^2 can be calculated according to Harrick expressions for thick films *(17,19,30)* using the appropriate incidence angle and refractive indexes for the IRE and the deposited layer. In practice, these expressions are good also for thin films *(30)*.

21. This requires that the distribution of the vector of the dipole moment variation has cylindrical symmetry around the long axis of the molecule, which is true in this case because the leukotoxin pore is an aggregate with cylindrical symmetry *(8)*.

22. Many materials used in FTIR, such as KBr, CsI, and KRS-5, are soluble in water and from moderately to highly hygroscopic. They have to be kept either inside the constantly purged instrument or, if not in use, in a desiccation cabinet.

23. If the surface of the IRE becomes cloudy in some places it should be repolished. We found that a diamond spray with particles of 0.1 μm diameter was suitable to clean it. However, polishing should not be repeated too often because this procedure would decrease the thickness of the crystal and compromise the geometry for the repeated reflections. Furthermore, for a correct interpretation of the polarization experiments the IRE surfaces should remain as flat, smooth, and parallel as possible.

24. Because nitrogen gas coming from most conventional sources may bring some water vapor with it, while bubbling in the D_2O it may contaminate it with H_2O. Hence, it is necessary to renew the D_2O in the reservoir from time to time, which can be decided from an incomplete exchange of the adsorbed H_2O appearing in the FTIR spectrum.

References

1. Freer, J. H. and Arbuthnott, J. P. (1983) Toxins of *Staphylococcus aureus*. *Pharmacol. Ther.* **19,** 55–106.

2. Bhakdi, S. and Tranum-Jensen, J. (1991) Alpha-toxin of *Staphylococcus aureus*. *Microbiol. Rev.* **55,** 733–751.

3. Couppié, P. and Prévost, G. (1997) Les leucotoxines staphylococciques. *Ann. Dermatol. Venereol.* **124,** 740–748.

4. Cooney, J., Kienle, Z., Foster, T. J., and O'Toole, P. W. (1993) The gamma-hemolysin locus of *Staphylococcus aureus* comprises three linked genes, two of which are identical to the genes of the F and S component of leukocidin. *Infect. Immun.* **61,** 768–771.

5. Ferreras, M., Höper, F., Dalla Serra, M., Colin, D. A., Prévost, G., and Menestrina, G. (1998) The interaction of *Staphylococcus aureus* bi-component gamma hemolysins and leucocidins with cells and model membranes. *Biochim. Biophys. Acta* **1414,** 108–126.

6. Gouaux, J. E., Hobaugh, M., and Song, L. Z. (1997) Alpha-hemolysin, gamma-hemolysin, and leukocidin from *Staphylococcus aureus*: distant in sequence but similar in structure. *Protein Sci.* **6,** 2631–2635.

7. Menestrina, G., Schiavo, G., and Montecucco, C. (1994) Molecular mechanism of action of bacterial protein toxins. *Mol. Asp. Med.* **15,** 79–193.

8. Song, L., Hobaugh, M. R., Shustak, C., Cheley, S., Bayley, H., and Gouaux, J. E. (1996) Structure of staphylococcal alpha-hemolysin, a heptameric transmembrane pore. *Science* **274,** 1859–1866.

9. Parker, M. W., Buckley, J. T., Postma, J. P. M., Tucker, A. D., Leonard, K., Pattus, F., and Tsernoglou, D. (1994) Structure of the *Aeromonas* toxin proaerolysin in its water-soluble and membrane-channel state. *Nature* **367,** 292–295.

10. Rossjohn, J., Feil, S. C., McKinstry, W. J., Tweten, R. K., and Parker, M. W. (1997) Structure of a cholesterol-binding, thiol-activated cytolysin and a model of its membrane form. *Cell* **89,** 685–692.

11. Petosa, C., Collier, R. J., Klimpel, K. R., Leppla, S. H., and Liddington, R. C. (1997) Crystal structure of the anthrax toxin protective antigen. *Nature* **385,** 833–838.

12. Arrondo, J. L., Muga, A., Castresana, J., and Goñi, F. M. (1993) Quantitative studies of the structure of proteins in solution by Fourier-transform infrared spectroscopy. *Prog. Biophys. Mol. Biol.* **59,** 23–56.

13. Susi, H. and Byler, D. M. (1986) Resolution enhanced Fourier transform infrared spectroscopy of enzymes. *Methods Enzymol.* **130,** 290–311.

14. Mayer, L. D., Hope, M. J., and Cullis, P. R. (1986) Vesicles of variable size produced by a rapid extrusion procedure. *Biochim. Biophys. Acta* **858,** 161–168.

15. Cauci, S., Monte, R., Ropele, M., Missero, C., Not, T., Quadrifoglio, F., and Menestrina, G. (1993) Pore-forming and hemolytic properties of the *Gardnerella vaginalis* cytolysin. *Mol. Microbiol.* **9,** 1143–1155.

16. Tatulian, S. A., Hinterdorfer, P., Baber, G., and Tamm, L. K. (1995) Influenza hemagglutinin assumes a tilted conformation during membrane fusion as determined by attenuated total reflection FTIR spectroscopy. *EMBO J.* **14,** 5514–5523.

17. Harrick, N. J. (1967) *Internal Reflection Spectroscopy*. Harrick Scientific Corporation, Ossining, NY.

18. Fringeli, U. P. and Günthard, H. H. (1981) Infrared membrane spectroscopy. *Mol. Biol. Biochem. Biophys.* **31,** 270–332.

19. Axelsen, P. H., Kaufman, B. K., McElhaney, R. N., and Lewis, R. N. A. H. (1995) The infrared dichroism of transmembrane helical polypeptides. *Biophys. J.* **69,** 2770–2781.

20. Mueller, E., Giehl, A., Schwarzmann, G., Sandhoff, K., and Blume, A. (1996) Oriented 1,2-dimyristoyl-*sn*-glycero–3-phosphorylcholine/ganglioside membranes: a Fourier transform infrared attenuated total reflection spectroscopic study. Band assignments; orientational, hydrational, and phase behaviour; and effects of Ca^{2+} binding. *Biophys. J.* **71,** 1400–1421.

21. Tamm, L. K. and Tatulian, S. A. (1993) Orientation of functional and nonfunctional PTS permease signal sequence in lipid bilayers. A polarized attenuated total reflection infrared study. *Biochemistry* **32,** 7720–7726.

22. Rodionova, N. A., Tatulian, S. A., Surrey, T., Jähnig, F., and Tamm, L. K. (1995) Characterization of two membrane-bound forms of OmpA. *Biochemistry* **34,** 1921–1929.

23. MacDonald, R. C., MacDonald, R. I., Menco, B. P. M., Takeshita, K., Subbarao, N. K., and Hu, L. (1991) Small-volume extrusion apparatus for preparation of large, unilamellar vesicles. *Biochim. Biophys. Acta* **1061,** 297–303.

24. Hope, M. J., Bally, M. B., Webb, G., and Cullis, P. R. (1985) Production of large unilamellar vesicles by a rapid extrusion procedure. Characterization of size distribution, trapped volume and ability to maintain a membrane potential. *Biochim. Biophys. Acta* **812,** 55–65.

25. Harrick, N. J. (1965) Electric field strength at totally reflecting interfaces. *J. Opt. Soc. America* **55,** 851–857.

26. Goormaghtigh, E., Cabiaux, V., and Ruysschaert, J.-M. (1994) Determination of soluble and membrane protein structure by Fourier transform infrared spectroscopy. II. Experimental aspects, side chain structure, and H/D exchange, in *Subcellular biochemistry,* Vol. 23: *Physicochemical methods in the study of biomembranes* (Hilderston, H. J. and Ralston, G. B., eds.), Plenum, New York, pp. 363–403.

27. Byler, D. M. and Susi, H. (1986) Examination of the secondary structure of proteins by deconvolved FTIR spectra. *Biopolymers* **25,** 469–487.

28. Goormaghtigh, E., Cabiaux, V., and Ruysschaert, J.-M. (1990) Secondary structure and dosage of soluble and membrane proteins by attenuated total reflection Fourier-transform infrared spectroscopy on hydrated films. *Eur. J. Biochem.* **193,** 409–420.

29. Mueller, E. and Blume, A. (1993) FTIR spectroscopic analysis of the amide and acid bands of ganglioside G_{M1}, in pure form and in mixtures with DMPC. *Biochim. Biophys. Acta* **1146,** 45–51.

30. Citra, M. J. and Axelsen, P. H. (1996) Determination of molecular order in supported lipid membranes by internal reflection Fourier transform infrared spectroscopy. *Biophys. J.* **71,** 1796–1805.

8

The Use of Fluorescence Resonance Energy Transfer to Detect Conformational Changes in Protein Toxins

William D. Picking

1. Introduction

1.1. An Introduction to Fluorescence

When a fluorophore absorbs a photon, an electron is excited to a higher energy level. This excited state electron returns to its ground state by one of two competing processes. In radiative de-excitation, a brief relaxation time (about 10^{-12} s) is followed by the electron's return to the ground state accompanied by the emission of a photon whose wavelength is longer than the one absorbed (1). Fluorescence emission competes with nonradiative processes that also allow the electron to return to the ground state. Because the competition between these processes is influenced by the fluorophore's surroundings, the nature of the emitted light provides information on the microenvironment of the probe. Therefore, when a fluorophore is part of a macromolecular structure (such as a protein or protein-containing complex), its fluorescence emission provides information on those events occurring within the structure.

Proteins can be studied using fluorescence spectroscopy by virtue of the intrinsic fluorescence of tryptophan residues that possess emission properties compatible with most fluorescence detection systems. Alternatively, proteins can be covalently labeled with extrinsic probes linked to protein functional groups such as primary amines or free sulfhydryls. The binding subunit monomers of important protein toxins such as shiga-like toxin (2), shiga toxin (3), cholera toxin (4), and heat-labile enterotoxin (4) each possess a single tryptophan residue that serves as a convenient site-specific probe within these proteins. For the B-subunit of cholera toxin (CTB), tryptophan 88 (W88) is found

From: *Methods in Molecular Biology, vol. 145: Bacterial Toxins: Methods and Protocols*
Edited by: O. Holst © Humana Press Inc., Totowa, NJ

as part of the GM1 binding site *(5)* and has been used by this and other laboratories as a convenient probe for investigating the interaction of this protein
with its ganglioside GM1 receptor *(6–9)*.

1.2. Fluorescence Resonance Energy Transfer

When a fluorophore's emission spectrum overlaps the absorption spectrum
of another molecule, its excitation can lead to the nonradiative transfer of its
excitation energy to the second molecule. This distance-dependent process is
fluorescence resonance energy transfer (FRET) and results from the dipole–
dipole coupling that occurs between the donor and acceptor molecules. FRET
is useful for studying distances within many proteins because it can be used to
obtain distance information in a range from 10 to 100 Å *(10,11)*.

The efficiency of FRET between a donor probe (**d**) and an acceptor probe
(**a**) as measured by steady-state and time-resolved fluorescence methods is
dependent upon three factors: (1) the degree of overlap between **d** emission
and **a** absorption (spectral overlap integral or J); (2) the relative orientation of
d and **a** dipoles (orientation factor or κ^2); and (3) the distance separating **d** and
a *(1,10,11)*. If J and κ^2 are known, FRET efficiency (E) can be used to calculate the distance separating the probes. More importantly, an inverse sixth-
power relationship between E and the distance from **d** to **a** makes this technique
sensitive to the movement of one fluorophore relative to the other.

J is easily calculated from spectral measurements of **d** and **a** according to
the relationship

$$J = F_D(\lambda)\varepsilon_A(\lambda)\lambda^4 d\lambda \tag{1}$$

where F_D is the normalized emission spectrum of **d**, ε_A is the molar absorption
coefficient of **a** (in $M^{-1}cm^{-1}$) and λ is the wavelength in nm *(1)*. Knowing J, it is
possible to calculate the Förster distance (that which gives 50% E for the **d-a**
pair) from

$$R_O^6 = 8.785 \times 10^{-5}(\kappa^2\phi_D J/n^2) \tag{2}$$

where ϕ_D is the quantum yield of **d** in the absence of **a**, and n is the refractive
index of the medium between **d** and **a** *(1,10)*. The importance of quantum yield in
monitoring the structural changes studied in our laboratory is discussed in **Note 5**.

The relative orientation of **d** and **a** dipoles is the most difficult factor to
assess in FRET measurements *(11)*. For probes linked to many biological components, the orientation is typically considered random ($\kappa^2 = 2/3$) which
assumes that each dipole adopts a dynamically averaged orientation over the
fluorescence lifetime of **d**. Different methods have been described to reduce
the uncertainty caused by assuming $\kappa^2 = 2/3$ *(12,13)*. In our laboratory, polarization measurements are used to estimate the limits of the uncertainty brought

about by assuming that **d** and **a** are oriented randomly *(11,12)*. More importantly, in the absence of precise knowledge of κ^2, FRET still provides one of the most sensitive methods for monitoring conformational changes within macromolecular complexes *(1)*.

FRET efficiency (*E*) is calculated from

$$E = 1 - (F_{da}/F_d) \tag{3}$$

where F_{da} is the fluorescent intensity of **d** in the presence of **a** and F_d is the fluorescence intensity of **d** in the absence of **a** *(1)*. Once *E* and R_o are known, they are used to calculate the distance between **d** and **a** from

$$E = R_o^6/(R_o^6 + r^6) \tag{4}$$

where *r* is the calculated distance between **d** and **a**.

2. Materials

2.1. Chemicals

1. Cholera toxin B subunit (CTB; List Biological Laboratories, Campbell, CA).
2. Lipids: Monosialoganglioside GM1 and bovine liver phosphatidylcholine (PC).
3. Fluorescent lipids: Pyrene–GM1 (Sigma Chemical, St. Louis, MO) and *N*-(7-nitrobenz-2-oxa-1,3-diazol-4-yl) dipalmitoyl-L-α-phosphatidylethanolamine (NBD–PE; Molecular Probes, Eugene, OR).
4. Reactive fluorescent probes: Dansylchloride (Dns), fluorescein isothiocyanate (FITC), 3-(4'-isothio-cyanatophenyl)-7-diethylamino-4-methylcoumarin (CPI) and 4-acetamido-4'-isothiocyanato-stilbene-2,2'-disulfonic acid (SITS).
5. Solvents: Dimethylformamide (DMF) and methanol.

2.2. Solutions and Buffers

1. All fluorescent probes are prepared freshly in DMF and kept out of direct light.
2. PC vesicles are prepared in 50 m*M* sodium citrate–50 m*M* sodium phosphate, pH 7.0; 0.15 *M* NaCl (citrate/phosphate buffer) and stored at 4°C for up to 5 d.
3. Gel filtration buffers:
   a. 0.1 *M* carbonate, pH 8.5 (to prepare CTB for fluorescence labeling).
   b. 50 m*M* phosphate, pH 7.5); 0.15 *M* NaCl (PBS) with 1 m*M* sodium azide (to separate labeled protein from unreacted probe).
3. Lipid stocks are suspended in methanol and stored at –20°C in the dark.
4. Protein solutions are stored at 4°C for a month or kept for longer times at –80°C.

3. Methods

3.1. Fluorescence Labeling

3.1.1. Preparation of PC Vesicles

1. Dry 1 mg of PC in chloroform–methanol under nitrogen with further removal of solvent in a vacuum for 1 h at 37°C.

**Table 1
Fluorescence Quenching of Labeled Lipids Present
on the Outer Leaflet vs the Inner and Outer Leaflets
of PC Vesicles**

Quenching agent[a]	K_{SV} (M^{-1}) for NBD–PE located on:	
	Inner and outer leaflets[b]	Outer leaflet only[c]
Acrylamide	0.209	0.439
Iodide	0.777	1.742
Methylviologen	1.768	5.054

[a]All three quenching agents are polar and do not cross phospholipid membranes. Acrylamide is a neutral quenching agent, iodide is anionic, and methylviologen is a cationic.

[b]To place fluorescent lipids on both the inner and outer leaflets of PC vesicles, they were included in the dried PC mixture which was sonicated in buffer containing 0.15 M NaCl.

[c]To incorporate fluorescent lipids into the outer leaflet of PC vesicles, they were added from a methanol stock to preformed vesicles in buffer containing 0.15 M NaCl. (Reprinted with permission from *Biochemistry* **36,** 9169–9178, 1997. Copyright 1997 American Chemical Society.)

2. Sonicate the dried PC in citrate–phosphate buffer, pH 7.0, at 20–22°C with three 30-s bursts at a medium setting.
3. PC vesicles form spontaneously when the sample is incubated 1 h at 37°C followed by incubation at 20–22°C for 18 h. Vesicles are used at 0.1 mg/mL in fluorescence experiments.
4. Incorporate GM1, pyrene–GM1, or NBD–PE into the outer leaflet of PC vesicles by adding to preformed PC vesicles directly from methanol stocks and incubating for 1 h at 37°C. Proper incorporation of fluorescent lipids into the vesicles can be monitored spectroscopically.
   a. Incorporation of pyrene–GM1 into membranes is observed spectroscopically by following the conversion of pyrene fluorescence from an excimer to a monomer state *(14)*.
   b. To confirm that the incorporation procedure has limited the introduction of fluorescent lipids to the outer leaflet of the PC vesicles, fluorescence quenching of probes present exclusively on the outer leaflet can be compared to that of lipids placed (by sonication) on both sides of the membrane (**Table 1**). The presence of a probe population on the inner leaflet causes a decrease in accessibility to polar quenching agents relative to vesicles with probes limited to the outer leaflet (*see* **Note 1**).

3.1.2. Fluorescence Labeling of CTB

1. To label CTB at primary amines, the protein is first passed over a Sephadex G25 (1 cm × 10 cm bed volume with gravity flow) equilibrated with 0.1 M carbonate, pH 8.5.

2. The protein is then labeled with 0.1 mg/mL of FITC or SITS, or made to 50% DMF for labeling with 0.05 mg/mL of CPI or Dns.

3. Incubate the CTB–probe mixture for 18 h at 4°C and remove any precipitate by centrifugation.

4. Separate the labeled CTB from unreacted dye by gel filtration on Sephadex G50 (1 cm × 15 cm bed volume with gravity flow) equilibrated with PBS containing 1 mM azide. Fluorescent fractions are collected manually with the aid of a hand-held UV light source.

5. CTB and probe concentrations are determined by absorbance at 280 nm and the peak of dye absorbance, respectively. Typically, CTB labels at a stoichiometry near one probe/monomer (FITC slightly exceeds this stoichiometry while CPI gives a lower labeling stoichiometry; *see* **Note 2**).

6. Sodium dodecyl sulfate-polyacrylamide gel electrophoresis (SDS-PAGE; 15% gels run until the dye front is ≤0.5 cm from the end of the gel) according to Laemmli *(15)* is used with UV detection to ensure that samples containing labeled protein are free of unreacted dye that travels at the dye front.

3.2. Fluorescence Methods

3.2.1. Fluorescence Detection

3.2.1.1. Fluorescence Measurements

Steady-state fluorescence is measured on a Spex (Edison, NJ) FluoroMax spectrofluorometer equipped with automatic blank subtraction, correction for wavelength dependence of lamp intensity, and excitation and emission polarizers (for anisotropy measurements). Wavelengths for fluorescence excitation are: 282 nm for W88, 330 nm for pyrene, 385 nm for CPI, and 480 nm for FITC. Fluorescence data are collected at the emission maximum for each probe with a sample absorbance <0.1 at the excitation and emission wavelengths.

3.2.1.2. Using FRET to Measure Structural Changes within CTB–GM1 Complexes

The transfer of excitation energy from a donor fluorophore to an acceptor probe results in a quenching of donor fluorescence. In **Fig. 1A**, the W88 donor is excited at 282 nm and emission scanned from 300 to 450 nm in the presence of either nonfluorescent GM1 (F_d) or pyrene–GM1 (F_{da}). Energy transfer causes a quantifiable decrease in the emission of W88 at 335 nm when the acceptor pyrene is present. **Figure 1B** shows the results of an energy transfer experiment with pyrene–GM1 as the donor and either CTB (F_d) or CPI-labeled CTB (F_{da}) as the acceptor. Energy transfer is seen as a quenching of pyrene emission in the presence of acceptor. Because the decrease in donor fluorescence is quantifiable, FRET is useful for determining the distance separating

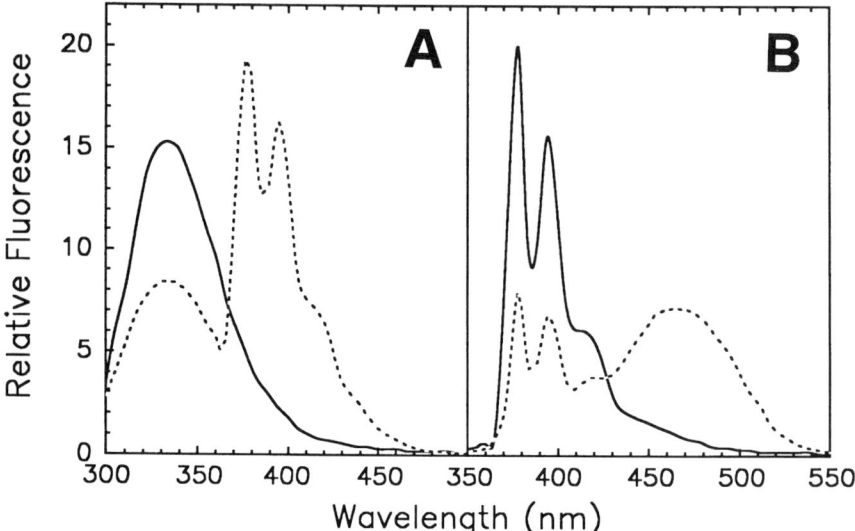

Fig. 1. FRET can be quantified from the quenching of donor fluorescence. In (**A**), the emission of W88 (0.75 μ*M* CTB in 0.6 mL) in the presence of excess nonfluorescent GM1 at pH 6.0 is shown by a *solid line*. The *dotted line* depicts the same concentration of CTB at pH 6.0 in the presence of excess pyrene–GM1 and shows a decrease in W88 emission at 334 nm. In (**B**), the emission of pyrene–GM1 incorporated into PC vesicles (at pH 7.0) is shown by a solid line. The same concentration of pyrene–GM1 in the presence of excess CPI-labeled CTB (at pH 7.0) is shown by the *dotted line* with a corresponding decrease in pyrene fluorescence at 380 and 400 nm.

the probes. Moreover, because of the inverse sixth-power relationship between FRET efficiency and the distance from **d** to **a**, events that alter the distance between them are readily detected.

To determine E from W88 to pyrene–GM1, the fluorescence intensity of W88 (using 150 n*M* CTB) is measured at the peak of W88 emission (338 nm) with F_d measured using 5 μ*M* nonfluorescent GM1 and F_{da} measured using 5 μ*M* pyrene–GM1. These FRET data can then be used to calculate the distance from W88 to the membrane-embedded pyrene on GM1, as a function of pH. When changes in the W88 quantum yield are considered (*see* **Note 5**), no change in the distance between the two sites is seen as a function of pH (**Fig. 2A**). This measurement provides a point of reference for further FRET experiments designed to test for pH-induced changes in CTB structure.

To determine E between membrane-embedded pyrene–GM1 and FITC, CPI, or SITS acceptors covalently linked to CTB, pyrene fluorescence intensity is most conveniently measured at its 398 nm emission peak. F_d is measured with

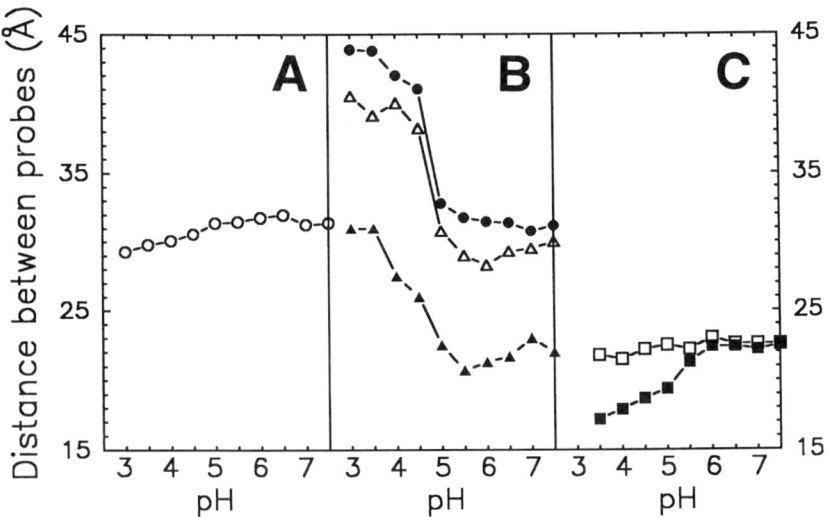

Fig. 2. FRET can be used to monitor distances between different sites within CTB–GM1 complexes as a function of pH. In (**A**), the distance between W88 and the probe on pyrene–GM1 does not change significantly as the pH is lowered. In contrast, (**B**) shows that changes in pH greatly influence the distance from the pyrene linked to the hydrophobic tail of GM1 to different acceptor probes attached to CTB. The acceptors used include CPI (*filled circles*), SITS (*open triangles*), and FITC (*closed triangles*). In (**C**), the distance between W88 and CPI (presumably linked to the same CTB monomer) does not change as a function of pH when GM1 is bound (*open squares*) but decreases at low pH in the absence of GM1 (*closed squares*). (Reprinted with permission from *Biochemistry 1997*, **36**, 9169–9178. Copyright 1997 American Chemical Society.)

150 nM pyrene–GM1 bound to 300 nM unlabeled CTB and F_{da} is measured with pyrene–GM1 bound to 300 nM fluorescent CTB. To give quantitative significance to these FRET measurements, it was important to determine the position of the labeled sites on CTB (*see* **Note 2**).

As shown in **Fig. 2B**, when CPI–CTB and SITS–CTB are bound to pyrene–GM1 in PC vesicles, the distance from the membrane-embedded pyrene to the probe linked to CTB increases by ≥10 Å as the pH is reduced to 4.5 or less. This distance change occurs without accompanying changes in the spectral properties of pyrene–GM1 (showing the results are not an artifact of pyrene–GM1 extraction from the membrane). Similar results are seen when FRET is carried out from pyrene–GM1 to FITC linked to CTB (**Fig. 2B**), except that a smaller change in the apparent distance (about 5 Å) from pyrene to the labeled site on CTB is observed.

For FRET from W88 to an extrinsic fluorophore (CPI, Dns, or SITS) linked to CTB, E is calculated in the presence and absence of 1 μM GM1. W88 emission was measured at its peak of fluorescence (350 nm without GM1 or 338 nm with GM1) with F_d measured using unlabeled CTB and F_{da} measured using the same concentration of labeled CTB. pH has little effect on the distance separating W88 and CPI in the presence of GM1 (**Fig. 1C**). Identical results obtained for each extrinsic probe tested (data not shown). In contrast, low pH caused a decrease in the distance between W88 and the CPI probe in the absence of GM1 (**Fig. 2C**).

The data in **Fig. 2C** support previous findings that GM1 binding stabilizes CTB structure against denaturation at low pH. Meanwhile, the data in **Fig. 2B** also suggest that a structural change in the CTB–GM1 complex is elicited at low pH. Because there is no change in the distance from W88 to extrinsic probes on CTB in the presence of GM1, it seems likely that a minor conformational change in CTB is eliciting a disturbance in membrane packing. If such a change in membrane structure does take place, it would have to occur without an obvious change in the distance from W88 to the tail of pyrene–GM1 (**Fig. 2A**). In any case, these data demonstrate the sensitivity of FRET analysis to detect movement between different points within a protein–lipid complex.

3.2.1.3. FRET FOR MONITORING MEMBRANE STRUCTURE

To determine distance changes from the nonpolar tail of pyrene–GM1 or sites on CTB, to acceptors on the surface of PC vesicles, either NBD–PE (an acceptor for CPI, SITS, pyrene, or W88) is preincubated with PC vesicles prior to incorporation of GM1 and the binding of CTB. Initial experiments should be designed to determine the amount of fluorescent phospholipid needed to obtain approx 50% E. That amount of lipid is then used in experiments where pH is the only variable. E is determined from emission spectra spanning the fluorescence of the donor and acceptor probes. No specific distances are calculated here because the observed FRET is an average value that depends upon acceptor density on the surface of the vesicles.

Data from this laboratory show that CTB binds to pyrene–GM1 incorporated into PC vesicles to disrupt local membrane structure (*8*). This phenomenon is diminished as the pH is lowered to 5 or less (*8*). These and FRET data suggest that low pH promotes conformational changes in CTB that influence membrane structure. To investigate these membrane effects, FRET can be used to monitor the position of the probe on pyrene–GM1 relative to the vesicle surface. NBD–PE is incorporated into PC vesicles and FRET used to look for changes in the distance between membrane-embedded pyrene and surface-associated NBD in the presence and absence of CTB. In the absence of CTB,

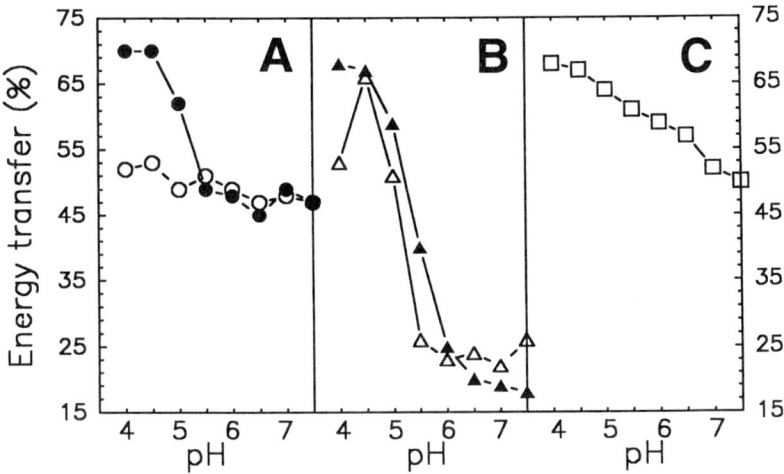

Fig. 3. FRET can be used to detect changes in the positions of probes on the CTB–GM1 complex relative to NBD–PE probes incorporated into the outer face of the PC vesicles. In (**A**), pH has no effect on FRET from the pyrene attached to GM1 in the absence of CTB (*open circles*), but has a significant effect when CTB is associated with the pyrene–GM1 (*closed circles*). In (**B**), pH also effects the relative FRET efficiency between the NBD–PE probes on the membrane surface relative to SITS (open triangles) and CPI (*closed triangles*) probes linked to CTB which is associated with GM1 that has also been incorporated into the outer face of the PC vesicles. Meanwhile, (**C**) shows that decreased pH causes a gradual change in the FRET from W88 to the membrane surface, but without the rapid pH-dependent effect seen in panels **A** and **B**. (Reprinted with permission from *Biochemistry* **36,** 9169–9178, 1997. Copyright 1997 American Chemical Society.)

pH has almost no effect on FRET between the pyrene and NBD probes; however, FRET increased substantially as the pH is lowered in the presence of CTB (**Fig. 3A**). These data show that CTB binding to GM1-containing membranes mediates pH-dependent changes in membrane structure and that the net movement of the embedded pyrene probe on GM1 is toward the surface of the membrane.

To look for pH-dependent changes in the distance between extrinsic probes on CTB and the membrane surface, NBD–PE is incorporated into PC vesicles as a FRET acceptor for CPI or SITS linked to CTB. In this case, a rapid and significant increase in FRET is observed at a pH of 5.0 or below (**Fig. 3B**), implying that pH conditions leading to movement of the extrinsic probes on CTB away from the pyrene on GM1 (and movement of the same pyrene toward the membrane surface) also result in the movement of extrinsic probes on CTB toward the membrane surface. In a related experiment, W88 can serve as a fluorophore to look for pH-dependent movement of this intrinsic probe relative to NBD–PE incorporated into the outer face of PC vesicles. In these experiments, FRET from W88 to the membrane surface increases as the pH is lowered (**Fig. 3C**); however, the observed change in FRET is gradual and occurs over a rather broad pH range. This may indicate that lowering the pH causes subtle changes in the spectral characteristics of one or both of the probes rather than an abrupt pH-dependent conformational change in the CTB–GM1 complex.

4. Notes

1. When excited fluorophores collide with small quenching agents like acrylamide, iodide, or methylviologen (1,1'-dimethyl-4,4'-bipyridinium), they return to their ground-state without emission. This quenching can be quantified from the relationship

$$F_o/F = 1 + K_{SV}[Q] \tag{5}$$

   as first described by Stern and Volmer *(16)* where F_o is the probe's fluorescence intensity in the absence of quenching agent, F is the intensity at a given concentration of quenching agent ($[Q]$), and K_{SV} is the Stern–Volmer quenching constant *(16)*. K_{SV} is the slope of F_o/F vs $[Q]$ and provides a measure of probe accessibility. Fluorescent lipids on the outer leaflet of PC vesicles are quenched more efficiently than those on the inner leaflet because of exposure to the external aqueous environment where the quenching agents are being added (*see* **Table 1**).

2. Protein and peptide analyses have shown that $\geq 80\%$ of the CTB was monolabeled with FITC with most of the labeling occurring at lysine 69 (K69) or 91 (K91). CPI labeling results in greater diversity in the sites modified. In all the cases described here, FRET results obtained using each extrinsic probe are similar. It is important to note that even without knowledge of the specific labeling sites on CTB, FRET is useful for detecting changes in structure. This is evident from the data in **Subheading 3.2.1.3.** (**Fig. 3**), where dispersion of acceptor probes within the PC vesicles allowed determination of only average FRET efficiencies. However, to use FRET analysis as a molecular ruler, precise knowledge of the position of donor and acceptor molecules is essential. For proteins with a single tryptophan, a convenient site-specific probe is already available. It is also possible to site-specifically label proteins containing single cysteine residues *(17)* or to design mutant proteins that possess a novel single cysteine residue to provide a site-specific anchor for a fluorescent label *(18,19)*. For proteins possessing multiple potential sites for labeling with extrinsic probes (such as CTB labeling at lysines), it is important to identify the positions at which actual labeling occurs.

   Proteolytic digestion and sequence analysis has been used to show that K69 and K91 dominate the labeling process. K91 is present at the edge of the GM1 binding site of CTB (about 10 Å from W88 on the same monomer) and may be involved in interactions with the terminal galactose moiety of GM1. Meanwhile, K69 is located at the middle of the large central helix of CTB but is relatively near the GM1 binding region (17 Å from W88 on the same CTB monomer). The ability of labeled CTB pentamers to bind pyrene–GM1 was indicated by: (1) observed FRET from pyrene to CPI–CTB, SITS–CTB, or FITC–CTB; and (2) changes in the fluorescence of the extrinsic probes upon incubation with unlabeled GM1. In fact, by monitoring changes in fluorescence as a function of GM1 concentration, the dissociation constant of labeled CTB for GM1 remains within an order of magnitude of that seen for unmodified CTB *(9)*.

3. A factor that complicates making distance determinations within protein homopolymers is the possibility that multiple donors are paired with multiple acceptors. When the distance between multiple acceptors and donors is asymmetric, the inverse sixth-power relationship between E and r ensures that the acceptor nearest each donor dominates the energy transfer process. In contrast, multiple equidistant acceptors require a modified analysis of FRET *(20)*. For CTB, it is known that the arrangement of donors relative to acceptors is asymmetric given the distances from W88 to the α-carbons of K69 and K91 in the crystal structure of CTB *(5,21)* and it is assumed that this asymmetry holds for fluorophores linked to these lysyl ε-amino groups. It is also assumed that acceptor probes at K69 and K91 are asymmetrically positioned relative to pyrene–GM1 that is approximately centered within the membrane beneath W88. These assumptions may introduce error into the determined distances; however, they do not lessen the significance of the conformational changes detected within the CTB–GM1 complex.

4. For probes attached to proteins or membrane components, the relative dipole orientation typically adopts a dynamically averaged orientation over the fluorescence lifetime of **d** ($\kappa^2 = 2/3$). Unfortunately, assuming a randomized orientation can introduce error into the resulting distance determinations as κ^2 can range from 0 to 4 *(1)*. Dos Remedios and Moens *(11)* present a case for confidently using FRET to measure protein conformational changes even without knowledge of κ^2 by showing that intraprotein distances determined in crystallographic studies agree with those determined in FRET studies using $\kappa^2 = 2/3$ when the sizes of the probes are taken into account *(11)*.

For FRET-based distance determinations, half-height limits of uncertainty can be estimated by determining the extent of isotropical randomization of **d** and **a** based on their fluorescence anisotropy values *(12)*. After excitation with polarized light, photons emitted by a fluorescent probe are also polarized *(1)*; however, rotational diffusion of a fluorophore causes depolarization of the emitted light over the probe's excited state lifetime. By using anisotropy to measure the rate of rotation for **d** and **a**, it is possible to estimate the average angular displacement of the probes over their excited state lifetimes. This angular displacement is affected by the rate of rotational diffusion during the fluorescence lifetime which is influenced inversely by the size of the diffusing molecule and regional flexibilities within the macromolecular complex.

Table 2
The Influence of Environmental pH
on the Quantum Yields of W88
and Pyrene–GM1

Donor	Φ^a	
	pH 7.5	pH 3.5
W88 (– GM1)	0.26	0.04
W88 (+ GM1)	0.28	0.22
pyrene–GM1	0.05	0.05

[a]The fluorescence quantum yield of the donor was calculated based on comparisons with a quinine sulfate standard *(22)*. (Reprinted with permission from *Biochemistry* **36,** 9169–9178, 1997. Copyright 1997 American Chemical Society.)

Fluorescence anisotropy data can be used according to the formula designed by Haas et al. *(12)* to estimate the degree of uncertainty introduced to a FRET-based distance determination by assuming $\kappa^2 = 2/3$. Anisotropy (r) is calculated from

$$r = (I_{vv} - I_{vh}) / (I_{vv} + 2I_{vh}) \tag{6}$$

where I_{vv} is the intensity when the excitation beam is polarized vertically and the excitation and emission polarizers are oriented parallel, and I_{vh} is the intensity when the excitation beam is polarized vertically and the excitation and emission polarizers are oriented perpendicular to one another (*see* **ref. [1]** for details).

5. Determining the distance (r) between a donor and acceptor probe based on FRET efficiency requires knowledge of the Forster distance (R_0). One of the factors that influences R_0 is the quantum yield of **d** which is affected, in turn, by the probe's microenvironment. Quantum yield is the ratio of photons emitted by a fluorophore to those absorbed and is determined by comparison with a quinine sulfate standard in 0.1 N H$_2$SO$_4$ *(22)*. In some cases, the dependence of donor quantum yield (or acceptor absorbance) on pH requires determining R_0 for each pH tested (**Table 2**). For example, the quantum yield of W88 is sensitive to pH; however, this sensitivity is diminished when GM1 is bound prior to lowering the pH (**Table 2**). It is therefore important to be aware of environmental changes that adversely affect the donor quantum yield. When quantum yield changes for W88 are seen, they are taken into account by calculating a new R_0 value for each change in the environmental conditions. In contrast, the quantum yield of pyrene–GM1 is not influenced by environmental pH once the lipid is incorporated into PC vesicles (**Table 2**). It is also important to correct for the influence that the environment has on the absorbance of the acceptor molecules like fluorescein.

References

1. Lakowicz, J. R., ed. (1983) *Principles of Fluorescence Spectroscopy,* Plenum, New York.
2. Calderwood, S. B., Auclair, F., Donohue-Rolfe, A., Keusch, G. T., and Mekalanos J. J. (1987) Nucleotide sequence of the shiga-like toxin genes of *Escherichia coli. Proc. Natl. Acad. Sci. USA* **84,** 4364–4368.
3. Surewicz, W. K., Surewicz, K., Mantsch, H. H., and Auclair, F. (1989) Interaction of shigella toxin with globotriaosyl ceramide receptor-containing membranes: a fluorescence study. *Biochem. Biophys. Res. Commun.* **160,** 126–132.
4. Spangler, B. D. (1992) Structure and function of cholera toxin and the related *E. coli* heat-labile enterotoxin. *M icrobiol. Rev.* **56,** 622–647.
5. Merritt, E. A., Sarfaty, S., van der Akker, R., L'Hoir, C., Martial, J. A., and Hol, W. G. J. (1994) Crystal structure of cholera toxin B-pentamer bound to receptor GM1 pentasaccharide. *Prot. Sci.* **3,** 166–175.
6. DeWolf, M. J. S., Fridkin, M., Epstein, M. M., and Kohn, L. D. (1981) Structure–function studies of cholera toxin and its A protomers and B protomers. *J. Biol. Chem.* **256,** 5481–5488.
7. DeWolf, M. S. J., Van Dessel, G. A. F., Lagrou, A. R., Hilderson, H. J. J., and Dierick, W. S. H. (1987) pH-induced transitions in cholera toxin conformation. *Biochemistry* **26,** 3799–3806.
8. Picking, W. L., Moon, H., Wu, H., and Picking, W. D. (1995) Fluorescence analysis of the interaction between ganglioside GM1-containing phospholipid vesicles and the B subunit of cholera toxin. *Biochim. Biophys. Acta* **1247,** 65–73.
9. McCann, J. A., Mertz, J. A., Czworkowski, J., and Picking, W. D. (1997) Conformational changes in cholera toxin B subunit–ganglioside GM1 complexes are elicited by environmental pH and evoke changes in membrane structure. *Biochemistry* **36,** 9169–9178.
10. Wu, P. and Brand, L. (1994) Resonance energy transfer: methods and applications. *Anal. Biochem.* **218,** 1–13.
11. dos Remedios, C. A. and Moens, P. D. (1995) Fluorescence resonance energy transfer spectroscopy is a reliable "ruler" for measuring structural changes in proteins: dispelling the problem of the unknown orientation factor. *J. Struct. Biol.* **115,** 175–185.
12. Haas, E., Katzir, E.-K., and Steinberg, I. Z. (1978) Effect of orientation of donor and acceptor on the probability of energy transfer involving electronic transitions of mixed polarization. *Biochemistry* **17,** 5064–5070.
13. Dale, R. E., Eisinger, J., and Blumberg, W. E. (1979) The orientational freedom of molecular probes: the orientation factor in intramolecular energy transfer. *Biophys. J.* **26,** 161–194.
14. Sonnino, S., Acquotti, D., Riboni, L., Giuliani, A., Kirschner, G., and Tettamanti, G. (1986) New chemical trends in ganglioside research. *Chem. Phys. Lipids* **42,** 3–26.
15. Laemmli, U. K. (1970) Cleavage of structural proteins during the assembly of the head of bacteriophage T4. *Nature* **227,** 680–685.

16. Stern, D. and Volmer, M. (1919) On the quenching-time of fluorescence. *Phys. Zeitschr.* **20,** 183–188.

17. Csortos, C., Matko, J., Erdodi, F., and Gergely, P. (1990) Interaction of the catalytic subunits of protein phosphatase-1 and 2A with inhibitor-1 and 2: a fluorescent study with sulfhydryl-specific pyrene maleimide. *Biochem. Biophys. Res. Commun.* **169,** 559–564.

18. Flitsch, S. L. and Khorana, H. G. (1989) Structural studies on transmembrane proteins. 1. Model study using bacteriorhodopsin mutants containing single cysteine residues. *Biochemistry* **28,** 7800–7805.

19. Picking, W. D., Kudlicki, W., Kramer, G., Hardesty, B., Vandenheede, J. R., Merlevede, W., Park, I., and DePaoli-Roach, A. (1991) Fluorescence studies on the interaction of inhibitor-2 and okadaic acid with the catalytic subunit of type 1 phosphoprotein phosphatase. *Biochemistry* **30,** 10,280–10,287.

20. Highsmith, S. and Murphy, A. J. (1984) Nd^{3+} and Co^{2+} binding to sarcoplasmic reticulum CaATPase: an estimation of the distance from the ATP binding site to the high-affinity calcium binding sites. *J. Biol. Chem.* **259,** 14,651–14,656.

21. Zhang, R.-G., Westbrook, M. L., Westbrook, E. M., Scott, D. L., Otwinowski, Z., Maulik, P. R., Reed, R. A., and Shipley, G. G. (1995) The 2.4Å crystal structure of cholera toxin B subunit pentamer: choleragenoid. *J. M ol. Biol.* **251,** 550–562.

22. Dawson, W. R. and Windsor, M. W. (1968) Fluorescence yields of aromatic compounds. *J. Phys. Chem.* **72,** 3251–3260.

9

Site-Directed Spin Labeling of Proteins

Applications to Diphtheria Toxin

Kyoung Joon Oh, Christian Altenbach, R. John Collier, and Wayne L. Hubbell

1. Introduction

1.1. Site-Directed Spin Labeling

Site-directed spin labeling (SDSL) has emerged as a powerful approach to study structure and dynamics of proteins that are not readily amenable to X-ray crystallography or nuclear magnetic resonance (NMR) spectroscopy *(1–3)*. SDSL involves the site-specific labeling of proteins with spin probes and the use of electron paramagnetic resonance (EPR) spectroscopy for analysis of the labeled proteins. Spin labeling is typically accomplished by cysteine-substitution mutagenesis followed by reaction with a sulfhydryl-specific nitroxide reagent *(4)*. The reagent most widely used is methanethiosulfonate spin label I *(5)*, which generates the nitroxide side chain designated R1, as shown in **Fig. 1**. Other spin label reagents are also used for specific purposes *(6)*, but examples in this chapter make use of R1 exclusively.

The paramagnetic resonance of a nitroxide side chain in a protein provides information on the mobility and solvent accessibility of the side chain. The mobility of the nitroxide determines the EPR spectral lineshape, and detailed information regarding the motional rate, anisotropy, and amplitude can in principle be obtained from spectral simulation *(7)*. However, owing to the complexity of the spectra of spin-labeled proteins, this approach has not yet been extensively exploited.

The inverse linewidth of the nitroxide central resonance ($m_I = 0$) and the spectral second moment can be used as qualitative measures of side chain

From: *Methods in Molecular Biology, vol. 145: Bacterial Toxins: Methods and Protocols*
Edited by: O. Holst © Humana Press Inc., Totowa, NJ

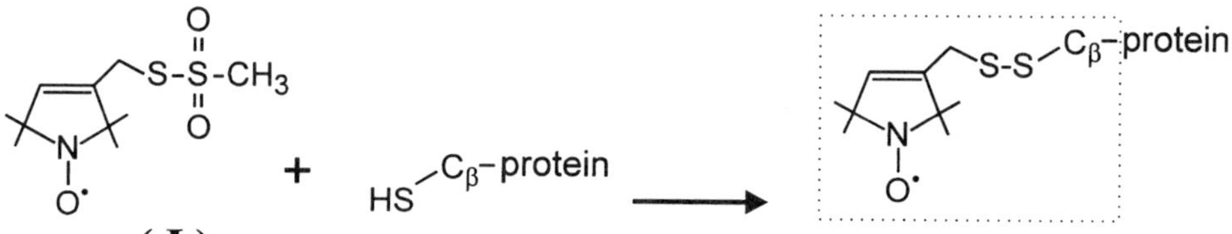

side-chain R1

Fig. 1. Reaction of the methanethiosulfonate spin label (I) with a cysteine residue to generate the side chain R1.

mobility (6). From a comprehensive study of many different spin-labeled T4 lysozyme mutants, it has been found that the mobility of R1 is strongly modulated by tertiary interactions and local segmental motion (6). Therefore, it is possible to determine from mobility alone whether the spin-labeled site is located in the protein interior, at the tertiary contact site, on the surface of a protein, or in a flexible loop (1,6).

Solvent accessibility is inferred from the collision frequency of the nitroxide side chain with paramagnetic reagents in solution (8–10). For static structure determination, the information content of side chain mobility and solvent accessibility are to a large extent redundant, and this chapter focuses on measurement and interpretation of R1 solvent accessibility in terms of protein structure. The most generally applicable method for determination of collision frequency is progressive power saturation. The theoretical basis and practical application of this methodology are described in detail in the following subsection.

In addition to mobility and solvent accessibility, introducing a pair of nitroxides (11–14), or a nitroxide and a paramagnetic metal ion, into the protein molecule can be used to estimate inter-residue distance (15). Although distance measurement between nitroxide side chains provides important structural information, this aspect of SDSL has not yet been exploited in the study of membrane-bound bacterial toxins. Lack of a theoretical framework prevents rigorous interpretation of data at room temperature. Although estimation of inter-residue distances is not described here, the reader may consult references (11–14) for examples of distance determination in low molecular mass proteins and for proteins at low temperature.

1.1.1. Spin Lattice Relaxation, Heisenberg Exchange, and Power Saturation EPR

The collision frequency of a nitroxide with a paramagnetic reagent is estimated from the change in spin-lattice relaxation time (T_1) of the spin label owing to the collision event. T_1 is a constant characterizing the relaxation of a spin system to thermal equilibrium. The reciprocal, $1/T_1$, is a measure of how rapidly a spin system equilibrates with its environment (historically called the "lattice"). Another relaxation time, T_2, related to interactions within the spin population, determines the shape of the EPR spectrum. An excellent introduction to spin relaxation is given by Poole *(16)*.

Whenever two paramagnetic molecules collide, they may exchange electrons in a process called Heisenberg exchange (HE). Thus, collision between a nitroxide and a paramagnetic reagent with a very short spin-lattice relaxation time effectively increases $1/T_1$ for the nitroxide. Excess energy in the nitroxide spin system is passed to the reagent by HE and then rapidly to the environment by spin-lattice relaxation of the reagent. For the nitroxide, HE is thus indistinguishable from a spin-lattice relaxation event, and collisions result in an increase in both $1/T_1$ and $1/T_2$ equal to the exchange frequency (W_{ex}). For typical EPR spectra of spin-labeled proteins, T_1 is a few microseconds and the effective T_2 (as measured from the central linewidth) is about 100× faster (10–30 ns). Thus a W_{ex} in the megahertz range will have a dramatic effect on T_1, but T_2 (and therefore lineshape) will only be affected at collision rates that are roughly 50–100× higher. For this reason T_2-based measurements are not sensitive enough to detect collisions with paramagnetic reagents in all cases, simply because the concentration of the paramagnetic reagent would need to be extremely high to have a measurable impact on lineshape. In summary, if T_1 can be measured in the absence and presence of a paramagnetic reagent, W_{ex} can be determined by the change in $1/T_1$.

Direct measurement of T_1 requires the use of a pulse saturation recovery EPR spectrometer. Although this specialized instrumentation is not generally available, it has been shown that indirect methods such as power saturation EPR spectroscopy using a loop-gap resonator can be equally effective for measuring relative changes in T_1 *(8)*. In a power-saturation EPR experiment, the amplitude of the EPR spectrum is measured as a function of incident microwave power. Under nonsaturating conditions (low microwave power), the amplitude of the EPR signal increases linearly with the square root of the incident microwave power. As the power is increased further, saturation of the EPR signal causes the signal to reach a maximum followed by a decrease with increasing power. Typically, the vertical peak-to-peak amplitude (*A*) of the central line of the first derivative EPR spectrum is measured as a function of incident microwave power (*P*). The resulting "power saturation curve" is described by

$$A = I \cdot \sqrt{P} \cdot [1 + (2^{1/\varepsilon} - 1) \cdot P / P_{1/2}]^{-\varepsilon} \qquad (1)$$

Equation (1) describes the saturation behavior of a first derivative EPR signal of unknown homogeneity and is sufficient to describe all data accurately. *I* is a scaling factor. $P_{1/2}$ is the power, where the first derivative amplitude is reduced to half of its theoretical unsaturated value determined by extrapolating the linear region (**Fig. 5C**). The parameter ε is a measure of the homogeneity of saturation of the resonance line *(8,9)*. For the homogeneous and inhomogeneous saturation limits, $\varepsilon = 1.5$ and $\varepsilon = 0.5$, respectively. The parameter $P_{1/2}$ is a function of T_1 and T_2 of the nitroxide spin label:

$$P_{1/2} \propto 1/(T_1 \cdot T_2) \tag{2}$$

Because the peak-to-peak linewidth of the first derivative spectrum (ΔH_{pp}) is inversely proportional to the effective T_2, dividing $P_{1/2}$ by ΔH_{pp} provides a parameter proportional to $1/T_1$ with a proportionality constant that is the same for all samples, irrespective of variations in nitroxide mobility (lineshape). This holds for paramagnetic reagent concentrations such that $W_{ex} \ll 1/T_2$, that is, at reagent concentrations in which the spectral lineshape is unchanged by the addition of reagent.

The change in $P_{1/2}$, $\Delta P_{1/2}$, owing to the presence of a paramagnetic reagent, is proportional to the exchange frequency between the spin label and the paramagnetic reagent according to

$$\Delta P_{1/2}/\Delta H_{pp} = (P_{1/2} - P_{1/2}{}^{\circ})/\Delta H_{pp} \propto W_{ex} \tag{3}$$

where $P_{1/2}$ and $P_{1/2}{}^{\circ}$ are the values in the presence and absence of a collision reagent, respectively. The parameter $\Delta P_{1/2}/\Delta H_{pp}$ is normalized to the same parameter for a reference sample to account for instrumental variations. This dimensionless parameter is an instrument and lineshape independent measure of W_{ex} and is referred to as Π, the accessibility parameter:

$$\Pi \equiv \{\Delta P_{1/2}/\Delta H_{pp}\}/\{P_{1/2}(\text{DPPH})/\Delta H_{pp}(\text{DPPH})\} \propto W_{ex} \tag{4}$$

where $P_{1/2}(\text{DPPH})$ and $\Delta H_{pp}(\text{DPPH})$ are the $P_{1/2}$ and linewidth values determined for a standard sample of crystalline 2,2-diphenyl-1-picrylhydrazyl (DPPH) in KCl, respectively *(10)*.

1.1.2. Residue Solvent Accessibility and Secondary Structure for Water-Soluble and Membrane Proteins

The collision frequency, and hence Π, is directly proportional to the solvent accessibility of the nitroxide residue R1 in the folded protein. For a membrane protein, there are effectively two solvents, the aqueous phase and the fluid lipid bilayer. To explore solvent accessibility for a membrane protein requires two paramagnetic reagents, one polar for the aqueous domain and the other nonpolar for the bilayer interior domain. Commonly used relaxation reagents are molecular oxygen (nonpolar) and the paramagnetic complexes Ni(II)ethylenediaminediacetate (NiEDDA) and Ni(II)acetylacetonate (NiAA), both polar. Note that each of these reagents is electrically neutral, and collision frequencies will be independent of local electrostatic potentials.

In a sequence of regular secondary structure, Π for a nitroxide side chain is a periodic function of sequence position, with a period that identifies the type of secondary structure and a phase that defines the orientation of the structural element in the protein. Thus, "nitroxide scanning," in which native side chains are sequentially replaced with a nitroxide side chain, can be used to determine the sequence-specific secondary structure in proteins *(1)*. The efficacy of the power saturation method of EPR for determining both secondary structure and topography was first demonstrated successfully in the helical membrane protein bacteriorhodopsin *(17)* and has been equally successful with water-soluble proteins *(1)*. The reagent selected for Π measurement is determined by the system. For a transmembrane sequence, the nonpolar reagent O_2 is employed. For a water-soluble protein, any of the above reagents suffice.

A particularly interesting situation arises when a secondary structural element is involved in asymmetric solvation, such as for an α-helix or β-strand lining a transmembrane aqueous pore. In this case, both $\Pi(O_2)$ and $\Pi(\text{NiEDDA})$ or $\Pi(\text{NiAA})$ will display a periodic dependence with sequence position, but $\Pi(O_2)$ will be 180° out-of-phase with $\Pi(\text{NiEDDA})$ *(18)*, diagnostic for asymmetrically solvated regular structures.

1.1.3. Estimation of Residue Immersion Depth in the Lipid Bilayer

The immersion depth of a nitroxide attached to a transmembrane polypeptide can be determined using the collision gradient method of Altenbach et al. *(19)*. This method relies on the observation that small molecules in a membrane–water system are partitioned between the water and the fluid hydrophobic phase of the bilayer according to their polarity, apparently forming gradients in concentration along the direction of bilayer normal. For example, the nonpolar molecule oxygen has a higher concentration in the center of the membrane than near the membrane–aqueous interface *(8)*. A polar reagent such as NiAA or NiEDDA has the opposite concentration profile. The immersion depth of a nitroxide from the bilayer surface, *D*, is given by *(19)*:

$$D = a \cdot \Phi + b \tag{5}$$

where

$$\Phi = \ln \left[\Pi(O_2) \, / \, \Pi(\text{NiEDDA}) \right] \tag{6}$$

and *a* and *b* are constants determined using nitroxide residues buried at known depths from the membrane surface *(19)*. After calibration, this relationship is used to estimate the immersion depth of the nitroxide side chain under investigation. The depth profile of a consecutive sequence can be used to determine the orientation of a given secondary structure in the lipid bilayer as shown in **Fig. 4C**. It is important to note that Eqs. (5) and (6) can be used only to determine depth for R1 residues fully exposed to the membrane environment. The Φ values for buried or partially buried R1 residues will be too large owing to molecular sieving effects caused by the size difference in the reagents.

1.2. Applications of SDSL to Diphtheria Toxin

Diphtheria toxin (DT) *(20–22)* belongs to a large class of toxic proteins that act by enzymatic modification of cytosolic substrates within eukaryotic cells *(23)*. Details of the process by which the catalytic moiety is transferred across a membrane lipid bilayer are not understood for any such toxin. Translocation of DT occurs only after the toxin has bound to its receptor at the cell surface and been delivered to the endosomal compartment by receptor-mediated endocytosis *(24,25)*. In the low pH environment, the toxin undergoes a conformational change that causes its transmembrane (T) domain to insert into the endosomal membrane *(26–34)*. The insertion event is known to induce the toxin's catalytic domain to cross the membrane to the cytoplasm *(35–37)*, where it catalyzes the adenosine diphosphate ribosylation of elongation factor-2, causing inhibition of protein synthesis and cell death *(38,39)*.

The T-domain (**Fig. 2**), which is situated between the toxin's N-terminal catalytic and C-terminal receptor domain, is composed of 10 α-helices *(42,43)*. Two long hydrophobic helices, TH8 and TH9, form the core of the T-domain, and are covered by two other "layers" of helices *(44)*. The holotoxin or the isolated T-domain forms voltage-dependent channels in planar bilayers under low pH conditions (pH ~5) *(45)*, and recent studies show that a toxin subfragment containing TH8 and TH9 as the only components of the T-domain is sufficient for channel formation *(46)*. Chemical modification studies indicate that the TH8–TH9 interhelical loop is located on the *trans* surface of the bilayer in the open channel state *(47)*. Thus it is postulated that the DT channel is formed by insertion of the TH8-TH9 helical hairpin into the membrane *(34,43,44,47)*. To explore the structure of the membrane-bound state in the isolated T-domain, SDSL has been applied to single-cysteine substitution mutants in the TH9 helix (*see* **Fig. 2**).

Figure 3 shows the EPR spectra of T-domain mutants containing a single R1 side chain at positions in the sequence 356–376 bound to phospholipid vesicles *(18)*. **Figure 4A** shows the corresponding values of $\Pi(O_2)$ and Π(NiEDDA) as a function of sequence position *(18)*. Π values for both reagents display a clear periodicity in sequence position, with a period approximately that of an α-helix. However, the oscillatory behavior for $\Pi(O_2)$ and Π(NiEDDA) are 180° out of phase. This indicates a helical structure throughout the entire sequence, with one face solvated by water and the other

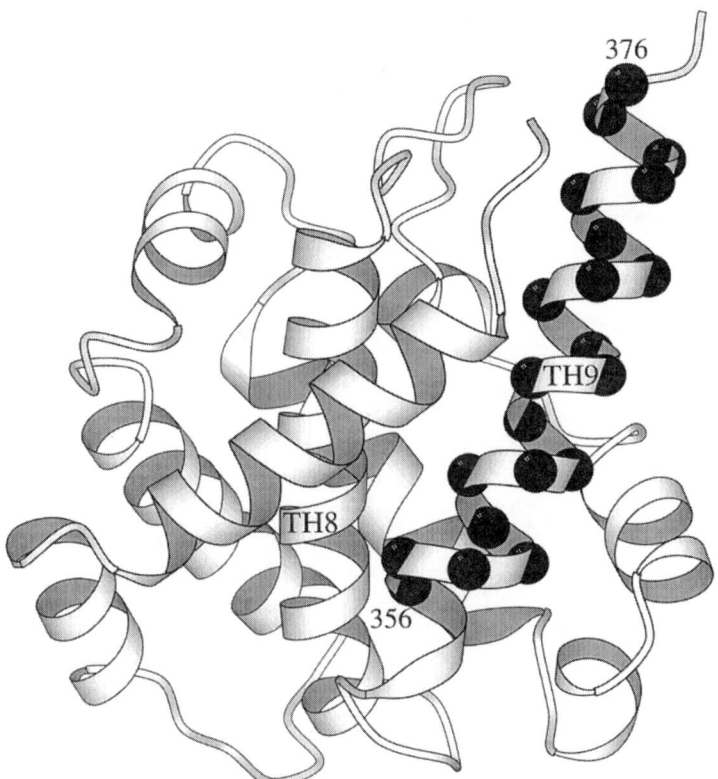

Fig. 2. Structure of the diphtheria toxin T-domain (residues 202–378). The locations of 21 residues selected for mutation to cysteine and attachment of spin probes are indicated as black spheres. Mutants of the isolated T-domain were prepared as described in the **Methods** *(40)*. The numbering of the mutant residues is that of the native DT. This ribbon diagram was generated using the MOLSCRIPT program *(41)* from the coordinates of diphtheria toxin provided by M. J. Bennett and D. Eisenberg *(42)*.

by the hydrophobic interior of the bilayer. Two structural models consistent with this result are shown in **Fig. 4D**.

The immersion depth measurement can be used to differentiate further between these models. **Figure 4C** shows a plot of depth vs amino acid number for the residues located on the surface of the helix facing the bilayer. Successive residues vary in depth by increments of about 5 Å, with residue 360 (**Fig. 4C**) near the center of the bilayer, consistent with a transmembrane orientation of the helix. These results strongly favor a transmembrane structure resembling a water-filled pore (**Fig. 4D**, model 2). An analysis of the nitroxide side chain

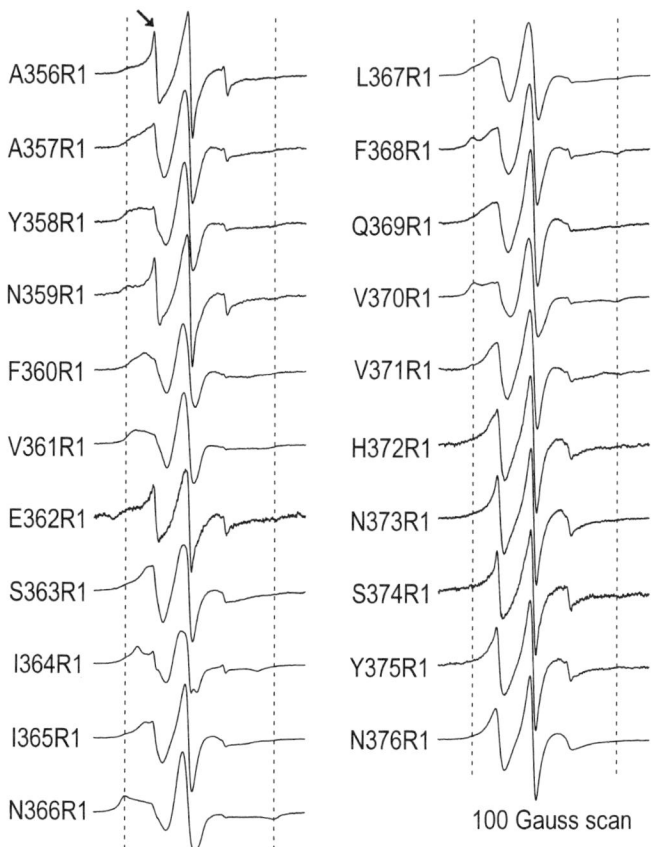

Fig. 3. EPR spectra of spin-labeled T-domain mutants in the presence of phospholipid vesicles at pH 4.6. The protein mutants used in this experiment have four additional residues at the N-terminus: Gly-Ser-His-Met. (The cysteine substitution mutant at residue 362 contained Met-Gly-Ser-Ser-His$_6$-Ser-Ser-Gly-Leu-Val-Pro-Arg-Gly-Ser-His-Met at the N-terminus.) Spin-labeled T-domain mutants bound to 17% POPG–POPC (mol/mol) vesicles were prepared as described in **Subheading 3.** *(40)*. The *vertical dashed lines* mark the locations of the outer hyperfine extrema. The sharp features in some spectra are the result of a small amount of unattached spin label (the arrow indicates one example). The EPR spectra shown are from a previous report *(18)*, except for F360R1 and V361R1. (To designate mutants with side chain R1, we use the notation XYR1, where X is the single-letter code for the original amino acid and Y is the position of the cysteine substitution. Single-letter codes: A, Ala; C, Cys; E, Glu; F, Phe; H, His; I, Ile; L, Leu; N, Asn; Q, Gln; S, Ser; V, Val; and Y, Tyr.) After **ref.** *(18)* was published, we found additional mutations in the nucleotide sequences for the original F360C and V361C mutants. We then prepared new constructs without such errors (*see* **Note 2**) and repeated the measurements, the results of which are shown. The overall conclusions remain unchanged from those in **ref.** *(18)*.

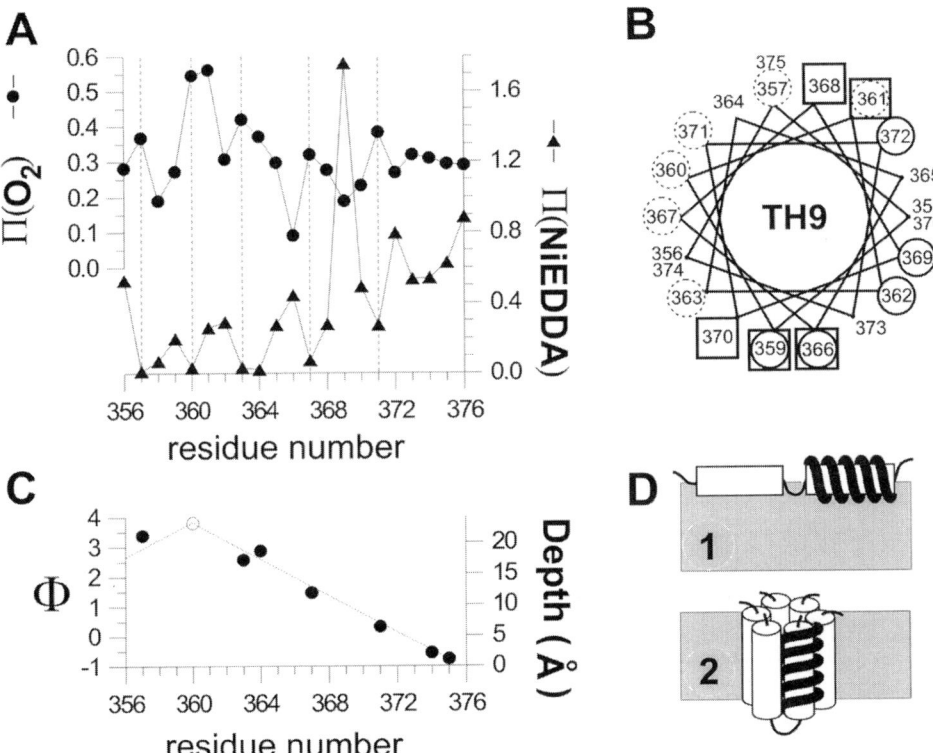

Fig. 4. **(A)** Accessibility parameters for oxygen, Π (O$_2$) (●) and for NiEDDA, Π(NiEDDA) (▲) vs residue number. The data shown are from a previous report *(18)* except F360R1 and V361R1, in which case the values are from newly made mutants (*see* **Note 2** and **Fig. 3** legend). Values of Π (O$_2$) and Π(NiEDDA) were measured for the spin-labeled T-domain mutants bound to vesicles containing POPG–POPC (lipid/protein molar ratio, 500:1) at pH 4.6, as described in **Subheading 3.** *(40)*. The O$_2$ concentration was that in equilibrium with air, and the concentration of NiEDDA was 200 m*M*. **(B)** Helical wheel showing locations of the spin-labeled residues. *Dashed* and *solid circles* identify sites of maxima in Π(O$_2$) and Π(NiEDDA), respectively. *Squares* indicate sites of immobilized residues. **(C)** Immersion depth of R1 chains vs residue number. The depth shown for S363R1 (○) is derived from a newly made construct in comparison with F360R1. **(D)** Two schematic models for the organization of the membrane-bound helix. Model 1: helices adsorbed on the surface of the lipid bilayer; model 2: a helical bundle forming a transmembrane pore. The *grey boxes* represent the lipid bilayer and the cylinders represent helices.

mobility (*see* **Fig. 3**) further supports this interpretation. Only sites 359, 361, 366, 368, and 370 have an immobilized lineshape (*see* **Fig. 3**), indicating sites of tertiary or quarternary interaction. As shown in **Fig. 4B**, these sites define contact zones on the helical surface that would be expected for a helical bundle arrangement.

The SDSL studies are in accordance with the postulate that the DT channel is formed by insertion of the TH8–TH9 helical hairpin into the membrane. This is also in accordance with the recent results by others *(33,34)*. The experimental details are described in the following section focusing on the SDSL method.

2. Materials

1. An EPR spectrometer fitted with a loop-gap resonator *(48)*. The resonator is commercially available from Medical Advances, Milwaukee, WI.
2. Gas-permeable TPX sample capillaries for the loop gap resonator (Medical Advances, Milwaukee, WI) and Pyrex and fused quartz sample capillaries of dimension 0.6 mm · 0.84 mm · 75 mm (i.d. · o.d. · length, VitroCom, Mt. Lakes, NJ).
3. (1-Oxyl-2,2,5,5-tetramethylpyrroline-3-methyl)-methanethiosulfonate (spin label I, **Fig. 1**). The reagent is available from REANAL Factory of Laboratory Chemicals, Budapest, Hungary or Toronto Research Chemicals, Ontario, Canada. A 10 mg/mL stock solution of the spin label in acetonitrile is prepared and stored at −80°C in the dark.
4. Spin-labeled phospholipids for calibration of depth measurements: 5-, 7-, and 12-doxyl phosphatidylcholine, obtained from Avanti Polar Lipids, Alabaster, AL.
5. Synthetic phospholipids for preparation of vesicles: 1-palmitoyl-2-oleoyl-*sn*-3-phospho-rac(1-glycerol) (POPG) and 1-palmitoyl-2-oleoyl-*sn*-glycero-3-phosphocholine (POPC), obtained from Avanti Polar Lipids, Alabaster, AL.
6. Extruder for preparation of vesicles (LiposoFast) and polycarbonate membrane (diameter = 19 mm, pore diameter = 100 nm) (both from Avestin, Ottawa, ON, Canada).
7. Ammonium molybdate tetrahydrate $((NH_4)_6Mo_7O_{24} \cdot 4H_2O)$ (Sigma).
8. Ampicillin (Sigma).
9. Bacto tryptone (Difco Laboratories, Detroit, MI).
10. Bacto yeast extract (Difco Laboratories, Detroit, MI).
11. Bovine serum albumin (fraction V, Sigma).
12. $CaCl_2 \cdot 2H_2O$ (J. T. Baker Chemical, Phillipsburg, NJ).
13. 2,2-Diphenyl-1-picrylhydrazyl (DPPH) (Aldrich, Milwaukee, WI).
14. Dithiothreitol (DTT) (ICN Biomedicals, OH).
15. Ethylenediaminetetraacetate (disodium salt) (Sigma).
16. Glutathione (Sigma).
17. Glycerol (Fisher Scientific, Fair Lawn, NJ).
18. Imidazole (Sigma).
19. Isopropyl-β-D-thiogalactopyranoside (IPTG) (Diagnostic Chemicals, Charlottetown, PEI, Canada).
20. NaCl (Mallinckrodt).

21. NaN$_3$ (Sigma).
22. Ni(II)acetylacetonate (NiAA) (Aldrich, Milwaukee, WI).
23. Ni(II)ethylenediaminediacetate (NiEDDA, for synthesis, *see* **Note 1**).
24. Perchloric acid (70%) (Mallinckrodt).
25. Sodium phosphate dibasic dihydrate (Mallinckrodt).
26. Thrombin (Novagen, Madison, WI).
27. Tris base (Boehringer Mannheim, Indianapolis, IN).
28. His-Bind resin (Novagen, Madison, WI).
29. Benzamidine Sepharose 6B resin (Amersham Pharmacia Biotech).
30. Glutathione Sepharose 4B (Amersham Pharmacia Biotech).
31. Gel filtration chromatography column (Superdex75, Amersham Pharmacia Biotech, Piscataway, NJ).
32. Column C 16/20 (Amersham Pharmacia Biotech).
33. A peristaltic pump (LABCONCO, Kansas City, MO).
34. Nitrogen or argon gas.
35. 1–10-µL Range gel loading tips; centriprep filter units (Amicon, Beverly, MA).
36. Syringe filters (0.45 µm) (Gelman Sciences, Ann Arbor, MI).
37. Buffer A (20 m*M* Tris, 150 m*M* NaCl, 1 m*M* EDTA and 0.02% [w/v] NaN$_3$, pH 8.0).
38. Binding buffer: 5 m*M* imidazole, 500 m*M* NaCl, 20 m*M* Tris-HCl, pH 7.9.
39. Wash buffer: 60 m*M* imidazole, 500 m*M* NaCl, 20 m*M* Tris-HCl, pH 7.9.
40. Elute buffer: 1 *M* imidazole, 500 m*M* NaCl, 20 m*M* Tris-HCl, pH 7.9.
41. Thrombin buffer: 2.5 m*M* CaCl$_2$, 150 m*M* NaCl, and 20 m*M* Tris-HCl, pH 8.4.
42. Sodium acetate buffer (100 m*M*, pH 4.6).

3. Methods

3.1. Preparation of Single Cysteine T-Domain Mutants

The *T*-domain is expressed as a fusion protein with an N-terminal hexa-histidine tag in the expression vector, pET15b (Novagen, Milwaukee, WI) as described *(40)*. The histidine tag is used for affinity purification of the protein and is later removed by digestion with thrombin. The T-domain mutants prepared this way contain residues 202–378, following the numbering scheme for the whole toxin, and four additional residues at the N-terminus: Gly-Ser-His-Met. All of the mutants except F360C and V361C (*see* **Note 2**) were prepared using this method. Following are the procedures for expression and purification of the T-domain.

1. The expression host, BL21(DE3) *E. coli* strain is transformed with pET15b expression vectors (Novagen, Milwaukee, WI) carrying genes for single cysteine T-domain mutants *(40)*.
2. About 200 mL of Luria broth (LB, 12 g/L of Bacto tryptone, 5 g/L of Bacto yeast extract, and 10 g/L of NaCl, pH 7.5) containing ampicillin (100 µg/mL) is inoculated with a single colony of the transformed cells and shaken vigorously for 15 h.

3. A total volume of 4 L of LB containing ampicillin (100 µg/mL), divided into four flasks, is inoculated with the 15-h culture. We use 25–50 mL of this culture to inoculate 1 L of LB. The flasks are shaken vigorously for 3–4 h until the optical density of the cells at 600 nm reaches 1.0–1.2.

4. The T-domain is expressed by induction with 1 mM IPTG for 2 h at 37°C, or 27°C (*see* **Note 3**).

5. The cells are harvested by centrifugation and stored at –20°C before lysis in a French Press.

6. The frozen cell pellets are resuspended in 40 mL of ice cold binding buffer (5 mM imidazole, 500 mM NaCl, 20 mM Tris-HCl, pH 7.9). The cell suspension is sonicated briefly on ice to disrupt the cell pellet. The cells are then ruptured in a French Press and sonicated once more to further break the cell membranes and the chromosomal DNA.

7. Cell debris is removed by centrifugation at 20,000g. The supernatant contains the T-domain.

8. The cell extracts are filtered through 0.45-µm pores (Gelman Sciences, Ann Arbor, MI), and the T-domain is purified with an Ni^{2+} affinity column (His-Bind Resin, Novagen, Milwaukee, WI). The Ni^{2+} affinity chromatography matrix is prepared with His-Bind Resin in a glass column (Column C 16/20, Amersham Pharmacia Biotech) connected to a peristaltic pump (LABCONCO, Kansas City, MO). We typically use 4 mL (bed volume) of the His-Bind resin for the cell extracts from a 4-L culture. The resin is washed with three column volumes of deionized water, charged with nickel ion with five column volumes of 50 mM NiSO$_4$, and equilibrated with three column volumes of the binding buffer. The cell extracts are then loaded onto the column at a flow rate of 1–2 mL/min. Nonspecifically bound materials are removed by washing the column with ten column volumes of the binding buffer, followed by six column volumes of wash buffer (60 mM imidazole, 500 mM NaCl, 20 mM Tris-HCl, pH 7.9). The T-domain is eluted with six column volumes of elute buffer (1 M imidazole, 500 mM NaCl, 20 mM Tris-HCl, pH 7.9).

9. The eluted protein is dialyzed against thrombin buffer (2.5 mM CaCl$_2$, 150 mM NaCl, 20 mM Tris-HCl, pH 8.4) for 18 h to remove the imidazole (*see* **Note 4**). The dialyzed sample is concentrated by centrifugation using a centriprep filter unit (Amicon, Beverly, MA).

10. The N-terminal hexahistidine tag sequence is cleaved from the T-domain by reaction with thrombin for 6–8 h *(40)*.

11. Benzamidine Sepharose 6B resin (Amersham Pharmacia Biotech, Piscataway, NJ) is added to the reaction mixture to remove thrombin and to stop the digestion reaction.

12. The reaction mixture is passed through the Ni^{2+} affinity column to remove the cleaved hexahistidine tag.

13. The purified T-domain is stored at –80°C in the presence of DTT (5–10 mM) and glycerol (15% or higher; *see* **Note 5**).

3.2. Spin Labeling of T-Domain Cysteine Mutants

1. T-Domain cysteine mutants are purified using gel filtration chromatography (Superdex75, Amersham Pharmacia Biotech), eluting with 20 mM Tris-HCl, pH 8.0, 150 mM NaCl, 1 mM EDTA, and 0.02% (w/v) NaN$_3$ (buffer A).
2. Immediately following purification, spin label stock solution (10 mg/mL, 37.8 mM) in acetonitrile is added to the fractions containing the T-domain to provide excess spin label (10-fold or higher molar excess over cysteine). The reaction mixture is shaken gently for 16 h at 20–22°C to generate the spin labeled side chain R1 (**Fig. 1**) (*see* **Note 6**).
3. Unreacted spin label is removed by gel filtration chromatography (Superdex75, Amersham Pharmacia Biotech). When the protein quantity is less than 0.5 mg, excess spin label is removed by dialysis against buffer A at 4°C.
4. The labeled protein is concentrated using a centrifugal concentrator with a molecular mass cutoff of 10 kDa (Microsep, Filtron Technology Corporation, Northborough, MA; or Microcon-10, Amicon, Beverly, MA).
5. The final concentration is determined by an appropriate protein assay using bovine serum albumin as standard.

3.3. Preparation of Large Unilamellar Vesicles (LUVs)

1. Chloroform solutions containing 20 mg of POPG and 100 mg of POPC are mixed and dried under vacuum for 18 h.
2. The phospholipids are resuspended in 1.2 mL of 100 mM sodium acetate buffer, pH 4.6, by brief vortex mixing.
3. Multilamellar vesicles (MLVs) are prepared from the lipid suspension using the reverse phase evaporation method *(49)*, or by repeated freeze-thawing of the lipid suspension *(50)*. The lipid suspension contained in a glass test tube or an Eppendorf tube, is frozen using liquid nitrogen, and then thawed at 20–22°C (>5×). LUVs are made by extruding the MLVs 10–15× through two sheets of polycarbonate membrane with a pore size of 100 nm *(51)*.
4. The Böttcher method *(52)* was used to determine phosphate content of the lipid vesicles. An aqueous phosphate solution, 0.3 mM Na$_2$HPO$_4$(dihydrate), is prepared as a standard. Phosphate standards (0–100 nmol phosphate) and samples of unknown concentration are prepared in equal volumes in pairs of glass test tubes. An aliquot of 0.4 mL of HClO$_4$ (70%) is added to the tubes and heated at 180–190°C for 30 min in a heating block. An aliquot of 4 mL of molybdate reagent (2.2 g of [NH$_4$]$_6$Mo$_7$O$_{24}$ · 4H$_2$O, 14.3 mL conc. sulfuric acid, dissolved in 1 L deionized water) is added after the heated solution is cooled. An aliquot of 10% (w/v) ascorbic acid water solution is added and heated for 10 min at 100°C in a water bath. The absorbance is measured at 812 nm using a spectrophotometer.

3.4. Preparation of Membrane-Bound T-Domain (40)

1. The pH of the LUV suspension is raised to pH 7–8 using 1 M NaOH just prior to addition of the T-domain.

2. The T-domain is added to the vesicles with an approximate 1:500 molar ratio of protein to phospholipid.

3. To induce binding, a predetermined amount (usually half the volume of protein–vesicle solution) of 100 mM sodium acetate solution is slowly added to the protein–vesicle mixture to a final pH of 4.6. The binding is allowed to proceed for about 30 min.

4. The membrane-bound T-domain is pelleted by centrifugation at 15,000g for 3 min at 20–22°C, and the pellet washed with 100 mM sodium acetate solution, pH 4.6.

3.5. Power Saturation Measurements and Determination of Π Values

1. Glass, quartz, or TPX sample capillaries are prepared with one end sealed. The TPX capillaries are used for power saturation measurements, while CW EPR spectra may be obtained in any of the sample capillaries. The capillaries are filled with 2–5 µL of sample in a concentration range from 10 to 500 µM in nitroxide (*see* **Notes 7** and **8**). For practical considerations in recording optimized CW EPR spectra, *see (53)*.

2. Determine ΔH_{pp} and $P_{1/2}$ of the saturation standard, DPPH. The EPR spectrum of a DPPH standard sample (*see* **Note 9**) is taken at a low incident microwave power to avoid saturation broadening, typically 2 mW. The linewidth of the spectrum (**Fig. 5A**), ΔH_{pp}(DPPH), can be conveniently measured on an expanded plot of the spectrum. The power saturation curve of the standard is measured as the vertical peak-to-peak amplitude as a function of incident microwave power in the range of 0.1–49 mW (*see* **Fig. 5B**). The scan width is typically 20 Gauss, which covers the entire DPPH spectrum, or the central line ($m_I = 0$) of the nitroxide EPR spectrum. The amplitude vs microwave power data are then fitted to Eq. (1) with a nonlinear least-squares algorithm to determine $P_{1/2}$ as well as the less useful fitting parameters I and ε (*see* **Note 10** and **Fig. 5C**). Power saturation measurements of a known reference sample such as DPPH should be done at regular intervals. This also provides a monitor for the "health" of the loop gap resonator and instrument in general. A sudden increase in $P_{1/2}$ of the reference indicates a hardware or resonator problem.

3. Determine $P_{1/2}$ of the samples in the presence of O$_2$. The sample prepared as described in **Subheadings 3.2.** and **3.4.** is loaded into a gas-permeable TPX sample capillary. It is convenient to use the concentration of O$_2$ in equilibrium with air. In this case the power saturation curve is measured directly and $P_{1/2}$ determined as described previously for the DPPH standard. Because the equilibrium concentration of oxygen is temperature dependent and significantly higher in a sample stored on ice, air should be blown through the resonator to ensure proper equilibration at the selected temperature.

4. Determine $P_{1/2}°$ (Eq. 3) of the samples under nitrogen or argon. After switching the gas flow from air to nitrogen or argon gas, the same sample is used to measure the power saturation curve in the absence of oxygen. It takes about 10 min to

remove the oxygen from the sample after initiation of gas flow (*see* **Note 11**). $P_{1/2}$ is determined from the saturation curve as described previously for DPPH.

5. Determine $P_{1/2}$ of the samples containing NiEDDA or NiAA. Samples containing appropriate concentrations of the reagents are prepared (*see* **Notes** 12 and 13), and the power saturation curve is measured after equilibration with nitrogen or argon gas as described previously. The $P_{1/2}$ values are again obtained by fitting the data to Eq. (1).

6. Measure the linewidths of the EPR spectra recorded at low incident microwave power from samples in TPX under nitrogen or argon flow as described previously for the DPPH standard.

7. Calculate $\Pi(O_2)$ and $\Pi(\text{NiEDDA})$ or $\Pi(\text{NiAA})$. The $P_{1/2}$ and $P_{1/2}^{\circ}$ values and linewidths obtained previously are used to calculate P values according to Eq. (4) (*see* **Note 14**).

3.6. The Depth Calibration Curve and Calculation of the Immersion Depths of Lipid-Facing Residues on Transmembrane Segments

1. Spin-labeled phospholipids (*see* **Note 15**) such as 7-doxyl, 10-doxyl, or 12-doxyl phosphatidylcholine (PC) are mixed with phospholipid–chloroform solutions containing 17% POPG–POPC (mol/mol), respectively. The molar ratio of the doxyl PC to host phospholipids (POPG–POPC) is 1:500.

2. LUVs are prepared as described under **Subheading 3.3.**

3. The T-domain (wild type) is added to the lipid vesicles with a molar ratio of 1:500 (T-domain/lipid).

4. Binding and insertion of the T-domain is induced as described in the procedures in section 3.4.

5. $\Pi(O_2)$ and $\Pi(\text{NiEDDA})$ values are measured for each sample, and the corresponding Φ values computed according to Eq. (6).

6. A calibration curve is made using the Φ values and the depth values of the doxyl nitroxides (*see* **Note 15**), and the values of the constants a and b determined.

7. Lipid-facing residues on the transmembrane segment are identified using the EPR lineshape, Π, and Φ values (*see* **Note 16**). The depths of the lipid-facing residues are calculated using the depth calibration curve.

4. Notes

1. Synthesis of NiEDDA: Ethylenediamine-*N,N'*-diacetic acid (Aldrich, Milwaukee, WI) (0.881 g, 0.005 mol) is added to ~300–400 mL of distilled water in a 1 L round flask fitted with a 24/40 ground glass joint and stirred until completely dissolved. To this is added 0.464 g (0.005 mol) of $Ni(OH)_2$ (Aldrich, Milwaukee, WI) and the turbid green solution is heated to 50–60°C. After a few hours all the insoluble green $Ni(OH)_2$ will be converted to the very soluble blue NiEDDA. Continue stirring at 20–22°C for 18 h and filter any undissolved material. The water is removed under vacuum, the solid residue washed with methanol to remove any remaining $Ni(OH)_2$ or EDDA, and the solid dried under vacuum for 18 h.

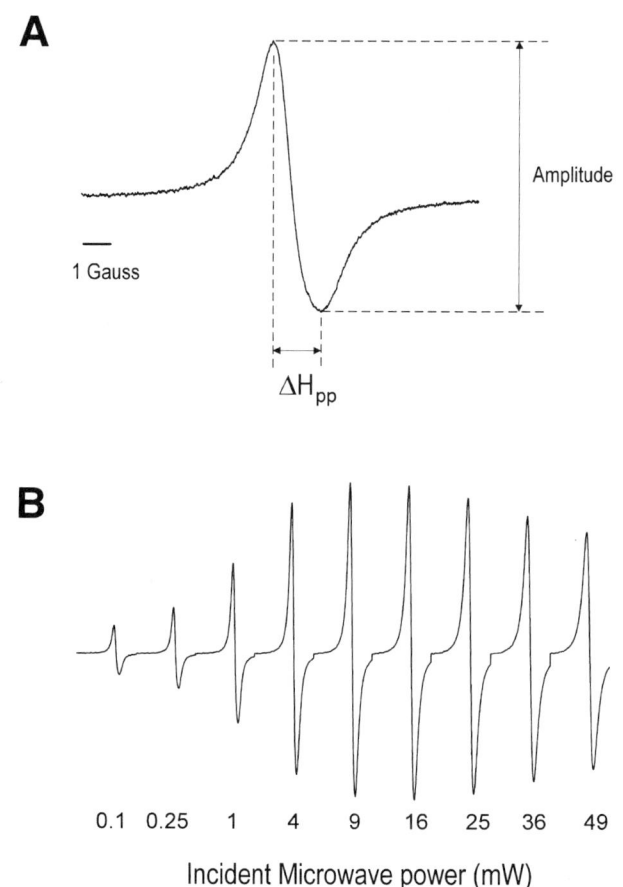

Fig. 5. Power saturation of the DPPH standard sample. **(A)** The EPR spectrum of DPPH with the amplitude and the linewidth indicated. Incident microwave power was 0.1 mW. **(B)** The EPR spectrum of DPPH at various incident microwave powers.

2. Alternative expression of the T-domain as a fusion protein to glutathione sulfur transferase (GST): The T-domain can be expressed as a fusion protein to the glutathione S-transferase in the pGEX-4T-1 vector (Amersham Pharmacia). In brief, a DNA fragment encoding the T-domain (residues 202–378 of DT) is incorporated into the expression vector pGEX-4T-1 (Amersham Pharmacia) using *Bam*HI and *Eco*RI restriction sites, which fuses the glutathione *S*-transferase (GST) to the N-terminus of the T-domain. The T-domain is cleaved by thrombin from the fusion protein. The T-domain mutants prepared this way have residues corresponding to 202–378 in the whole diphtheria toxin and additional Gly-Ser-His-Met residues at the N-terminus. F360C and V361C mutants were prepared using this method.

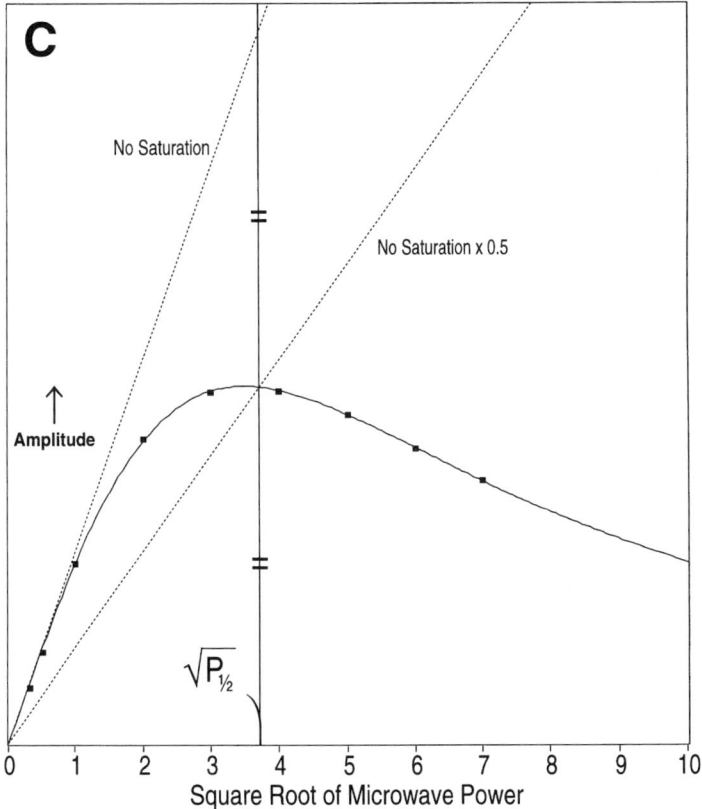

Fig. 5C. Fitting of the data of (B) by a program developed by C. Altenbach (*see* **Note 10**). A dotted line (no saturation) extrapolating the initial slope shows the theoretical behavior in the absence of saturation. The other dotted line with half this slope intersects the power saturation where $\sqrt{P} = \sqrt{P_{1/2}}$ as indicated.

3. Some mutants gave higher yield when expressed at a temperature lower than 37°C. To express the proteins at lower temperature, cells are chilled to ~27°C with tap water before addition of the IPTG.
4. In the presence of high concentration of imidazole, it is difficult to concentrate the protein eluate owing to the poor hydration of the concentrator membrane.
5. To avoid the precipitation of the T-domain after freeze and thawing, glycerol is added to the protein solution to a final concentration of 15% (v/v) or higher.
6. Other satisfactory solvents for spin label stock solution are dimethylsulfoxide and ethanol. The methanethiosulfonate spin label cannot be stored in water for prolonged periods because it forms relatively unreactive disulfide-linked homodimers. The appropriate pH range for spin labeling with the methanethiosulfonate label is pH 6–8.

7. Quartz and glass capillaries: Some quartz capillaries contain background EPR signals. It is recommended to test each batch of capillaries for background signal. Pyrex tubes can be used for measurements at 20–22°C. Quartz capillaries are required at low temperatures because Pyrex exhibits a signal under these conditions. Glass and quartz capillaries are sealed on one end with a gas–oxygen torch. Care must be taken that the sealed end of the capillary does not increase in diameter as a result of drop formation. Samples can be easily loaded through the open end into the capillary tubes using thin gel loading tips (Geloader Tips, 1–10 µL range, Eppendorf). The sample can be conveniently moved toward the sealed end of the capillary by centrifugation at low speed in a clinical centrifuge, holding the capillary in a conical centrifuge tube.

8. TPX capillary tubes are temporarily sealed with wax before sample loading. It is convenient to load the sample into the TPX capillary by the same method described for the quartz and glass capillaries, except that centrifugation is carried out by supporting the small TPX capillary in an Eppendorf tube. The TPX is brittle and care must be taken in handling it. To prevent contamination of the resonator by the wax used to seal the capillary, the tip of the TPX must be wiped clean on the outside.

9. The DPPH standard is prepared by mixing crystalline DPPH and KCl, and grinding the mixture to a fine powder containing 1.5×10^{18} spins/g. The powder is placed in a quartz capillary sealed on both ends and stored in the dark.

10. Power saturation data (amplitude vs P), can be fit to Eq. (1) using a wide variety of commercially available software packages. A nonlinear Levenberg–Marquardt algorithm typically works well with the following initial parameter estimates: I = (slope of line going through the origin and points at low power), $P_{1/2}$ = (power of data point with highest signal), $\varepsilon = 1$. A fitting program is available from C. Altenbach upon request. The exponent ε is typically close to 1.5 for a nitroxide under nitrogen, but lower in the presence of paramagnetic reagents. Many different factors can contribute to this apparent inhomogeneity in saturation (distribution of states, multicomponent spectra, etc.). Nonlinear least-squares fit of power saturation data to equation (1), however is typically excellent, and $P_{1/2}$ provides a good estimation for $1/T_1$, irrespective of differences in ε. While there is certainly information contained in the exponent, we have so far exclusively focused on $P_{1/2}$. The saturation curve is relatively featureless and it is thus not recommended to try to fit with more elaborate models, such as multiple components, each with its own $P_{1/2}$ and ε. An increase in parameters typically does not improve the fit quality, but leads to unwarranted parameter inflation, interacting parameters, and possibly nonconvergence of the fitting routine. To obtain good estimates for $P_{1/2}$ it is important to measure the signal amplitude over a range of powers that exceed the value of $P_{1/2}$. Obviously, if all points are well below saturation, the power saturation curve (amplitude vs P) is nearly linear and only I (Eq. 1) can be determined accurately. Even at the highest power setting of a typical microwave bridge, saturation is difficult to achieve in conventional microwave cavities because of the low H_1, the microwave magnetic field, at the

sample. Loop gap resonators are much more efficient and saturation occurs usually below 30 mW of incident microwave power, well within the capabilities of most spectrometers. Typically, the centerline is recorded at eight different power settings. A convenient set of powers is [0.1, 0.25, 1, 4, 9, 16, 25, 36] mW (the last six points are spaced equally in H_1, which is proportional to the square root of incident microwave power). If needed, additional points are measured until the amplitude decreases with increasing power, [... 49, 64, 81, 100...] mW. If it is not possible to reach that point, the paramagnetic reagent concentration is too high and needs to be reduced. Even with excellent data, $P_{1/2}$ cannot be determined reliably if it is higher than twice the highest power measured.

11. Oxygen is dissolved in the sample medium in equilibrium with air. When the power saturation curve is measured in the presence of NiEDDA, or in the absence of any collision reagent, it is necessary to remove oxygen from the sample. This is achieved by blowing nitrogen or argon gas over the sample in a gas-permeable TPX capillary before and during the measurement. The bottom of the Medical Advances Loop Gap Resonator has an inlet for gas flow.

12. For spin-labeled proteins in solution, a NiEDDA concentration of 3 mM is optimal for power saturation measurement of $P_{1/2}$. For spin-labels buried in the bilayer, concentrations of 100–200 mM are needed, which is high enough to differentiate depths of the residues in the core of the bilayer. To investigate residues near the membrane–water interface, 20 mM NiEDDA is a good compromise.

13. To prepare NiEDDA containing samples for power saturation, a stock solution of 200 mM NiEDDA (for synthesis, *see* **Note 1**), is dissolved in distilled water. The solution is diluted with distilled water to desired concentrations for subsequent use. For power saturation experiments, 3 µL of the stock or diluted solution is dispensed into small (0.5 mL) Eppendorf tubes, and dried under vacuum. An equal volume of the spin labeled sample is then added to the dry NiEDDA powder to obtain the desired concentration of NiEDDA without diluting the sample. When the protein sample is sufficiently concentrated and sample dilution tolerable, the stock solution can be directly added to the protein sample. For samples containing lipid membranes, care must be taken that NiEDDA is distributed equally in all compartments of the system. Under normal conditions, it is not membrane permeable, and several freeze–thaw cycles are required to obtain meaningful results.

14. Measurement of the linewidth is a major source of error in the determination of Π for samples with poor signal-to-noise ratio. Spectra containing multiple components also make it difficult to obtain accurate linewidths. In addition, multicomponent spectra are problematic for power saturation, as each component tends to have different saturation behavior.

15. The calibration curve was obtained by using 5-doxyl, 7-doxyl, 10-doxyl, and 12-doxyl PCs *(19)* and *N*-tempoyl palmitamide *(54)* in the host phospholipid vesicles containing 17% POPG–POPC (mol/mol) vesicles *(40)* at pH 4.6 without the bound T-domain. The depths taken for *N*-tempoyl palmitamide, 5-doxyl, 7-doxyl, 10-doxyl, or 12-doxyl groups are 0, 8.1, 10.5, 14, and 16 Å, respectively *(10)*.

The plot of this depth vs Φ was fit by the linear equation depth $(\text{Å}) = 4.81\Phi + 4.9$, which was used for the depth estimation in **Fig. 4C**. The data obtained in the presence of bound T-domain were fit by the equation depth $(\text{Å}) = (4.81\Phi + 11.6)$. This has the same slope as in the absence of the T-domain, but a different intercept. The lack of a spin label standard for the depth near the center of the bilayer makes it difficult to estimate accurately the immersion depth of very deeply buried residues. It is recommended that a separate calibration curve be constructed for each lipid composition.

16. It is expected that the lipid-facing residues have relatively large $\Pi(O_2)$ and small $\Pi(\text{NiEDDA})$. In addition, they would have EPR lineshapes with features showing high mobility. For these residues, we calculate the Φ values using Eq. (6) and calculate the immersion depth using Eq. (5). It should be noted that not all sites are suitable for depth calculation. The residues in protein contact sites have limited access to the relatively bulky NiEDDA in contrast to the small oxygen molecule, resulting in the large value of Φ owing to the steric effect rather than concentration gradient effect in the bilayer. This is why residues that have mobile line shape are used for depth calculation.

References

1. Hubbell, W. L., Mchaourab, H. S., Altenbach, C., and Lietzow, M. A. (1996) Watching proteins move using site-directed spin labeling. *Structure* **4,** 779–783.
2. Hubbell, W. L. and Altenbach, C. (1994) Site-directed spin labeling of membrane proteins, in *Membrane Protein Structure: Experimental Approaches* (White, S. H., ed.), Oxford University Press, London, pp. 224–248.
3. Hubbell, W. L. and Altenbach, C. (1994) Investigation of structure and dynamics in membrane proteins using site-directed spin labeling. *Curr. Opin. Struct. Biol.* **4,** 566–573.
4. Todd, A., Cong, J., Levinthal, F., Levinthal, C., and Hubbell, W. L. (1989) Site-directed mutagenesis of colicin E1 provides specific attachment of spin labels whose spectra are sensitive to local conformation. *Proteins* **6,** 294–305.
5. Berliner, L. J., Grunwald, J., Hankovszky, H. O., and Hideg., K. (1982) A novel reversible thiol-specific spin label: papain active site labeling and inhibition. *Anal. Biochem.* **119,** 450–455.
6. Mchaourab, H. S., Lietzow, M. A., Hideg, K., and Hubbell, W. L. (1996) Motion of spin-labeled side chains in T4 lysozyme. Correlation with protein structure and dynamics. *Biochemistry* **35,** 7692–7704.
7. Budil, D. E., Lee, S., Saxena, S., and Freed, J. H. (1996) Nonlinear-least-squares analysis of slow-motion EPR spectra in one and two dimensions using a modified Levenberg–Marquardt algorithm. *J. Magn. Res.* **120,** 155–189.
8. Subczynski, W. K. and Hyde, J. S. (1981) The diffusion concentration product of oxygen in lipid bilayers using the spin label T_1 method. *Biochim. Biophys. Acta* **643,** 283–291.
9. Altenbach, C., Flitsch, S. L., Khorana, H. G., and Hubbell, W. L. (1989) Structural studies on transmembrane proteins. 2. Spin labeling of bacteriorhodopsin mutants at unique cysteines. *Biochemistry* **28,** 7806–7812.

10. Farahbakhsh, Z. T., Altenbach, C., and Hubbell, W. L. (1992) Spin labeled cysteines as sensors for protein–lipid interaction and conformation in rhodopsin. *Photochem. Photobiol.* **56,** 1019–1033.

11. Rabenstein, M. D. and Shin, Y. K. (1995) Determination of the distance between two spin labels attached to a macromolecule. *Proc. Natl. Acad. Sci. USA* **92,** 823–943.

12. Mchaourab, H. S., Oh, K. J., Fang, C. J., and Hubbell, W. L. (1997) Conformation of T4 lysozyme in solution. Hinge-bending motion and the substrate-induced conformational transition studied by site-directed spin labeling. *Biochemistry* **36,** 307–316.

13. Steinhoff, H. J., Radzwill, N., Thevis, W., Lenz, V., Brandenburg, D., Antson, A., Dodson, G., and Wollmer, A. (1997) Determination of interspin distances between spin labels attached to insulin: comparison of electron paramagnetic resonance data with the X-ray structure. *Biophys. J.* **73,** 3287–3298.

14. Hustedt, E. J., Smirnov, A. I., Laub, C. F., Cobb, C. E., and Beth, A. H. (1997) Molecular distances from dipolar coupled spin-labels: the global analysis of multifrequency continuous wave electron paramagnetic resonance data. *Biophys. J.* **74,** 1861–1877.

15. Voss, J., Salwinski, L., Kaback, H. R., and Hubbell, W. L. (1995) A method for distance determination in proteins using a designed metal ion binding site and site-directed spin labeling: evaluation with T4 lysozyme. *Proc. Natl. Acad. Sci. USA* **92,** 12,295–12,299.

16. Poole, C. P. (1983) *Electron Spin Resonance.* Dover Publications, Mineola, NY.

17. Altenbach, C., Marti, T., Khorana, H. G., and Hubbell, W. L. (1990) Transmembrane protein structure: spin labeling of bacteriorhodopsin mutants. *Science* **248,** 1088–1092.

18. Oh, K. J., Zhan, H., Cui, C., Hideg, K., Collier, R. J., and Hubbell, W. L. (1996) Organization of diphtheria toxin T domain in bilayers: a site-directed spin labeling study. *Science* **273,** 810–812.

19. Altenbach, C., Greenhalgh, D., Khorana, H. G., and Hubbell, W. L. (1994) A collision gradient method to determine the immersion depth of nitroxides in lipid bilayers: application to spin-labeled mutants of bacteriorhodopsin. *Proc. Natl. Acad. Sci. USA* **91,** 1667–1671.

20. Collier, R. J. (1975) Diphtheria toxin: mode of action and structure. *Bacteriol. Rev.* **39,** 54–85.

21. Pappenheimer, A. M., Jr. (1977) Diphtheria toxin. *Annu. Rev. Biochem.* **46,** 69–94.

22. Greenfield, L., Bjorn, M. J., Horn, G., Fong, D., Buck, G. A., Collier, R. J., and Kaplan, D. A. (1983) Nucleotide sequence of the structural gene for diphtheria toxin carried by corynebacteriophage beta. *Proc. Natl. Acad. Sci. USA* **80,** 6853–6857.

23. Moss, J., Iglewski, B., Vaughan, M., and Tu, A. T., eds. (1995) *Bacterial Toxins and Virulence Factors in Disease: Handbook of Natural Toxins,* Vol. 8. Marcel Dekker, New York.

24. Morris, R. E., Gerstein, A. S., Bonventre, P. F., and Saelinger, C. B. (1985) Receptor-mediated entry of diphtheria toxin into monkey kidney (Vero) cells: electron microscopic evaluation. *Infect. Immun.* **50,** 721–727.

25. Naglich, J. G., Metherall, J. E., Russell, D. W., and Eidels, L. (1992) Expression cloning of a diphtheria toxin receptor: identity with a heparin-binding EGF-like growth factor precursor. *Cell* **69,** 1051–1061.

26. London, E. (1992) Diphtheria toxin: membrane interaction and membrane translocation. *Biochim. Biophys. Acta* **1113,** 25–51.

27. Tortorella, D., Sesardic, D., Dawes, C. S., and London, E. (1995) Immunochemical analysis shows all three domains of diphtheria toxin penetrate across model membranes. *J. Biol. Chem.* **270,** 27,446–27,452.

28. Cabiaux, V., Quertenmont, P., Conrath, K., Brasseur, R., Capiau, C., and Ruysschaert, J.-M. (1994) Topology of diphtheria toxin B fragment inserted in lipid vesicles. *Mol. Microbiol.* **11, 43–50.**

29. Moskaug, J. O., Stenmark, H., and Olsnes, S. (1991) Insertion of diphtheria toxin B-fragment into the plasma membrane at low pH. *J. Biol. Chem.* **266,** 2652–2659.

30. Sandvig, K. and Olsnes, S. (1980) Diphtheria toxin entry into cells is facilitated by low pH. *J. Cell Biol.* **87,** 828–832.

31. Draper, R. K. and Simon, M. I. (1980) The entry of diphtheria toxin into the mammalian cell cytoplasm: evidence for lysosomal involvement. *J. Cell Biol.* **87,** 849–854.

32. Kagan, B. L., Finkelstein, A., and Colombini, M. (1981) Diphtheria toxin fragment forms large pores in phospholipid bilayer mambranes. *Proc. Natl. Acad. Sci. USA* **78,** 4950–4954.

33. Wang, Y., Malenbaum, S. E., Kachel, K., Zhan, H., Collier, R. J., and London, E. (1997) Identification of shallow and deep membrane-penetrating forms of diphtheria toxin T domain that are regulated by protein concentration and bilayer width. *J. Biol. Chem.* **272,** 25,091–25,098.

34. Kachel, K., Ren, J., Collier, R. J., and London, E. (1998) Identifying transmembrane states and defining the membrane insertion boundaries of hydrophobic helices in membrane-inserted diphtheria toxin T domain. *J. Biol. Chem.* **273,** 22,950–22,956.

35. Montecucco, C., Papini, E., Schiavo, G., Padovan, E., and Rossetto, O. (1992) Ion channel and membrane translocation of diphtheria toxin. *FEMS Microbiol. Immun.* **105,** 101–112.

36. Madshus, I. H., Wiedlocha, A., and Sandvig, K. (1994) Intermediates in translocation of diphtheria toxin across the plasma membrane. *J. Biol. Chem.* **269,** 4648–4652.

37. O´Keefe, D. O., Cabiaux, V., Choe, S., Eisenberg, D., and Collier, R. J. (1992) pH-Dependent insertion of proteins into membranes: B-chain mutation of diphtheria toxin that inhibits membrane translocation, Glu–349-Lys. *Proc. Natl. Acad. Sci. USA* **89,** 6202–6206.

38. Honjo, T., Nishizuka, Y., and Hayaishi, O. (1968) Diphtheria toxin-dependent adenosine diphosphate ribosylation of aminocyl transferase II and inhibition of protein synthesis. *J. Biol. Chem.* **243,** 3553–3555.

39. Van Ness, B. G., Howard, J. B., and Bodley, J. W. (1980) ADP-ribosylation of elongation factor 2 by diphtheria toxin. Isolation and properties of the novel ribosyl-amino acid and its hydrolysis products. *J. Biol. Chem.* **255,** 10,717–10,720.

40. Zhan, H., Oh, K. J., Shin, Y. K., Hubbell, W. L., and Collier, R. J. (1995) Interaction of the isolated transmembrane domain of diphtheria toxin with membranes. *Biochemistry* **34,** 4856–4863.

41. Kraulis, P. J. (1991) MOLSCRIPT: a program to produce both detailed and schematic plots of protein structures. *J. Appl. Crystallogr.* **24,** 946–950.

42. Bennett, M. J. and Eisenberg, D. (1994) Refined structure of monomeric diphtheria toxin at 2.3 Å resolution. *Protein Sci.* **3,** 1464–1475.

43. Choe, S., Bennett, M. J., Fujii, G., Curmi, P. M. G., Kantardjieff, K. A., Collier, R. J., and Eisenberg, D. (1992) The crystal structure of diphtheria toxin. *Nature* **357,** 216–222.

44. Silverman, J. A., Mindell, J. A., Finkelstein, A., Shen, W. H., and Collier, R. J. (1994) Mutational analysis of the helical hairpin region of diphtheria toxin transmembrane domain. *J. Biol. Chem.* **269,** 22,524–22,532.

45. Mindell, J. A., Zhan, H., Huynh, P. D., Collier, R. J., and Finkelstein, A. (1994) Reaction of diphtheria toxin channels with sulfhydryl-specific reagents: observation of chemical reactions at the single molecule level. *Proc. Natl. Acad. Sci. USA* **91,** 5272–5276.

46. Silverman, J. A., Mindell, J. A., Zhan, H., Finkelstein, A., and Collier, R. J. (1994) Structure–function relationships in diphtheria toxin channels: I. Determining a minimal channel-forming domain. *J. Membr. Biol.* **137,** 17–28.

47. Mindell, J. A., Silverman, J. A., Collier, R. J., and Finkelstein, A. (1994) Structure–function relationships in diphtheria toxin channels: II. A residue responsible for the channel's dependence on *trans* pH. *J. Membr. Biol.* **137,** 29–44.

48. Hubbell, W. L., Froncisz, W., and Hyde, J. S. (1987) Continuous and stopped flow EPR spectrometer based on a loop gap resonator. *Rev. Sci. Instrum.* **58,** 1879–1886.

49. Szoka, F., Jr. and Papahadjopoulos, D. (1980) Comparative properties and methods of preparation of lipid vesicles (liposomes). *Annu. Rev. Biophysics. Bioeng.* **9,** 467–508.

50. Mayer, L. D., Hope, M. J., Cullis, P. R., and Janoff, A. S. (1985) Solute distributions and trapping efficiencies observed in freeze-thawed multilamellar vesicles. *Biochim. Biophys. Acta* **817,** 193–196.

51. Mayer, L. D., Hope, M. J., and Cullis, P. R. (1986) Vesicles of variable sizes produced by a rapid extrusion procedure. *Biochim. Biophys. Acta* **858,** 161–168.

52. Böttcher, C. J. F., Van Gent, C. M., and Pries, C. (1961) A rapid and sensitive sub-micro phosphorus determination. *Anal. Chim. Acta* **24,** 203–204.

53. Jost, P. and Griffith, O. H. (1976) Instrumental aspects of spin labeling, in *Spin Labeling: Theory and Applications* (Berliner, L. J., ed.), Academic Press, New York, pp. 251–272.

54. Shin, Y. K. and Hubbell, W. L. (1992) Determination of electrostatic potentials at biological interfaces using electron-electron double resonance. *Biophys. J.* **61,** 1443–1453.

10

Characterization of Molecular Properties of Pore-Forming Toxins with Planar Lipid Bilayers

Mauro Dalla Serra and Gianfranco Menestrina

1. Introduction

Pore-forming toxins (PFTs) belong to the membrane-damaging toxins supergroup, a family of proteins already quite large but still quickly growing *(1)*. PFTs are used by the bacteria to attack potentially harmful cells of the host, for example cells of the immune system, or to obtain nutrients, for example by lysing red blood cells *(2)*. They produce well-defined pores in the plasma membrane of attacked cells, thus increasing their permeability to ions and small molecules. These exogenous channels can rupture small non-nucleated cells (e.g., red blood cells and platelets) via the so-called colloid-osmotic shock. In the case of large nucleated cells, for example, the leukocytes, they may trigger Ca^{2+} entry followed by a number of secondary effects elicited by this messenger, leading finally to apoptosis and cell death. Besides being a very important pathogenic factor in a number of widespread diseases, bacterial PFTs are also finding new, and sometimes unexpected, applications. For example, they can be used to selectively permeabilize cells to molecules up to a certain size (the pore cutoff) for in vitro studies of cell functions *(3,4)*, or to build immunotoxins or mitotoxins specifically directed against cancer cells *(5,6)*, or even to construct new engineered single molecule biosensors for electronic devices *(7)*.

Planar lipid membranes (PLMs) are stable portions of a lipid bilayer supported on a hole in a Teflon septum between two aqueous solutions *(8,9)*. They are made of purified lipids and may be quite useful in studying protein channels opened by PFTs in a simple model membrane environment *(10)*. Among the advantages they offer are: (1) easy control of all the physicochemical parameters, for example, lipid composition of the membrane and chemical

From: *Methods in Molecular Biology, vol. 145: Bacterial Toxins: Methods and Protocols*
Edited by: O. Holst © Humana Press Inc., Totowa, NJ

composition and temperature of the water phase; (2) full control of the trans-membrane voltage, which is not readily offered by other systems; (3) current resolution that allows the detection of single-channel events. Among the drawbacks are: (1) sensitivity to small amounts of impurities; (2) necessity that the channel self-incorporates into the lipid film from the water phase; (3) inefficiency of channel incorporation, typically less than 1 molecule in 10^{10} inserts into the bilayer; and (4) lack of semiautomatic procedures for rapid screening of different conditions. With the majority of PFTs, at least the spontaneous incorporation is not a problem, because it is their inherent property to be able to cross a water phase before reaching and integrating into their final target membrane.

PLMs can provide information at a molecular level on PFT properties. For example, one can determine the size of the pore, and its selectivity, that is, the preference for either anions or cations, as well as the permeability of the different ions according to their size. Selectivity and voltage dependence of ion current may be correlated with the presence and distribution of fixed charges in the pore lumen *(11)*. For example, we have used them in a study of the effects of modification of charged lysine residues on the conductance of the *Staphylococcus aureus* α-toxin pore *(12)*, arriving at a model of the distribution of charges on the pore's mouths fairly consistent with the recently determined three-dimensional structure *(13)*.

In this chapter we describe how to prepare PLMs and to perform and interpret experiments aimed at obtaining molecular information on PFT channels. Because the setup is not commercially available and should be self-made, we include also instructions on how to assemble an efficient, yet rather inexpensive, PLM setup.

2. Materials

2.1. Reagents and Buffers

1. Tridistilled H_2O (*see* **Note 1**).
2. Chloroform.
3. Pentane.
4. *n*-Hexane.
5. *n*-Hexadecane.
6. Ethanol.
7. Acetone.
8. Lipids useful for PLM preparation (*see* **Note 2**), which can be obtained for example from Avanti Polar Lipids (Pelham, AL) are:
   a. Egg phosphatidylcholine (PC).
   b. Palmitoyl-oleoyl-phosphatidylcholine (POPC).
   c. Diphytanoyl-phosphatidylcholine (DPhPC).

    d. Egg phosphatidylethanolamine (PE).

    e. Brain phosphatidylserine (PS).

9. High-vacuum source to dry the lipids (Teflon diaphragm pump by Vacuubrand, Wertheim, Germany).

10. Buffer A: 100 mM NaCl (or other concentrations as specified), 20 mM Tris-HCl, 1 mM EDTA, pH 7.0.

11. Electrode chloriding solution: 0.1 M HCl.

12. Clorox bleach.

13. Electrode equilibrating solution: 3 M KCl.

2.2. Electrical Equipment

1. Ultralow noise operational amplifier based on a field-effect transistor (FET), for example, OPA 104C of Burr-Brown (Tucson, AZ) and a feedback resistor of 10^8 or 10^9 Ω (Eltec Instruments, Daytona Beach, FL).

2. Waveform generator, either homemade *(8)* or commercial (e.g., from Wavetek, San Diego, CA).

3. Low-pass analog Bessel filter, at least 24 dB/octave (e.g., model 900 of Frequency Devices, Haverhill, MT, or 4302 Itacho, Itacha, NY).

4. Digital storage oscilloscope (e.g., from Tektronix, Beaverton, OR).

5. Fast-response potentiometric X–t chart recorder (e.g., from Yokogawa, Newnan, GA).

6. Two-channel digital tape recorder (DAT), with typical bandwidth between DC and 20 kHz per channel, minimum 12-bit resolution and S/N > 92 dB (e.g., DTR 1200 of Bio-Logic, Claix, France, or DAS-75 of Dagan Corp., Minneapolis, MN).

7. Digital audiocassettes of 1 or 2 h duration (Sony, Tokyo, Japan).

8. Computer acquisition board with 12 or, if possible, 16-bit resolution, at least two analog inputs and one analog output, sampling rate 100 kHz, complete with a suitable software; for example, a ready-to-use interface plus software (from Axon, Foster City, CA, or Bio-Logic or Dagan) or a separate card (e.g., LabVIEW, from National Instruments, Austin, TX) for assembling a self-developed platform.

9. BNC cables of different lengths (*e.g.,* from ITT, Pomona, CA).

10. BNC and coaxial adapters to connect all the components of the system (e.g., from World Precision Instruments, Sarasota, FL).

11. Other electronic supplies, which can be found in most specialized shops and should be chosen of the highest possible quality, particularly for low noise and high precision, are:

    a. Connectors.

    b. Resistors.

    c. Capacitors.

    d. Potentiometers.

12. Digital multimeter for checking voltage and resistance system (e.g., from World Precision Instruments).

2.3. Set-Up Conditioning

1. Vibration isolation table with best possible characteristics at low frequencies (in the range 10–1000Hz), for example, from Technical Manufacturing Corporation (Peabody, MA).
2. Thermostatic bath circulator (Techne, Princeton, NJ or Neslab, Portsmouth, NH) or Peltier temperature control (World Precision Instruments).
3. Perfusion system: Automatic device consisting of two identical syringes mechanically coupled and driven by a stepped motor (e.g., SP260P of World Precision Instruments) (*see* **Note 3**).
4. Battery driven magnetic stirrer (e.g., from Prolabo, Fontenay Sous Bois, France).

2.4. Chamber and Electrode Preparation

1. Bulk Teflon block for machining the chambers.
2. Teflon foil of 12, 25, and 100 µm thickness (e.g., Goodfellow Corp., Berwyn, PA).
3. High-voltage spark device, generating single-voltage pulses of a few hundred volts between two sharpened electrodes. The length of the pulse (of the order of milliseconds) should be variable. It is used for puncturing the septum.
4. Low-magnification (×20–×40) optical microscope with long focal distance, for checking the quality and diameter of the hole into the septum (Nikon, Tokyo, Japan).
5. High-vacuum silicon grease for assembling the chamber (e.g., Apiezon N from Fluka, Buchs, Switzerland).
6. Magnetic microspin bars (e.g., microfleas of Bel-Art, Pequannock, NJ, or Sigma, St. Louis, MO).
7. 99.99% Purity silver wire, 1 mm diameter for the electrodes (e.g., from Aldrich, Milwaukee, WI).
8. Tin soldering (for electrode connections).
9. Agarose, electrophoresis grade, for salt bridges (e.g., from Sigma).

3. Methods

3.1. Description of the Setup

3.1.1. Electrical Equipment

3.1.1.1. HEADSTAGE

The heart of the acquisition system is the headstage, a high-impedance current/voltage (*I/V*) converter that transforms the current flowing through the membrane into a voltage that can be analyzed (*see* **Fig. 1** and *[8]*). For a basic inexpensive setup one can build its own *I/V* converter (*see* **Note 4***)*.

1. Use an ultra-low noise FET operational amplifier in the virtual ground configuration (i.e., signal wired to the inverting input and non-inverting input wired to ground). Input impedance should be $>10^{12}\ \Omega$ (typically $10^{14}\ \Omega$), input capacitance as low as possible (typically <2 pF).

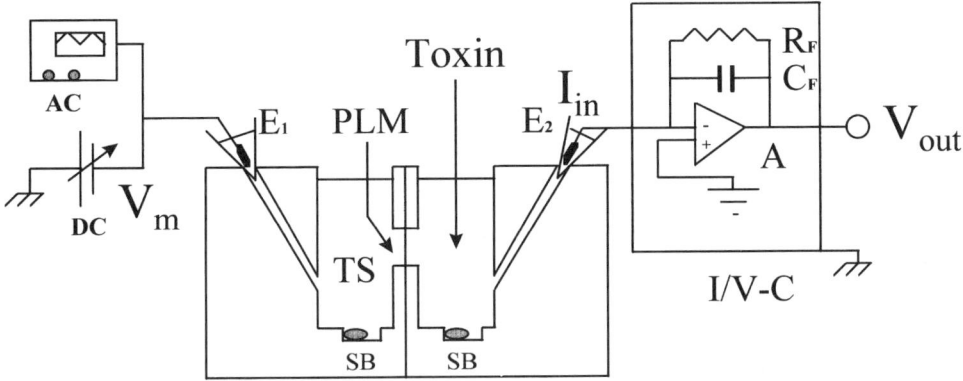

Fig. 1. Chamber for planar lipid membranes used in the study of pore-forming toxins. Schematic representation of a cell for the preparation of PLMs. A supported bilayer is prepared by apposing two lipid monolayers onto a small (0.1–0.2 mm) hole, punched in a thin (12 µm) Teflon septum (TS) separating two water filled compartments. Two Ag–AgCl electrodes (E_1 and E_2), immersed in agarose bridges, are shown. One is used to apply to the membrane the voltage (V_m) which is either a steady voltage generated by a DC battery supply or a periodic signal created by a waveform generator (AC). The other electrode drives the membrane current (I_{in}) into the current/voltage converter (I/V-C). This is a virtual grounded operational amplifier (A) with a feedback resistance, R_F (usually 10^9 Ω), and a feedback capacitance, C_F (usually 1 pF), that provide a time resolution of 1 ms. The output voltage (V_{out}) is fed to an oscilloscope, an X–t chart recorder, and a digital cassette recorder (not shown). Buffer solutions are stirred with magnetic spin-bars (SB) and the toxin is applied in the *cis* compartment (at virtual ground). The various elements are not drawn to scale.

2. Mount it on a Teflon board (which preserves maximal isolation).
3. Place a suitable resistor (R_F) and capacitor (C_F), in parallel configuration, in the feedback loop. Resistors should be chosen for low noise and low parasite capacity (*see* **Note 5**). The time constant of the converter is given by $R_F \times C_F$ and is normally chosen as approx 1 ms (e.g., 10^8 Ω with 10 pS); *see* **Note 6**.
4. Supply with two DC batteries of 9 V in series (*see* **Note 7**).
5. Enclose everything into a small metal box with two connectors (one for the input current and one for the output voltage) and connect the box to ground.
6. Place it next to the PLM chamber and connect the input to the current electrode (keeping the connection as short as possible).

3.1.1.2. Connecting Electrodes

For measuring the electrical parameters of the lipid bilayer this should be connected to the electronics via suitable electrochemical electrodes. The most diffuse are Ag–AgCl electrodes, which may be prepared as follows:

1. Clean two pieces of silver wire down to the bare metal with ultra-fine sandpaper and fine steel wool pad (*see* **Note 1**).
2. Immerse the two electrodes in a 0.1 M HCl solution and connect them together to the positive terminal of a 4.5 V battery. Place also a Pt (or graphite) electrode, the anode, in the solution and connect to the negative terminal (*see* **Note 8**). Note that hydrogen bubbles form at the anode and the Ag electrodes start darkening.
3. Continue the Cl⁻ plating for 2–3 h. Reverse the battery's polarity for 30 s, then disconnect the electrodes and remove them from the solution.
4. Before use, equilibrate the new electrodes by short-circuiting them in 3 M KCl for some hours.
5. Alternatively, the chloriding procedure (**steps 2** and **3**) could be performed by immersing the clean Ag electrodes in Clorox bleach for 14 h (at a bleach-to-water ratio of 1:5).

3.1.1.3. SALT BRIDGES

In experiments in which the solutions at the two sides of the membrane have different Cl⁻ concentrations, or when the solutions do not contain Cl⁻ at all, it is recommended to use agar bridges to avoid the formation of a junction potential between the electrodes and the solution. They may be prepared as follows:

1. Put 300 mg of agarose in 20 mL of 3 M KCl. Stir and gently heat the mixture until the agarose dissolves and melts (the solution becomes clear).
2. Remove it from the bath and aliquot in Eppendorf tubes (≈1 mL of solution per tube). Store the aliquots at 20–22C°.
3. For making agarose bridges melt one of those tubes and fill a 200-μL pipet tip with the solution. Pay particular attention in avoiding the formation of air bubbles.
4. Insert the Ag–AgCl electrode into the pipet tip filled with agarose. The electrode should be completely surrounded by agarose, and never contact the tip walls directly (*see* **Note 9**).
5. Store the salt bridges in 3 M KCl and short-circuit the electrodes together.

3.1.1.4. OTHER ELECTRONIC ELEMENTS IN THE PRINCIPAL CIRCUIT

1. Connect a DC voltage supply to the voltage electrode. Use batteries and a potentiometric voltage divider to produce an output potential spanning ±200 mV (*see* **Note 7**).
2. Put in parallel a waveform generator. This is used to generate a triangular wave suitable for monitoring membrane formation or for measuring continuous I–V curves. The generator can also be used to apply step-like voltage changes to the membrane.
3. Feed the output of the *I/V* converter to a low-pass analog filter which is used for narrowing the signal bandwidth (normally from DC to a few kHz or less). Even in the case of using a commercial headstage and amplifier with built-in low-pass filter, a separated low-pass filter is useful in reanalyzing experimental data recorded with a high-frequency bandwidth on the recording device.

4. Close the circuit in a storage oscilloscope that provides visualization of the output current (for clarity about the signal source we will use the term current even if it is converted to voltage). The oscilloscope has a high input impedance and an extremely fast response that ensures nothing of the signal is lost due to deficient instrument specifications (this is not true for other recording devices with reduced input impedance or narrower bandwidth).

5. Prepare a star-shaped ground connection at which to link *all* metal items surrounding the experiment (either the instruments' boxes, or the thermostatting jacket, or the Faraday cage or any other mechanical part). Connections should be done with copper wire of large section (dia. approx 5 mm). Pay attention that they may transmit mechanical vibrations (*see* **Note 10**).

3.1.1.5. OTHER RECORDING DEVICES

Before inserting any new instrument listed in this section, please *see* **Note 10**.

1. To visualize the current signals in real time, and permanently reproduce on paper, connect a fast X–t chart recorder in parallel to the oscilloscope (*see* **Note 11**).

2. For subsequent analysis, store the whole experiment on a two-channel DAT recorder (typical bandwidth DC–20 kHz) using 2-h audio cassettes. Note that although the analog signal is digitized before being stored, it can be reproduced only in analog form.

3. During the experiment, or during later analysis, transfer the pieces of current trace that contain the events of interest to a computer via an acquisition board. Because of the fast rate of data transfer (usually approx 100 kHz) these boards are excellent to reproduce fast events, but they load the computer with a mass of data and therefore cannot be routinely used to store the whole experiment.

3.1.2. Mechanical Setup

3.1.2.1. ISOLATION AND CONDITIONING OF THE SETUP (*SEE* **NOTE 10**)

1. The membrane is quite a delicate object and can easily be broken by mechanical shocks. In addition, mechanical vibrations cause it to behave as a vibrating capacitor that generates an undesired fluctuating current. To increase mechanical stability and to lower the noise, place the setup on a vibration isolation table. A cheap, reasonably efficient alternative to expensive commercial instruments is to use a heavy stone or metal platform (e.g., $40 \times 40 \times 5$ cm) placed on four tennis balls (or an inflated inner tube) partially immersed in a sand box.

2. Surround the PLM chamber with a Faraday cage of generous dimensions to eliminate electrical interference from the environment on the very small currents measured (from pA to nA) (*see* **Note 12**).

3. Eliminate acoustic vibrations that induce membrane vibration causing noise as described in **step 1**. If the acoustical insulation offered by the cage is not good enough cover the inside with a suitable plastic foam.

4. Apply to one or both the chambers a perfusion system consisting of two identical syringes mechanically coupled back to back, either hand- or motor-driven. This

is used for changing the water solution on one or both sides of the bilayer. The volume of the syringes should be at least 5× that of the chamber.

5. Install a temperature control system that either *(10)* surrounds the chamber with a jacket in which a temperature-controlled fluid circulates or establishes a good contact with a metal block thermostatted with a Peltier system (*see* **Note 13**).

6. Position a magnetic stirring device close to the bottom of the chamber (but not in contact with it). A common stirrer with a battery driven DC motor (*see* **Note 7**) and a 2-cm magnet fixed on the rotor is normally sufficient to stir both chambers. Use a switch and a potentiometer (preferably removed from the antivibration platform) to turn it on/off and to regulate the stirring speed (*see* **Note 14**).

3.1.2.2. Preparation of Home-Made Chamber for Membrane Formation

1. Prepare a number of identical blocks of typical size 2.5 × 2.5 × 2 cm (base × height × width) machined out of a Teflon bar (the complete chamber consists of two of these elements, called cuvets, separated by a partition septum holding the membrane).

2. From the top drill into each cuvet an elliptical hole of approx 4.5 mL volume that will constitute the main chamber for the bathing solution.

3. Parallel to this, drill another cylindrical hole for placing the electrode (dia. approx 4 mm) and a smaller one ($\varnothing \approx 1.5$ mm) for adding the toxin or for use with the perfusion system. These holes must be connected to the main chamber through a passage near its bottom.

4. Approximately in the center of one face drill a round aperture (dia. approx 6 mm) where the partition will be applied.

5. At the bottom of the main chamber carve a site (dia. approx 10 mm) for the magnetic stirring bar.

6. Prepare a chamber holder made of metal for clamping the two chamber elements together *(10)*. This should leave the top free and should fit rather precisely into the thermostatic element.

3.1.2.3. Preparation of the Septum Supporting the Bilayer

1. Cut a round piece (dia. approx 14 mm) from a Teflon foil 12 or 25 μm thick.

2. Cut two crown-shaped pieces (outer diameter $\approx$ 14 mm, inner diameter $\approx$ 5 mm) from a Teflon foil 100 μm thick.

3. Prepare a sandwich with the thin film between the two thick pieces.

4. Place between two clean glass slides and put onto a hot plate until the three pieces melt together (avoid excessive heating).

5. Make a hole of the desired diameter (typically approx 0.1 mm) using the high-voltage spark generated between two electrodes. The diameter of the hole will be smaller with a higher distance between the electrodes; more sparks will enlarge the hole (*see* **Note 15**).

6. Use a low-power optical microscope to check the diameter and the quality of the hole into the septum. Give particular attention to the rim, which should be as smooth and regular as possible.

7. Clean the new septum with solvents and store it in a clean space, attached to a card, with the indication of its diameter.

3.2. Preparation of the Planar Bilayer

3.2.1. Assembling the Chamber

1. With a 200-μL pipet tip, spread a thin layer of silicon grease on the two chambers (if necessary remove the excess grease with a clean paper).
2. Pretreat the septum with 5 μL of pentane (or hexane)–hexadecane (10:1) on each side of the septum, and wait until the short-chain alkane evaporates.
3. Prepare a sandwich with the septum between the two chambers, matching the inner portion of the septum with the apertures on the lateral sides of the cuvets.
4. Clamp tightly together in the chamber holder (*see* **Note 16**).

3.2.2. Membrane Formation

1. Lodge the assembled chamber into its stand (with the temperature control if necessary).
2. Insert the magnetic bars into each chamber (sometimes only into the *cis* side where the toxin will be added).
3. Add 1 mL of aqueous solution (e.g., buffer A) into each chamber through the electrode hole using a micropipet. Add the solution through the electrode passages to ensure they are filled and in contact with the main chamber. The water level should be at least 5 mm below the hole.
4. Place the electrodes in their cavities and connect to the electronic circuit.
5. Add 10 μL of the lipid–pentane (or lipid–hexane) solution (10 mg/mL) on top of each bathing solution (*see* **Note 17**).
6. Apply a triangular wave of 100 mV$_{p-p}$ and 25 Hz frequency to monitor the capacitance of the septum. A capacitance of 100 pF should respond with a square wave of current of 1 nA$_{p-p}$ (this is easy to check by placing a capacitor of 100 pF between the two electrode connections) (*see* **Note 18**).
7. Wait 5 min until the solvent has evaporated and then raise sequentially the water level in both chambers to above the hole (adding approx 2 mL of water solution on each side).
8. Monitor on the oscilloscope the membrane formation process by the ensuing sharp increase in capacitance. In fact, the capacitance of the 25 μm thick septum with the hole clogged is about 15 pF (30 pF for a thickness of 12 μm), whereas a membrane with a diameter of 100 μm should have a capacitance of 65 pF, assuming a specific capacitance of 0.8 μF/cm^2 (*see* **Note 19**).
9. Remove the triangular waveform.
10. Test the membrane stability by applying DC voltages of ±100 mV and ±140 mV for 5–10 min before adding the toxin. A membrane with a diameter of 100 μm should have a conductance $G < 10$ pS, that is, it should respond with a stable current $I < 1$ pA to the application of 100 mV DC (*see* **Note 20**).

3.3. Detection of Single Events and Determination of Pore Properties

3.3.1. Observing Channels Induced by PFT

1. Reduce the DC voltage applied to the membrane.
2. Add a few nanograms per milliliter of PFT to one side of the membrane (this is the cis side, normally the virtual-grounded compartment).
3. Turn the stirrer on.
4. Observe, after sometime, stepwise increases of the membrane current.
5. Turn the stirrer off to reduce the vibrational noise (*see* **Note 21**).
6. Analyze the current steps: usually one (or in some cases a few) typical amplitudes may be recognized. Divide this current value by the applied voltage to obtain the pore conductance G. With different PFT, G may range between 10 pS and several hundred pS, under physiological conditions. Some characteristic pore properties can be determined from the channel conductance by varying the experimental conditions, for example, selectivity between anions and cations, influence of lipid composition, effect of voltage (gating), size and geometry of the channel, molecularity (i.e., the number of monomeric units involved in forming the conductive pathway), and the presence of water in its lumen.

3.3.2. Estimate of Pore Size

1. Measure the pore conductance at different salt concentrations.
2. Plot the single-pore conductance (at various voltages) vs solution conductivity (or salt activity). If a linear relationship is observed, the pore is probably filled with water.
3. Estimate pore size according to the following equation:

$$r = \sqrt{\frac{Gl}{\sigma\pi}} \qquad (1)$$

where r is the pore radius, σ the conductivity of the solution, l the length of the pore, and G its conductance (this assumes that the pore is simply a cylindrical hole filled with water, where the mobility of ions is similar to that in the bulk aqueous solution; *see* **Note 22**).

3.3.3. Determining the Presence of Fixed Charges on the Pore

3.3.3.1. SATURATION

1. Measure the pore conductance at different salt concentrations.
2. Plot the single-pore conductance (at various voltages) vs salt activity. If the conductance increases sublinearly with the ion concentration of the solution (i.e., it shows a saturation) there is probably an excess of charged residues of one sign at the entrance of the pore. An indication of this effect is that the conductance depends on the square root of the salt concentration *(14)*.

3.3.3.2. SELECTIVITY

1. Prepare a membrane under asymmetrical conditions, using different concentrations of a monovalent salt on the *cis* and *trans* sides (C^I and C^{II}).
2. Measure the current flowing through the pore at different applied voltages (e.g., from -140 mV to $+140$ mV in steps of 10 mV).
3. Plot the single-pore current vs voltage. If the curve intercepts the voltage axis out of the origin (at a point called the reversal voltage, V_{rev}), the pore is selective, that is, anions and cations have different permeabilities (*see* **Note 23**).
4. Use the Goldman–Hodgkin–Katz *(15)* equation for correlating V_{rev} to the pore selectivity:

$$V_{rev} = \frac{RT}{zF} \ln \frac{P_+ + C^{II} + P_- C^I_-}{P_+ C^I_+ + P_- C^{II}_-} \quad \text{or} \quad \frac{P_+}{P_-} = \frac{C^{II}/C^I\, e^{\frac{zFV_{rev}}{RT}} - 1}{C^{II}/C^I - e^{\frac{zFV_{rev}}{RT}}} \tag{2}$$

where R is the gas constant, T the absolute temperature, F the Faraday constant, z the valence (at 23°C RT/zF is 25 mV); P_- and P_+ are the permeability of the anion and the cation respectively.

5. Estimate the presence of a potential Ψ_{pore} at the pore entrance (generated by uncompensated fixed charges), from the following rule *(12)*:

$$\frac{P_-}{P_+} = \frac{u_-}{u_+}\, e^{\Psi'_{pore}} \tag{3}$$

where Ψ'_{pore} is the reduced potential (i.e., Ψ_{pore} /25 mV) and u_- and u_+ are the mobility of the anion and the cation respectively. As an example an entrance potential of $+25$ mV would imply that anions are $\sim 3\times$ more permeant than cations. Because this electrostatic filter can discriminate ions only on the basis of their charge it is rather poor. However, this is typical for toxin pores that are presumed to inflict a rather unselective damage to target cells.

3.3.3.3. NONLINEAR I/V CHARACTERISTIC

1. Measure the current flowing through the pore at different applied voltages (e.g., from -140 mV to $+ 140$ mV in steps of 10 mV).
2. Plot the single-pore current vs voltage. If the current has a nonlinear hyperbolic shape there is probably an excess of charges of one sign at one entrance of the pore generating a potential Ψ_{pore}.
3. Calculate the ratio between the current at the highest positive voltage (I_+) and at the corresponding negative voltage (I_-). Obtain Ψ_{pore} from

$$\frac{I_-}{I_+} = e^{\Psi'_{pore}} \tag{4}$$

where Y' is again the reduced potential.

3.3.4. Channel Gating

1. Obtain one single channel in the membrane.
2. Keep a constant applied voltage and record the fluctuations arising from the opening and closing of the channel, or, more in general, from the fluctuations of the pore conductance between a high and a low value.
3. Measure all the intervals during which the pore is open (or closed).
4. Plot them in a dwell-time histogram, reporting the number of events longer than a given time vs time.
5. Fit this with a single exponential and use the time constant as the lifetime of the open (or closed) state at that voltage.
6. Change the applied voltage (e.g., from –140 mV to + 140 mV in steps of 10 mV) and repeat **steps 2–6**.
7. Plot the logarithm of the lifetime of the open (or closed) state vs applied voltage. A linear dependence suggests that channel opening occurs by a conformational change promoted by the movement of a charged part of the pore through the membrane *(16,17)*.
8. Derive this charge, measured in electronic units, from

$$Q = (m - n) \cdot RT/F \qquad (5)$$

where m and n are the slope of the closing and of the opening rate, respectively.

3.4. Detection of Macroscopic Currents and Determination of Pore Properties

PLMs are stable enough to allow studying also currents deriving from the simultaneous presence of several thousand pores in a so-called multichannel experiment. The information that can be drawn from such experiments is very often the same as that from the single channel, with the advantage that it provides values that are averages of many individual contributions (*see* **Note 24**).

3.4.1. Ion Selectivity

1. Prepare a membrane as usual in symmetric solutions.
2. Add a few micrograms per milliliter of PFT to obtain many channels inserted.
3. Add to one side the necessary amount of salt (either as a concentrated solution or even as a powder) to establish the desired gradient across the bilayer.
4. Record some *I–V* curves, applying a triangular wave (±140 mV, 0.01 Hz) while the number of inserted channels is still growing.
5. All the curves intercept each other and the voltage axis at one point, the reversal potential (V_{rev}) (*see* **Note 23**).
6. Proceed as in **steps 4** and **5** of **Subheading 3.3.3.2.**

3.4.2. Molecularity

1. Prepare a membrane as usual in symmetric solutions.
2. Add some PFT (typically 0.1 µg/mL) and estimate the number of channels inserted.

3. Double the dose of PFT and repeat **steps 2** and **3** until a final dose of at least 10 µg/mL is reached.
4. Plot the number of pores observed vs the concentration of the PFT, in a double logarithmic scale.
5. Estimate the number of monomers involved in the formation of an active unit from the slope of this plot.

3.4.3. Conductance and Lifetime (Noise Analysis Measurements)

1. Prepare a membrane as usual in symmetric solution.
2. Add a few micrograms per milliliter of PFT and wait the time necessary to obtain a large and constant number of channels inserted.
3. Apply a constant voltage, for example, +100 mV.
4. Record the current fluctuations arising from the fact that, as pores are discrete molecules that open and close stochastically, the number of open channels in a voltage-clamped membrane fluctuates even at equilibrium.
5. Apply spectral (Fourier) analysis to this extra-noise (easily distinguished from common sources of electrical noise such as thermal, shot, or $1/f$). Assuming that the channels present are constant in number, all equal and independent, and have only two possible states (open and closed), estimate their conductance and life time *(18)*.
6. Change the applied potential (e.g., from –140 mV to +140 mV in steps of 10 mV) and repeat **steps 4–6**.
7. Analyze conductance and lifetime data as in the case of single channels (**Subheadings 3.3.3.3. and 3.3.4.**) and compare the results.

4. Notes

1. Many of the artefacts and problems arising with PLMs come often from minute amounts of contaminants present as impurities in the chemicals used, and/or from aged, possibly oxidized, lipids. Therefore, one should use MilliQ grade water, organic solvents of the purest grade available, HPLC grade reagents and lipids at least 98% pure by TLC.
2. Store lipids at –80°C in glass vials with polytetrafluoroethylene (PTFE)-silicon septa, under nitrogen or argon if opened. They should be used within few months from purchase. Remember that oxidized lipids could produce leakage and membrane instability and even the formation of purely lipidic channel-like structures (*see also* **Note 20**).
3. Alternatively one could also use a peristaltic pump, with two channels per chamber; however, this device, owing to the fluctuating pressure it applies, introduces more unwanted mechanical noise.
4. Alternatively it is possible to purchase an integrated PLM module that normally comprises a ready-to-use headstage, a variable gain amplifier with built-in low-pass Bessel filter, a low-noise voltage supply, a module for capacity measurement (e.g., Axopatch 200B with the headstage CV203BU from Axon Instruments, or Dagan 3900A with the 3910 expander module for PLM and headstages 3901 for single-channel or 3902 for multichannel measurements from Dagan Corp.).

5. It may be convenient to install a couple of connectors that allow the possibility of changing the feedback elements in relation to the type of experiment performed (e.g., single-channel vs multichannel experiment).

6. With resistors $\geq 10^9 \, \Omega$ there is no need of feedback capacitance because they possess their own input capacitance. For the same reason, with resistors $\geq 10^{10} \, \Omega$ there is the need of a capacitance compensation circuit, to improve the time response; *see (10)*. Take care that the feedback elements are held tightly and do not vibrate, producing noise.

7. Avoid using a power supply based on an AC/DC transformer, because this introduces a powerful source of 50-Hz noise (60 Hz in the US) into the system.

8. Insert a potentiometer between the battery terminal and the anode to limit the current to approx 1 mA/cm^2 of silver surface area.

9. Drill a 1.5-mm hole in a rubber cork, so that it can hold the Ag–AgCl electrode in place in the pipet tip. The cork prevents the electrode from separating from the gel.

10. Avoid making ground loops. These are formed when a single item is connected two or more times to the ground, for example, when two instruments that are separately grounded are connected with a Bayonet Neill Concellman (BNC) cable that links also the two shieldings. This generates a closed wire that will pick up an electronmotive force from the oscillating magnetic fields present in the environment (the main sources of these fields are transformers and AC-powered motors which should be avoided, or at least shielded, whenever possible).

11. Typically, such recorders have a bandwidth between DC and 3–5 Hz which will cut all high-frequency noise, but also fast transient signals. This should be kept in mind while analyzing recorded traces that might look different from what was seen on the oscilloscope.

12. For a good electrostatic screening the Faraday cage can be made of any metal (steel, aluminium, copper); however, to screen from magnetic fields one must cover it with μ-metal, for example, from Amuneal, Philadelphia, PA *(19)*. It can be made of a metal net or from solid plates. In the last case these should be at least 3 mm thick to stop the transmission of acoustic vibrations.

13. The second alternative introduces less mechanical noise. In both cases one should consider that, because of the strong isolating properties of Teflon, the temperature of the membrane bath may be a few degrees different from that of the jacket, and variations will occur more slowly. When using higher temperatures remember also that, owing to the small volume of the chambers, evaporation may considerably change the volume and concentration of the two baths.

14. Stirring the solution is necessary to obtain a good diffusion of the peptides through the solution to the membrane; however, it introduces a large noise. Very often the stirrer is on during incorporation but is then switched off when the data are acquired.

15. So that the sparkle perforates the septum right in the middle, before applying it, gently engrave the thin Teflon foil near its center, with a small metal needle. Alternatively, tightly join the new sandwich to an old, already punctured, septum

(e.g., holding them between two bored Plexiglas stands), and proceed to **Subheading 3.1.2.3., step 5**.

16. It is not always necessary to disassemble and reassemble the chamber after an experiment. In some cases one can carefully wash the chamber with the septum already mounted with tap water and MilliQ water (to remove salt solutions), and then with ethanol and acetone (to remove lipids and excess of silicon grease). However, if a toxin (or other surface-active peptide) was added during the experiment, it is always better to dismantle the chamber and proceed to a thorough cleaning.

17. If possible dissolve lyophilized lipids directly in pentane or hexane. If they are already dissolved in chloroform or other solvents, first dry them under nitrogen or argon, then expose for 18 h to high vacuum (e.g., in a desiccator connected to a Teflon diaphragm pump) to remove any residual trace of the original organic solvent and finally add pentane or hexane. Some lipids are incompletely dissolved in these alkanes; normally, this can be circumvented by adding small amounts (5–10% v/v) of ethanol.

18. Dagan amplifier with PLM extension module displays directly the capacitance of the membrane in pF.

19. Sometimes the capacitance of the membrane is smaller than expected, which might indicate that the membrane does not occupy all the hole or that it is not completely solvent free. Very often, it is sufficient to wait a few minutes and the capacitance will reach the correct value. Avoid starting experiments with membranes that present a capacitance too large or too small in comparison with what is expected. A too large value may indicate that the membrane is inflated, and projects out on one side of the septum. It is possible to drive it back in place by increasing the hydrostatic pressure on that side. This is done by gently rising the solution level in one or the other of the two chambers. A too low capacitance is an indication of a clogged hole (by hexadecane or a multilayer) or of a membrane containing too much solvent.

20. Avoid starting experiments with membranes that present a conductance that is too large or unstable. Unstructured current fluctuations, especially if present at high potential, may indicate the formation of lipid or water channels *(10)*. These may be due to insufficient quality of the lipid (impurities, oxidized lipids, etc.); residual traces of organic solvents such as chloroform, or the presence of surface-active molecules, such as detergents. It should be emphasized that, although normally they appear as abrupt noisy fluctuations, in some cases they are more channel-like. For example, beautiful channels have been observed with Triton X–100 *(20)* and with asolectin alone (a mixture of poorly purified lipids of plant origin).

21. Vibrational noise can be distinguished from the conductance fluctuations discussed in **Note 20**, by the fact that the first is symmetrically distributed around the current trace, whereas the second corresponds to transitions from a low to a high conductance state and always has the same sign, or direction, as the applied voltage.

22. An improved method was recently introduced *(21–23)*, based on measuring G in the presence of sugars of different size. If the sugars are able to enter the pore, G is decreased; if they are too big, G remains unchanged. The size of the largest molecule allowed to pass inside the pore is thus easily determined. If the shape of the pore is not that of a regular cylinder this method provides essentially the radius of its entrance.

23. V_{rev} should be corrected for some small offsets coming from the electronics (amplifiers) and from the polarization of the electrodes. A way to estimate this, after having measured V_{rev}, is to break the membrane and to determine again the value of potential corresponding to zero current. This value (which should be small, only a few millivolts) has to be subtracted from the measured V_{rev}.

24. In general, a good rule, which may be helpful in avoiding artefacts, is that single-channel and multichannel experiments, when aimed at determining the same property, give the same result.

25. If you have already tried to obtain a membrane without success, never try to add more lipid solution when the water level is below the hole. You will clog the hole. If you need to add more lipid, first increase the water level, add lipid, wait 10 min, decrease the water level, and try again to obtain the membrane.

26. Ground loops are easily formed when many electronic instruments are attached to the setup. They are normally detected from the presence of a line frequency noise (50 or 60 Hz). When this is observed, one should check the grounding connection pattern again. This is done by connecting a membrane simulator (e.g., a parallel of a 10 $G\,\Omega$ resistor and a 100 pF capacitor) in place of the membrane, applying a small voltage (e.g., 20 mV) with the DC generator, and disconnecting all other instrumentations but the oscilloscope, where the 50- or 60-Hz noise level is observed. Starting from this minimal configuration verify that each metal item present is connected to the ground star by just one single wire. To do this, remove that single wire and check with a digital multimeter that that item is now indeed disconnected from the ground. After verifying that the basic configuration is correct and the noise level is low (it can be kept to less than 0.1 pA_{p-p}); increase the complexity of the connections by adding one instrument at a time and repeating previous controls.

27. Every 6 mo (or when needed) check calibration and tuning of all the electronics devices, in particular headstage, DC offset of the amplifier, analog Bessel filter offset, DC offset and gain of the tape and X–t recorder (follow, when available, the vendor's instruction).

28. Cleaning procedures (for both chambers and partition). Wash with tap water and MilliQ water (to remove salt solutions), then with ethanol and acetone (to remove lipids and excess silicon grease). To remove any residual of the organic molecules used, especially if very hydrophobic peptides are used, from time to time the cuvets should be cleaned by 24-h exposure to the chromic acid mixture. (**Caution:** Chromic mixture is very corrosive.) Thereafter, they should be rinsed extensively with distilled water and finally with methanol.

References

1. Menestrina, G., Schiavo, G., and Montecucco, C. (1994) Molecular mechanism of action of bacterial protein toxins. *Mol. Asp. Med.* **15,** 79–193.
2. Sato, N., Kurotaki, H., Watanabe, T., Mikami, T., and Matsumoto, T. (1998) Use of hemoglobin as an iron source by *Bacillus cereus. Biol. Pharmacol. Bull.* **21,** 311–314.
3. Ahnert-Hilger, G. and Gratzl, M. (1988) Controlled manipulation of the cell interior by pore-forming proteins. *Trends Pharmacol. Sci.* **9,** 195–197.
4. Ahnert-Hilger, G., Bader, M. F., Bhakdi, S., and Gratzl, M. (1989) Introduction of macromolecules into bovine adrenal medullary chromaffin cells and rat pheochromocytoma cells by permeabilization with streptolysin O: inhibitory effect of tetanus toxin on catecholamine secretion. *J. Neurochem.* **52,** 1752–1758.
5. Panchal, R. G., Cusak, E., Cheley, S., and Bayley, H. (1996) Tumor protease-activated, pore-forming toxins from a combinatorial library. *Nat. Biotechnol.* **14,** 852–856.
6. Al-yahyaee, S. A. S. and Ellar, D. J. (1996) Cell targeting of a pore-forming toxin, CytA δ-endotoxin from *Bacillus thuringiensis* subspecies *israelensis*, by conjugating CytA with anti-thy 1 monoclonal antibodies and insulin. *Bioconjugate Chem.* **7,** 451–460.
7. Braha, O., Walker, B., Cheley, S., Kasianowicz, J. J., Song, L., Gouaux, J. E., and Bayley, H. (1997) Designed protein pores as components for biosensors. *Chem. Biol.* **4,** 497–505.
8. Alvarez, O. (1986) How to set up a bilayer system, in *Ion channel reconstitution* (Miller, C., ed.), Plenum, New York, pp. 115–131.
9. Kagan, B. L. and Sokolov, Y. (1997) Use of lipid bilayer membranes to detect pore formation by toxins, in *Bacterial Pathogenesis* (Clark, V. L. and Bavoil, P. M., eds.), Academic Press, San Diego, CA, pp. 395–409.
10. Hanke, W. and Schlue, W.-R., (1993) *Planar Lipid Bilayers. Methods and Applications.* Academic Press, London, UK.
11. Ropele, M. and Menestrina, G. (1989) Electrical properties and molecular architecture of the channel formed by *E. coli* hemolysin in planar lipid membranes. *Biochim. Biophys. Acta* **985,** 9–18.
12. Cescatti, L., Pederzolli, C., and Menestrina, G. (1991) Modification of lysine residues of *S. aureus* α-toxin: effects on its channel forming properties. *J. Membr. Biol.* **119,** 53–64.
13. Song, L., Hobaugh, M. R., Shustak, C., Cheley, S., Bayley, H., and Gouaux, J. E. (1996) Structure of staphylococcal alpha-hemolysin, a heptameric transmembrane pore. *Science* **274,** 1859–1866.
14. Menestrina, G. and Antolini, R. (1981) Ion transport through hemocyanin channels in oxidized cholesterol artificial bilayer membranes. *Biochim. Biophys. Acta* **643,** 616–625.
15. Stein, W. D. (1990) *Channels, carriers and pumps. An introduction to membrane transport.* Academic Press, San Diego, CA.

16. Ehrenstein, G. and Lecar, H. (1977) Electrically gated ionic channels in lipid bilayers. *Q. Rev. Biophys.* **10,** 1–34.
17. Schwarz, G. (1978) On the physico-chemical basis of voltage-dependent molecular gating mechanisms in biological membranes. *J. Membrane Biol.* **43,** 127–148.
18. Neher, E. and Stevens, C. F. (1977) Conductance fluctuations and ionic pores in membranes. *Annu. Rev. Biophys. Bioeng.* **6,** 345–381.
19. Kasianowicz, J. J., Brandin, E., Branton, D., and Deamer, D. W. (1996) Characterization of individual polynucleotide molecules using a membrane channel. *Proc. Natl. Acad. Sci. USA* **93,** 13,770–13,773.
20. Schlieper, P. and De Robertis, E. (1977) Triton-X100 as a channel-forming substance in artificial lipid bilayer membranes. *Arch. Biochem. Biophys.* **184,** 204–208.
21. Krasilnikov, O. V., Sabirov, R. Z., Ternovsky, O. V., Merzlyak, P. G., and Muratkhodjaev, J. N. (1992) A simple method for the determination of the pore radius of channels in planar lipid bilayer membranes. *FEMS Microbiol. Immunol.* **105,** 93–100.
22. Parsegian, V. A., Bezrukov, S. M., and Vodyanoy, I. (1995) Watching small molecules move: interrogating ion channels using neutral solutes. *Biosci. Rep.* **15,** 503–514.
23. Krasilnikov, O. V., DaCruz, J. B., Yuldasheva, L. N., Varanda, W. A., and Nogueira, R. A. (1998) A novel approach to study the geometry of the water lumen of ion channels: Colicin Ia channels in planar lipid bilayers. *J. Membr. Biol.* **161,** 83–92.

11

Determination of Affinity and Kinetic Rate Constants Using Surface Plasmon Resonance

Luke Masson, Alberto Mazza, and Gregory De Crescenzo

1. Introduction

Surface plasmon resonance (SPR) is a relatively new technique extremely useful for studying macromolecular interactions between proteins, proteins and DNA, or proteins and lipids. Biospecific interaction analyses using SPR provides valuable information about the strength, speed, and stoichiometry of the interaction in real time and without the use of labels. An excellent review on the commercial SPR instrument called BIAcore™ has been published recently *(1)*. In general the device is a biosensor that measures mass accretion/loss within a finite surface volume as a function of time. The initial step is to stably immobilize a known quantity of one of the interactants (ligand) on the surface. Following this, the second or free-flowing interactant (analyte) is made available for binding to its putative surface-linked homologue through diffusion from a pool or source that steadily flows over the immobilized ligand surface. By replacing analyte solution with buffer the source becomes a sink taking away complex-dissociated analyte from the ligand surface, resulting in a detectable loss of mass.

Informative data come from the rate at which complexes assemble/disassemble, the net magnitude of complex formation, and the effects on these two parameters of modulating analyte availability. **Figure 1** shows a typical graphical representation (sensorgram) of data generated by BIAcore. The vertical axis represents the variation in mass at the surface described by resonance units (RUs) with 1 kRU equaling 1 ng/mm^2 of protein. The horizontal axis of course represents time in seconds. The initial baseline corresponds to a ligand-coupled surface of known quantity referred to as the matrix. As the injection of sample

From: *Methods in Molecular Biology, vol. 145: Bacterial Toxins: Methods and Protocols*
Edited by: O. Holst © Humana Press Inc., Totowa, NJ

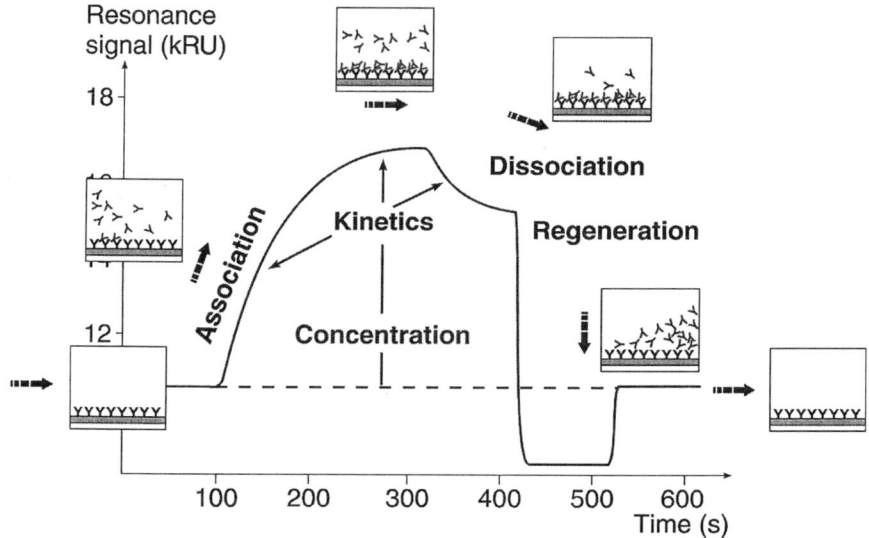

Fig. 1. Interpretation of a generalized sensorgram. This figure shows the injection of an analyte at $T = 100$ s to initiate the association phase of the curve. The injection is switched over to buffer alone at $T = 325$ s to initiate the start of the dissociation phase of the curve. After 100 s of complex dissociation, the immobilized ligand surface is regenerated and is ready for a new binding reaction. (This figure is reproduced with permission of Biacore AB, Uppsala, Sweden.)

begins analyte rapidly diffuses from the source into the matrix, whereby it freely interacts with ligand at a rate determined by such intrinsic properties as affinity (K_d), kinetics (k_{assoc}, k_{dissoc}), and stoichiometry (molar ratio of analyte to ligand), producing a sensorgram curve that plateaus at either a steady state level (R_{eq}) or at saturation (R_{max}). At the end of this associative phase, sample (analyte) is replaced by buffer, resulting in unbound analyte diffusing away from the matrix at a rate determined by the same intrinsic properties described previously. The dissociative phase is followed by a mild regeneration in which the matrix is chemically treated to remove all bound analyte without significantly damaging the ligand surface, thus enabling the experiment to be repeated any number of times.

The **Notes** section of this chapter deals with the practical aspects of experimental design as well as those related to other technical limitations that affect the precision of describing macromolecular interactions using BIAcore. We take you through an experiment in which the binding of a bacterial insecticidal protein toxin called Cry1Ac *(2)* to its purified glycoprotein receptor was assayed. Furthermore, by comparing the experimental sensorgrams with a related aminopeptidase N receptor, we provide additional information that can be extracted about the stoichiometry and the subsequent nature of the receptor.

2. Materials
2.1. Equipment

1. BIAcore (BIAcoreAB, Uppsala, Sweden) upgraded to 1000 or 2000 level specifications. The system described here is a 1000, which includes reagent delivery robotics and is PC-driven (NEC 466 ES, Windows 3.1®).
2. Miniprotein II slab gel apparatus (Bio-Rad).
3. A Pentium-based desktop computer using BIAevaluation™ 2.1 data analysis software (BIAcoreAB Uppsala, Sweden) as well as Microsoft Excel and global analysis software, for example, SP*Revolution*©.
4. 650E Advanced Protein Purification System (Waters).
5. Q-Sepharose anion-exchange resin (Pharmacia, Uppsala, Sweden).
6. Reagent-grade CM5 sensor chips (Pharmacia Biosensor).

2.2. Reagents
2.2.1. BIAcore™ Analysis: Solutions and Buffers

1. Running buffer: N-2-Hydroxyethylpiperazine-N'-2-ethanesulfonic acid (HEPES) buffered saline (HBS, 10 mM HEPES, pH 7.4, 150 mM NaCl, 3.4 mM EDTA) containing 0.05% BIAcore™ Surfactant P20.
2. Analyte buffer: Surfactant-free HBS.
3. Regeneration buffer: 5 mM NaOH.
4. Amine coupling solutions: activation solution (NHS = 0.1M N-hydroxy-succinimide, EDC = 0.1 M N-ethyl-N'-[3-diethylaminopropyl]carbodiimide).
5. Coupling buffer (20 mM ammonium acetate, pH 4.5).
6. Deactivation solution (1 M ethanolamine, pH 8.5).
7. Bovine serum albumin (BSA) blocking solution: HBS+ BSA (2.0 mg/mL–fraction 5, radioimmunoassay (RIA) grade, Sigma).

2.2.2. Protein Quantitation Solutions

1. Bio-Rad protein assay dye reagent (cat. no. 500-0006).
2. BSA (2.0 mg/mL) in distilled water (stock solution for standard curve production).

2.2.3. Protein Gel Electrophoresis Solutions

1. *Bis*-acrylamide stock solution: 40% w/v acrylamide/N,N'-methylene *bis*-acrylamide, 29:1, w/w (Bio-Rad).
2. Running buffer: 0.375 M Tris-HCl, pH 8.8, 0.1% w/v sodium dodecyl sulfate (SDS).
3. Stacking buffer: 0.15 M Tris-HCl, pH 6.8, 0.1% w/v SDS.
4. Electrophoresis buffer: 25 mM Tris, 250 mM glycine, pH 8.3, 0.1% w/v SDS.
5. Coomassie stain solution: 2.5 μL/mL in glacial acetic acid–water–methanol 10:40:50, by vol.

6. Destain solution: Glacial acetic acid–water–methanol 10:30:60, by vol.
7. Molecular mass markers: Standard broad range (Bio-Rad).
8. Sample buffer: 50 mM Tris-HCl, pH 6.8, 100 mM dithiothreitol, 2% w/v SDS, 10% v/v glycerol, 0.1% w/v bromophenol blue.

2.2.4. Receptor-Toxin Purification: Buffers and Reagents

1. Trypsin stock solution: 150 mM NaCl, 10 mM phosphate pH 7.4, 0.2% w/v EDTA, 0.5% w/v trypsin (1:250) (Boehringer-Mannheim).
2. Toxin solubilization and chromatography buffer A: 40 mM Na$_2$CO$_3$ buffer, pH 10.5.
3. Chromatography elution buffer B: buffer A + 1.0 M NaCl.

3. Methods
3.1. Protein Determination

1. Total protein quantification is carried out using the micro dye protocol of **ref. *(3)*** as per reagent manufacturer instructions (Bio-Rad).
2. Activated toxin or receptor in HBS is diluted in HBS to 800 μL to which 200 μL of dye is added.
3. The readings taken at an optical density of 595 nm are blanked against HBS.

3.2. Toxin-Receptor Purification

It is essential for subsequent kinetic SPR analyses that highly pure reagents are utilized during the experiments.

3.2.1. Purification of Analyte (Cry1Ac Toxin)

1. Recombinant Cry1Ac protoxin is expressed in *Escherichia coli* and isolated as insoluble inclusion bodies (*see* Chapter 6 by Schwartz and Masson for an in-depth description of how to purify trypsin-activated Cry toxins).
2. All activated toxins are quantitated by Bradford assay *(3)* and the purity verified by SDS-polyacrylamide gel electrophoresis (SDS-PAGE).

3.2.2. Purification of Ligand (Aminopeptidase N Receptor)

Two different aminopeptidase N (APN) preparations are described in these experiments and were obtained from two external sources *(4,5)*. Both APNs are ectoenzymes that are found attached to larval brush border epithelial cell membranes by a glycosylphosphatidylinositol (GPI) anchor. These membrane-bound receptors are easily solubilized by cleaving this GPI anchor with PIPLC (phosphatidylinositol-specific phospholipase C). Both APN preparations used in this study were isolated by a combination of Cry1Ac affinity chromatography and anion-exchange chromatography. The isolated APNs are judged pure as determined by the presence of a single band on an SDS-PAGE gel and are stored in HBS at –80°C.

1. One APN preparation is a 105-kDa glycoprotein isolated from the gypsy moth *Lymantria dispar L (4)*.
2. One APN preparation is a 115-kDa glycoprotein isolated from the tobacco hornworm *Manduca sexta (5)*.

3.3. SDS-PAGE

SDS-PAGE was carried out according to a standard procedure *(6)* using a 7 cm × 10 cm gel of 0.75 mm thickness.

1. A 10% polyacrylamide resolving gel is poured and allowed to solidify.
2. A 5% stacking gel is poured and solidified on top of the 10% gel.
3. Because these gels are used to verify quality and quantity of purified toxins and receptors, the total protein loaded per lane should not exceed 2–3 µg.
4. Samples were electrophoresed at 200 V (constant) for 1 h. Initial amperage should not be higher than 70 mA.
5. The gel is stained for 2 h in staining solution and destained until all the blue background disappears.

3.4. BIAcore™ Binding Experiments

When planning binding experiments, there are certain conditions that should be considered to eliminate extraneous factors such as mass transport and rebinding, among others, that can interfere with subsequent curve fitting (**Table 1**).

3.4.1. Coupling/Immobilization

1. The purified 105-kDa *L. dispar* APN is immobilized by covalently attaching it to the carboxymethylated dextran surface of a CM5 sensor chip using the instrument's manual injection mode and using the standard amine coupling procedure and reagents supplied by the manufacturer.
2. Carboxyl groups along the CM-dextran chains of the sensor chip surface are activated by exposure (35 µL at 5 µL/min) to a mixture of NHS (0.1 M N-hydroxysuccinimide)–EDC (0.1 M N-ethyl-N'-[3-diethylaminopropyl] carbodiimide) 1:1, v/v. The resulting succinimidyl ester groups are highly reactive with the free amine group of the N-terminal residue and the nonburied lysine or arginine residues arrayed along the polypeptide to be immobilized.
3. APN is injected over the surface at a concentration of 0.1 mg/mL in coupling buffer with the contact time (i.e., flow rate and injection volume) controlled so as to immobilize the precise quantity desired. Coupling solution must be relatively salt free (less than 50 mM) and at a pH below the isoelectric point of the sample to minimize charge repulsion that would inhibit covalent attachment.
4. After coupling, unreacted surface ester groups are blocked by exposure (35 µL at 5 µL/min) to an amine-rich deactivation solution. The net quantity of immobilized receptor was determined from the difference in baseline signal strength (RU) precoupling and post-deactivation—bearing in mind that any additional baseline

Table 1
Considerations for Experimental Design and Data Analysis

Potential Problems	Remedy/Solution
Surface heterogeneity Standard amine coupling can be very nonspecific, resulting in heterogeneous surfaces with variable access to ligand binding sites and subsequent poor kinetic data fit.	Alternate coupling strategies such as thiol coupling, antibody capture, His tag capture etc., can be employed to produce a homogeneously oriented population of ligand molecules.
Sample heterogeneity Chemical and structural inhomogeneity of ligand and/or analyte resulting in poor data fit due to the possible presence multiple affinity variants.	Extensive sample purification. If unavoidably impure must be considered critically during data analysis.
Steric hindrance Physical proximity of binding complexes limits access of analyte to unbound ligand sites on the surface. Distorts stoichiometric evaluation because binding sites are never truly saturated *(9)*. Results in poor kinetic data fit.	Use low surface densities of ligand. Manual injection mode permits the user to pause injections in mid-course so as to evaluate the quantity of ligand coupled to the surface. Once the desired level of immobilization has been achieved the coupling injection can be aborted.
Mass transport limitation Analyte binding rate exceeds its diffusion rate, creating a localized depletion within analyte source that in turn distorts concentration-dependent analyses *(10)*.	Can be experimentally minimized using high analyte flow rates (20 µL/min or higher) as well as low surface densities of ligand. Best approach is to mathematically (data analysis) correct for mass transport limitation.
Protein aggregation or valency The system relies on a mass-based detection principle unable to discriminate self-association from other macromolecular interactions. Aggregation can interfere with ligand binding directly and mask stoichiometric relationships.	Check using other techniques such as light scattering or analytical ultracentrifugation. Determine handling and solution conditions that minimize aggregation.
Avidity The propensity of a bound analyte to influence, cooperatively, additional binding events either specifically or nonspecifically distorting both affinity and kinetics.	Can be identified and corrected for with proper data analysis *(7)*. Experimentally difficult to control.

Table 1 (continued)

Nonspecific binding	
Biologically irrelevant macromolecular analyte association with surface-bound ligand or other matrix components.	Blank injections (samples injected over a nonligand surface). Inclusion of BSA in samples.
	Precondition the surface prior to actual experiment by a sample injection and regeneration.
Artifacts	
Instrument effects such noise, baseline drift, or sample dilution.	Blank injections and high analyte flow rates. Use KINJECT command (injection noise reduction feature on BIAcore 1000 to 3000). Wash cycles between injections. Randomize sample injection order between replicates. Condition the surface.

decay downstream will have to be incorporated into surface quantity estimates (*see* **Table 1**). For this experiment, surface densities of approx 250 RU were used.

3.4.2. Binding

All binding experiments are carried out using the methods automation feature available with BIAcore.

1. Toxins are injected at a flow rate of 5 μL/min for all experiments. Sample injection volumes were held constant at 25 μL. Higher flow rates should be considered when mass transport is a factor (*see* **Table 1**).

2. Inject different sample concentration ranges (300, 500, 1000, 1500 and 2000 n*M* were used in **Fig. 2**). Toxins are dissolved and diluted in the same HBS solution used with BIAcore to minimize refractive index shifts between sample and running buffer (*see* **Table 1**). Centrifugation to remove insoluble particulate matter is performed prior to sample quantification and use. This is vital to prevent blockages of the instrument's fluidics and also to improve the precision of sample quantity estimates. Automation enables the user to generate a superimposable series of data curves that have similar injection start/stop points and more importantly, identical data collection rates.

3. Blank injections are incorporated into these methods and consist of analogous toxin injections (*i.e.*, 300, 500, 1000, 1500, and 2000 n*M*) over nonreceptor surfaces activated and deactivated in an identical fashion as the surface where the receptor is bound.

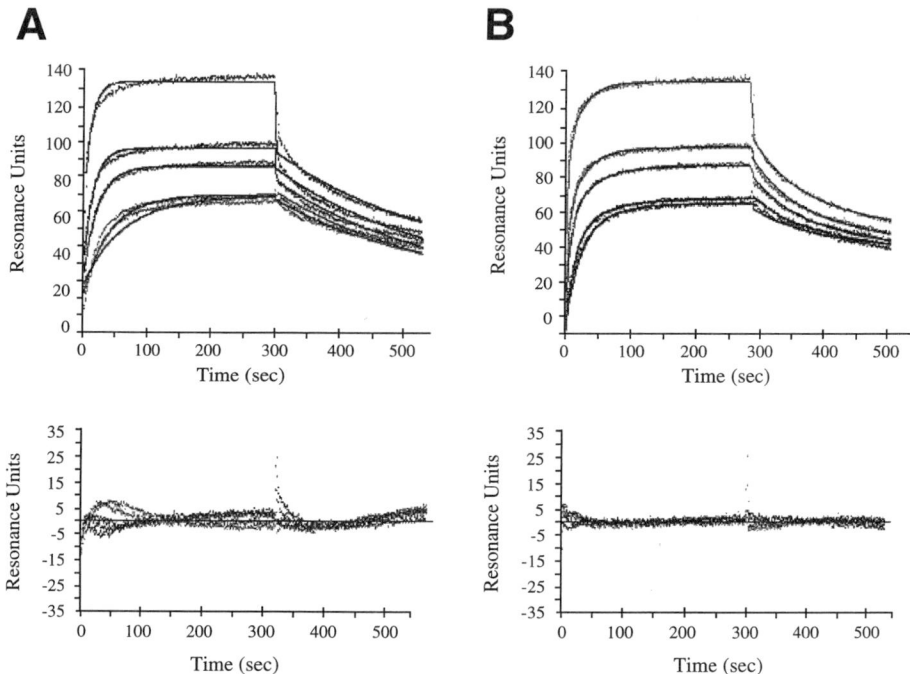

Fig. 2. Global fitting of toxin–receptor interactions. Binding data (*dotted curves*) from five different concentrations of Cry1Ac (300, 500, 1000, 1500, and 2000 n*M*) were fitted (*solid lines*) either to a simple model (A + B ⇔ AB) (**A**) or to a heterogeneous surface model (A + B ⇔ AB and A + B* ⇔ AB*) (**B**). A plot of the residuals that represents the difference between the experimental data points and the corresponding fitted data points is presented immediately below the fitted sensorgrams.

3.4.3. Regeneration

These surfaces are regenerated using a single 2 min pulse of 5 m*M* NaOH at 5 µL/min and are reusable for at least five experiments without any overt loss of binding capacity. Regenerations are also incorporated into the automated methods for reasons described previously.

3.5. BIAcore™ Data Analysis

To analyze the sensorgrams generated in **Fig. 2**, we use a software program developed in our laboratory called SP*Revolution*© (*7*) that is freely available for general use (Internet address: http://www.bri.nrc.ca/csrg/equip.htm#biacore). Unlike in the integrated rate equation method, the solution of the differential rate equation sets describing the kinetics is not analytically calculated but approximated by using numerical integration algorithms (*see* **Note 1**).

SP*Revolution*© uses an iterative process based on a Levenberg–Marquardt algorithm to manage the search of the apparent kinetic parameters. The integration of the differential set of equations describing the time dependence of the various species concentrations, for one given set of kinetic parameters, is then performed by an adaptive step-size Runge–Kutta algorithm. The criterion used to determine the best set of kinetic parameters for a given model is the minimization of the sum of squared residuals (difference between the fit and the experimental values).

3.5.1. Curve Fitting

3.5.1.1. Data Preparation

The data must be prepared by subtracting the individual toxin curves that were injected over a blank surface (absence of immobilized ligand on the CM-5 sensor chip) from the identical toxin concentration injected over a receptor surface. This can be done using the BIAevaluation™ 2.1 software that comes with the BIAcore apparatus, or alternatively, directly in the SP*Revolution*© software.

3.5.1.2. Model Selection

1. Select the simple A + B ⇔ AB model. The SP*Revolution*© software will automatically fit the experimental data to this model and will present the simulated curves fitted onto the data curves. This output shown in **Fig. 2A** clearly shows that the data describes an interaction more complex than a simple model, as the simulated curves do not fit the actual experimental data. A plot of the residual curves, which represents the difference between the calculated and experimental data vs time, is shown directly below the sensorgram. It is clear from the residuals produced by the simple A + B ⇔ AB fitting that they do not fall within the range of the noise generated by the BIAcore apparatus (±2 RU) and presents clear trends in the dissociation phase.
2. Other factors such as mass transport or avidity (*see* **Table 1**) may influence the shape of the binding curve. Select these different models available in the software and refit the curves.
3. Select the heterologous model that assumes the presence of two different sites (i.e., a mixed population of receptors on the surface). The data presented in **Fig. 2B** can be nicely fitted to this model, which indicates that there are two populations of receptors on the surface each with a single, but kinetically distinct, binding site.

3.5.1.3. Determination of Affinity and Rate Constants

The SP*Revolution*© software outputs a table that summarizes all of the kinetic parameters of the reaction (**Table 2**).

1. k_{assoc}. This kinetic rate constant is an indicator of how fast the analyte recognizes and binds to the receptor to form a complex. It is described as a function of con-

Table 2
Global Analysis Summary of the *L. dispar* Sensorgrams

	Model	
	Two receptor populations	Simple interaction
k_{assoc1} (M^{-1}s^{-1})	$(139 \pm 6) \times 10^3$	$(38 \pm 2) \times 10^3$
$k_{dissoc1}$ (s^{-1})	$(4.4 \pm 0.1) \times 10^{-3}$	$(4.0 \pm 0.1) \times 10^{-3}$
k_{assoc2} (M^{-1}s^{-1})	$(13 \pm 1) \times 10^3$	n/a
$k_{dissoc2}$ (s^{-1})	$(1.2 \pm 0.5) \times 10^{-3}$	n/a
Total receptor (RU)	199 ± 3	113 ± 2
[Receptor 1]/[Receptor 2]	0.86 ± 0.02	n/a
K_{d1} (M)	$(3.6 \pm 0.2) \times 10^{-9}$	$(1.05 \pm 0.07) \times 10^{-7}$
K_{d2} (M)	$(9.2 \pm 0.9) \times 10^{-7}$	n/a
χ^2	3.27	9.38

centration and time. For example, $[M^{-1}s^{-1}]$ defines the number of AB complexes formed per second in a 1 M solution containing both A and B.

2. k_{dissoc}. This kinetic rate constant is an indicator of stability or how fast the formed complex dissociates into its original components. It is described solely as a function of time. For example, $[s^{-1}]$ defines the fraction of complexes that dissociates or decays per second. A k_{dissoc} of $10^{-2}s^{-1}$ indicates that 0.01 or 1% of the AB complexes dissociate per second.

3. Total receptor (RU). The SP*Revolution*© software will automatically calculate the maximal amount of functional binding surface (R_{max}). This is useful as it provides an indicator of how much of the immobilized receptor is inactive. In the two receptor populations that were immobilized in our experiment, although 250 RU was initially immobilized, only 200 RU was determined to be functional.

4. [Receptor 1]/[Receptor 2]. This term represents the ratio of the concentration of the first receptor to the second receptor in terms of RU.

5. K_d. This is the dissociation constant (more commonly known as the affinity constant) is a descriptor of the overall strength of the binding reaction. It is expressed as a measure of concentration and is calculated from the ratio of k_{dissoc}/k_{assoc}.

6. χ^2. This (Chi^2) value indicates the goodness-of-fit; the lower the χ^2 value, the better the fit.

3.5.2. Binding Stoichiometry

By simply injecting a high concentration of analyte over a low density receptor surface, one can extract a valuable piece of information concerning the toxin–receptor interaction, namely the stoichiometry of the binding interaction.

1. Inject a high concentration of Cry1Ac analyte (= 1 μM) over a 200 RU surface of two relatively similar APNs isolated from two different larval species *(4,5)*.

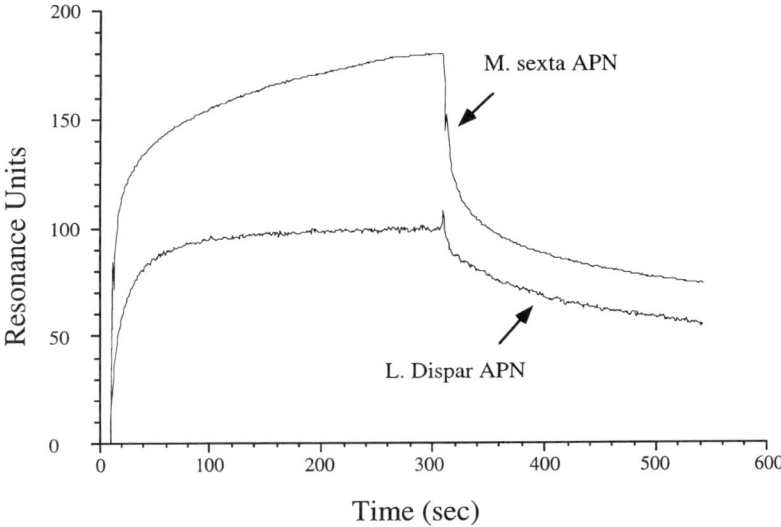

Fig. 3. Composite curve showing stoichiometric differences in total binding. An injection of Cry1Ac analyte (1000 n*M*) was passed over a 200 RU surface of the 115-kDa APN from *Manduca sexta* (*curve A*). For comparative purposes, we injected a 1500nM solution of Cry1Ac over a 200 RU surface of the 105-kDa APN from *Lymantria dispar* (*curve B*).

2. With SPR, the molecular mass of any protein is proportional to the response (RU). In general, it has been determined that 1 ng/mm^2 of protein represents 1000 RU. Therefore 200 RU of an immobilized 105-kDa protein or a 115-kDa protein represents approx 1.9 or 1.7 fmol/mm^2, respectively.
3. Determine the R_{max} of a relatively saturating level of analyte (**Fig. 3**). Using the RU-protein conversion factor, 100 RU of Cry1Ac toxin (65 kDa) is equivalent to 1.54 fmol/mm^2 of toxin.
4. **Figure 3** shows that Cry1Ac toxin can create more toxin–receptor complexes when it binds to the *M. sexta* APN than when it binds to the *L. dispar* APN.
5. By comparing the immobilized surface density to the Rmax of toxin binding one can calculate molar ratios for Cry1Ac binding to *L. dispar* (1.5 fmol/mm^2 of toxin bound to 1.9 fmol/mm^2 of receptor). It is clear that the toxin is binding to the receptor in a 1:1 molar ratio further supporting a model of two receptor populations with each population containing a single binding site for the toxin (*see* **Note 2**). In contrast, it is known that the *M. sexta* APN has the ability to bind two molecules of Cry1Ac (*8*). This stoichiometry is reflected in **Fig. 3** where 180 RU of toxin (approx 2.8 fmol/mm^2) can bind to 200 RU (1.7 fmol/mm^2) of *M. sexta* receptor, which surpasses a 1:1 molar ratio. In summary, this simple experiment provides important additional information concerning the nature of the ligand and can provide additional support for the chosen binding model.

4. Notes

1. The main advantage of using global analysis, in addition to the capability of using more complex kinetic models than the ones previously available using software based on linearization or integrated rate equation methods, is the ability to apply the desired model to every part of a sensorgram (numerical analysis) or even to many sensorgrams at the same time (global analysis). When the kinetic models consist of two or more successive steps (e.g., a conformational change model: A + B ⟺ AB and AB ⟺ AB*), the use of linear transformations technique or the use of the analytical integration of the differential rate equations describing the model chosen is impossible without ignoring parts of the sensorgrams or without doing any approximation, respectively. Data similar to that shown in **Fig. 2** for the *L. dispar* APN has been fitted to a simple A (toxin) + B (receptor) ⟺AB model *(4)* by integrated rate equations using the first 100 s of the dissociation phase of the curve and using the determined k_{dissoc} rate constant to constrain the k_{assoc} rate constant assessment of the initial part of the binding curve. Surface heterogeneity was not evident in those analyses. The global integration of a whole curve (association and dissociation) or even many curves simultaneously, which is possible with numerical integration, also eliminates the need of self-consistency tests that are highly recommended when the kinetic parameters are obtained by analyzing association and dissociation data separately using older techniques *(11)*.

2. Because the APN receptor is a glycoprotein and it is known that the carbohydrate *N*-acetylgalactosamine forms part of the binding site, it is entirely feasible that the two different populations, determined here by global analysis, differ only in the carbohydrate structure and not in the protein itself.

Acknowledgments

We wish to acknowledge Drs. Sangadala, Adang, and Valaitis for access to the isolated aminopeptidases.

References

1. Fivash, M., Towler, E. M., and Fisher, R. J. (1998) BIAcore for macromolecular interaction. *Curr. Opin. Biotechnol.* **9,** 97–101.
2. Schnepf, E., Crickmore, N., van Rie, J., Lereclus, D., Baum, J., Feitelson, J., Zeigler, D. R., and Dean, D. H. (1998) *Bacillus thuringiensis* and its pesticidal crystal proteins. *Microbiol. Mol. Biol. Rev.* **62,** 775–806.
3. Bradford, M. M. (1976) A rapid and sensitive method for the quantitation of microgram quantities of protein utilizing the principle of protein-dye binding. *Anal. Biochem.* **72,** 248–254.
4. Valaitis, A. P., Mazza, A., Brousseau, R., and Masson, L. (1997) Interaction analyses of *Bacillus thuringiensis* Cry1A toxins with two aminopeptidases from gypsy moth midgut brush border membranes. *Insect Biochem. Mol. Biol.* **27,** 529–539.
5. Sangadala, S., Walters, F. W., English, L. H., and Adang, M. J. (1994) A mixture of *Manduca sexta* aminopeptidase and phosphatase enhances *Bacillus thuringiensis*

insecticidal CryIA(c) toxin binding and 86Rb⁺-K⁺ efflux *in vitro. J. Biol. Chem.* **269,** 10,088–10,092.

6. Laemmli, U. K. (1970) Cleavage of structural proteins during the assembly of the head of bacteriophage T4. *Nature* **227,** 680–685.

7. O'Connor-McCourt, M. D., De Crescenzo, G., Lortie, R., Lenferink, A., and Grothe, S. (1998) The analysis of surface plasmon resonance-based biosensor data using numerical integration, in *Quantitative Analysis of Biospecific Interactions.* (Lundahl, P., Lundqvist , A., and Greijer, E., eds.), Harwood, GmbH, Chur, Switzerland.

8. Masson, L., Lu, Y. J., Mazza, A., Brousseau, R., and Adang, M. J. (1995) The CryIA(c) receptor purified from *Manduca sexta* displays multiple specificities. *J. Biol. Chem.* **270,** 20,309–20,315.

9. O'Shannessy, D. J. (1994) Determination of kinetic rate and equilibrium binding constants for macromolecular interactions: a critique of the surface plasmon resonance literature. *Curr. Opin. Biotechnol.* **5,** 65–71.

10. Myszka, D. G. (1997) Kinetic analysis of macromolecular interactions using surface plasmon resonance biosensor. *Curr. Opin. Biotechnol.* **8,** 50–57.

11. Schuck, P. and Minton, A. P. (1996) Kinetic analysis of biosensor data: elementary tests for self-consistency. *Trends Biochem. Sci.* **21,** 458–460.

12

ADP-Ribosylation of α-G$_i$ Proteins by Pertussis Toxin

Positional Dissection of Acceptor Sites Using Membrane Anchored Synthetic Peptides

Lars von Olleschik-Elbheim, Ali el Bayâ, and M. Alexander Schmidt

1. Introduction

Bordetella pertussis, the causative agent of whooping cough, secretes a plethora of virulence factors contributing to the onset and characteristic of the disease. The exotoxin pertussis toxin (PT) is regarded as particularly important during pertussis infection, as it is involved in both the adhesion of the pathogen and the initiation of systemic disease *(1,2)*. A number of symptoms observed during pertussis infections, such as, for example, lymphocytosis, islet-cell activation, and histamine sensitization have been associated with the activity of pertussis toxin. Like cholera toxin or the *Escherichia coli* heat-labile enterotoxin LT, PT is an A–B type bacterial toxin that is composed of the enzymatically active S1 subunit (A-protomer) and the binding B-oligomer consisting of the S2, S3, S5, and two S4 subunits. The observed pleiotropic effects are the result of the ability of S1 to ADP-ribosylate certain inhibitory α-subunits of heterotrimeric GTP-binding proteins (G-proteins) involved in a variety of signaling pathways. ADP-ribosylation not only uncouples these G-proteins from their receptors but in addition also deranges the adenylate cyclase system and in this way interferes with second-messenger pathways *(3–7)*. The modification is introduced at a distinct cysteine residue at position –4 of the C-terminus *(3,4)*. For uncoupling signal transduction this seems to be an ideal site as the C-terminus of the α-G$_i$ proteins has been shown to be necessary for interactions with the receptor(s) *(8,9)*. Especially the last four amino acids seem to

From: *Methods in Molecular Biology, vol. 145: Bacterial Toxins: Methods and Protocols*
Edited by: O. Holst © Humana Press Inc., Totowa, NJ

influence receptor selectivity *(9)*. PT-sensitive and -insensitive α-subunits of
G-proteins were thought to differ by the presence or absence of the target
cysteine residue near the C-terminus. However, as C-terminal sequences of
PT-sensitive G-protein α-subunits are highly conserved *(10)* protein regions
distant to the –4 cysteine residue also seemed to play a role *(11,12)*. For further
analysis of PT interactions with α-G$_i$ proteins and in particular of the requirements of the acceptor site for ADP-ribosylation already several molecular
approaches have been investigated.

As native α-subunits are difficult to isolate as single species and are furthermore often contaminated by βγ-subunits, Vaughan and Wolf and their colleagues employed oligonucleotide-directed mutagenesis using the polymerase
chain reaction (PCR) technique to address sequence and/or structural substrate
requirements *(13,14)*. Using this technique a set of 10 amino acid point mutations or minimal deletions involving the last five C-terminal positions was generated. Based on theses studies it was concluded that three of the four terminal
amino acids seem to be critical for pertussis toxin-mediated ADP-ribosylation.

Synthetic peptides have been frequently shown in various systems to be able
to mimic certain aspects of antigenicity of even complex proteins *(15,16)*. In
recent years it became apparent that besides immunological characteristics also
other functional aspects of proteins can be addressed by synthetic peptides and
especially so when the corresponding segments are involved in protein-protein
interactions. Thus, in another approach Graf and colleagues *(17)* recently demonstrated that PT is able to ADP-ribosylate a cysteine at position –4 of free
soluble 15-mer and 20-mer peptides encompassing the C-terminal amino acid
sequence. The K_m of PT for the ADP-ribosylation of free peptides was shown
to be even 10-fold higher than that determined for the heterotrimeric G-proteins. To our knowledge thus far this is the only study on ADP-ribosylation
employing synthetic peptides as substrates. That functional aspects of native
α-subunits could be mimicked by synthetic peptides was further emphasized
by the study of Dratz et al., who reported that an 11-amino-acid peptide derived
from the C-terminal sequence of the α-subunit of the heterotrimeric G-protein
transducin (G$_t$) bound to rhodopsin and was able to stabilize its active form *(8)*.

With the method described in this chapter the synthetic peptide approach
was taken even further as the possibility of employing synthetic peptides was
combined with the potential, versatility, and speed of peptide synthesis via a
C-terminal anchor on cellulose membranes, the "spot synthesis" recently
developed by Ronald Frank *(18)*. This technique, which is already widely
applied for the delineation of epitopes recognized by, for example, monoclonal
antibodies had previously been shown to be also applicable for the analysis of
protein–protein interactions *(19)*. Thus, we were interested to see whether
membrane-anchored peptides might also be suited to serve as substrates for

enzyme reactions and in this way could be employed for the dissection of acceptor sites. In particular, we were interested in the analysis of sequence requirements for the PT-mediated ADP-ribosylation of G_i α-subunits *(21)*.

Spot synthesis is based on the development of specific "anchor" moieties used to covalently modify the surface of a cellulose filter sheet with amino functions. With the introduction of amino functions the "activated" cellulose membrane can be used for the synthesis of C-terminally anchored peptides in defined "spots" basically according to the procedures of Merrifield synthesis using Fmoc-amino acid derivatives. Depending on the experimental set up filter membranes of the size of an ordinary postcard (about 9 × 13 cm) are easily derivatized with 425 C-terminally anchored peptides following a preselected array. The critical issue of this method—the stability of the linkage between anchor and cellulose membrane during peptide synthesis—has been solved by Frank by the introduction of β-alanine spacer moieties *(18)*. Thus, a rather large number of peptide sequences—using just one or potentially also several filters - can subsequently be assessed under identical assay conditions for their particular activity, such as, for example, their activity as substrates for toxin binding or ADP-ribosylation.

2. Material

2.1. Equipment

Although the spot synthesis can in principle also be performed manually, precision and ease of handling is greatly enhanced if an automated and programmable pipetting robot adapted to the small volumes involved in the synthesis would be available. An example for a possible automated pipetting system (ABIMED Auto-Spot Robot ASP 222; Abimed Analysentechnik, Langenfeld, Germany) is shown in **Fig. 1**. With the help of an automated device the activated amino acid derivatives can be applied in volumes of 0.2 μL to the "spot" area on cellulose membranes (Whatman 540, Whatman, Kent, England) with an accuracy of 0.1 mm. Peptide synthesis was routinely performed in "spots" with a diameter of 2–3 mm and generated approx 20 nmol of peptide per spot. Owing to the restricted diameter of the synthesis area up to 425 different peptide sequences on a filter of the given size can be synthesized.

For critical equipment and the necessary modifications introduced in the system, *see* **Note 1**.

For quantifying the incorporation of ^{32}P-labeled ADP-ribose into PT target proteins, a Fuji BAS 1000 Bioimager was used.

2.2. Chemicals and Buffers

1. ABTS-substrate solution (Boehringer, Mannheim, Germany).
2. Binding buffer: 5 m*M* $MgCl_2$, 50 m*M* Tris-HCl, pH 7.5.

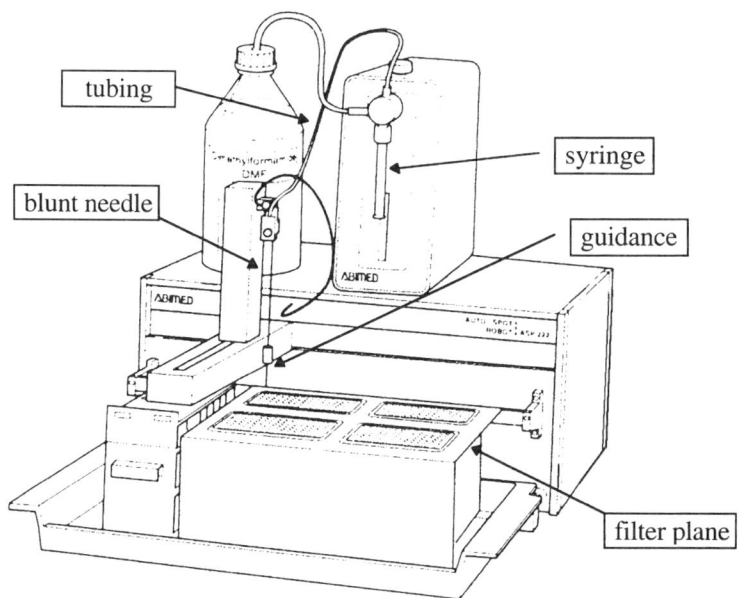

Fig. 1. Schematic representation of an automated pipetting robot employed in membrane-anchored synthesis of peptides (ABIMED, Langenfeld, Germany). *Arrows* indicate equipment parts critical for the performance of the synthesis.

3. Blocking solution: 0.05% Tween-20 (Pharmacia Biotech, Uppsala, Sweden), 5% sucrose, 20% blocking buffer (Genosys, Ismaning, Germany), 80% Tris-buffered saline (TBS), pH 7.0.
4. Bromphenol blue POD substrate, precipitating (Boehringer, Mannheim, Germany).
5. BPB solution: 1% bromphenol blue (Merck, Darmstadt, Germany) in $N,N,$-dimethylformamide (DMF, Fluka, Neu-Ulm, Germany).
6. Buffer A (peptide synthesis): 8 M urea, 1% sodium dodecyl sulfate (SDS), 0.5% mercaptoethanol, pH 7.0.
7. Buffer B (peptide synthesis): 10% acetic acid, 50% ethanol, 40% H_2O.
8. Capping solution: 2% acetic acid anhydride (Ac$_3$OH, Merck, Darmstadt, Germany) in DMF.
9. Citrate-buffered saline (CBS): 0.8% NaCl, 0.02% KCl, 10 mM citric acid-1-hydrate, pH 7.0.
10. Coomassie Brilliant Blue staining solution (Sigma, Deisenhofen, Germany), 2.75 mg of Coomassie Brilliant Blue R250 dissolved in 500 mL of ethanol, 100 mL of acetic acid, and 400 mL of H_2O. The gels are destained in 10% ethanol–7% acetic acid.
11. Cova-buffer: 2 M NaCl, 40 mM MgSO$_4$, 0.05% Tween-20 in phosphate-buffered saline (PBS).

12. Derivatizing solution (3 mL per cellulose sheet): 140 mg Fmoc-β-Alα-OH (Novabiochem, Bad Soden, Germany) in 2.245 mL of DMF, 105 μL of *N,N'*-diisopropylcarbodiimide (DICD, Fluka, Neu-Ulm, Germany). The reaction time is 10 min at 20–22°C. Afterwards 45 μL of 1-methylimidazole (MELM, Merck, Darmstadt, Germany) are added.

13. Diethanolamine buffer: 10% diethanolamine, 5 μM $MgCl_2$, 0.02% NaN_3 in H_2O, pH 9.8.

14. Dimethyl sulfoxide (DMSO, Fluka, Neu-Ulm, Germany).

15. EDC solution: 6.5 mM 1-ethyl-3-(3-diethylaminopropyl)-carbodiimide (Sigma, Deisenhofen, Germany) in H_2O.

16. Hydroxybenzotriazole (Novabiochem, Bad Soden, Germany).

17. Instant Scint-Gel Plus (Packard, Meriden, CT).

18. 1-Methyl-2-pyrrolidone (NMP, Fluka, Neu-Ulm, Germany).

19. *N*-α-carboanhydride (Neosystems, Strasbourg, France).

20. PBS: 8 mM Na_2HPO_4, 2 mM NaH_2PO_4, 140 mM NaCl, pH 7.2.

21. Piperidine solution: 20% piperidine (Fluka, Neu-Ulm, Germany) in DMF.

22. SDS-polyacrylamide gel (SDS-PAGE): The separating gel consists of 15% acrylamide–*bis*-acrylamide (29:1, Pharmacia Biotech, Uppsala, Sweden); 0.375 M Tris-HCl, pH 8.8, 2 mM Na_2EDTA; 0.1% SDS. The stacking gel consists of 5% acrylamide–*bis*-acrylamid, 0.125 M Tris-HCl, pH 6.8, 2 mM Na_2EDTA, 0.1% SDS. To induce polymerization of the solutions 0.1% ammonium persulfate (Sigma, Deisenhofen, Germany) and 0.0625% *N,N,N',N'*-tetramethylethylendiamine (TEMED, Sigma, Deisenhofen, Germany) is added. The gels are run in 25 mM Tris-base, 192 mM glycine, 0.1% SDS at pH 8.3.

23. Staining solution (AP): 5 mM $MgCl_2$, 0.4% bromo-4-chloro-3-indolylphosphate-toluidine salt (BCIP, Biomol, Hamburg, Germany), 0.6% 3-[4,5-dimethylthiazol-2-yl]-2,5-diphenyl-tetrazolium bromide (MTT, Sigma, Deisenhofen, Germany) in CBS.

24. Staining solution (AP2): 100 mM NaCl, 50 mM $MgCl_2$, 100 mM Tris-base, 0.3% nitroblue tetrazolium (NBT, Sigma, Deisenhofen, Germany), 0.3% BCIP.

25. Staining solution-AP: 1 mg/mL of *p*-nitrophenyl phosphate (Sigma, Deisenhofen, Germany) in diethanolamine buffer.

26. Stock solution BCIP: 6% BCIP in DMF (store at –20°C).

27. Stock solution MTT: 5% MTT in DMF (70%)–H_2O (30%) (store at –20°C).

28. Stock solution NBT: 5% NBT in 70% DMF, 30% H_2O (store at 4°C).

29. TBS: 0.8% NaCl, 0.02% KCl, 50 mM Tris-HCl, pH 7.0.

30. T-TBS: 0.05% Tween-20 in TBS.

31. TFE-cleavage solution: 45% $CHCl_2$, 3% triisobutylsilane (TIBS, Aldrich, Steinheim, Germany), 50% trifluoroacetic acid (TFA, Fluka, Neu-Ulm, Germany), 2% H_2O.

32. Wash buffer: 0.06% Brij 35 solution (Sigma, Deisenhofen, Germany) in PBS.

2.3. Special Reagents

1. Pertussis toxin was obtained as a kind gift from the Institute Pasteur Mérieux Connaught (Lyon, France).

2. Transducin and α-G_i subunits were obtained from P. Gierschick (Institute for Pharmacology and Toxicology, University of Ulm, Germany).
3. [^{32}P]NAD (1000 Ci/mmol, Amersham Buchler, Braunschweig, Germany).
4. Activated and protected amino acid derivatives were from Novabiochem (Bad Soden, Germany) or Neosystems (Strasbourg, France). The following amino acid derivatives are used: moc-Alα-OPfp-ester, Fmoc-Cys (Acm)-OPfp-ester, Fmoc-Asp (OtBu)-OPfp-ester, Fmoc-Glu (OtBu)-OPfp-ester, Fmoc-Phe-OPfp-ester, Fmoc-Gly-OH, Fmoc-His (Trt)-OPfp-ester, Fmoc-Ile-NCA, Fmoc-Lys (Boc)-OPfp-ester, Fmoc-Leu-NCA, Fmoc-Met-OPfp-ester, Fmoc-Asn (Trt)-OPfp-ester, Fmoc-Pro-OPfp-ester, Fmoc-Gln (Trt)-OPfp-ester, Fmoc-Arg-(Pmc)OH, Fmoc-Ser(tBu)-ODhbt, Fmoc-Thr-OPfp-ODhbt, Fmoc-Val-NCA, Fmoc-Trp (Boc)-OPfp-ester, Fmoc-Tyr (tBu)-OPfp-ester (*see* **Note 1**).

3. Methods

3.1. Peptide Synthesis on Cellulose Membranes

Although the synthesis of peptides on cellulose membranes has been modified and optimized to be used with a pipetting robot (**Fig. 1**, *see* **Notes 2, 3, and 4**) the principal reactions involved are essentially performed as developed by Frank *(18)*.

In the following paragraphs the first appearance of particular solutions and reagents described in **Subheading 2.** is indicated in *italics*.

3.1.1. Introduction of Free Amino Functions in Cellulose Membranes by Derivatization with β-Alanine

For the introduction of free amino groups in cellulose membranes the hydroxyl groups of cellulose are esterified with β-alanines to provide nonimmunogenic linear spacers serving as anchors for the growing peptide chains. The reaction has to take place under absolutely anhydrous conditions (*see* **Note 5**). Thus, the generated water has to be constantly removed to shift the equilibrium towards the synthesis of the esters and—after completion of the coupling step—to avoid their hydrolysis.

1. For this the cellulose membranes (Whatman 540) are cut to a convenient "postcard" size of 9×13 cm and are dried with silica gel for 24 h in a desiccator.
2. For derivatization the filters are incubated for 3 h at 20–22°C in an air-tight glass Petri dish with 3 mL *derivatizing solution* for every 9×13 cm cellulose sheet.

3.1.1.1. Efficiency of the Reaction Step

1. To control success and efficiency of the derivatization about 1 cm^2 of cellulose is carried along in the reaction. After 2 h the small piece of cellulose is removed and transferred to a glass vial with seal.
2. The cellulose filter is treated 4× with 2 mL of DMF for 1 min with shaking.

3. Subsequently, for deprotection of the amino groups the cellulose is incubated for 5 min with shaking in 1 mL of *piperidine solution* and again washed four times with 2 mL of DMF.
4. To assay for the introduction of free amino groups 50 μL of *BPB solution* (bromphenol blue solution) are added to 2 mL of DMF and the control filter is incubated until a deep blue color develops.
5. The cellulose sheet is air dried and after transfer to a new glass vial completely decolorized with 2 mL of piperidine solution.
6. For quantification 100 μL of the blue supernatant are added to 2 mL of piperidine solution. The optical density ($\lambda = 605$ nm) of this solution is determined using piperidine solution as control. The derivatization of the cellulose with free amino groups is calculated according to Lambert–Beer's law ($E = \varepsilon c d$) using the extinction coefficient for bromphenol blue of $\varepsilon_{605} = 95{,}000$ mol^{-1} cm^{-1}.

Thus, the derivatization with amino groups is calculated according to:

$$\text{amino group [mmol/cm}^2] = \frac{\text{OD}_{605\,\text{nm}} \times \text{factor of dilution} \times \text{volume of decolorizing solution (mL)}}{95 \times \text{area of cellulose filter (cm}^2)}$$

The synthesis should be continued only if the Whatman 540 cellulose filter has been derivatized with amino groups with a minimum density of 0.2 μmol/cm^2.

3.1.1.2. DEPROTECTION OF β-ALANINE AMINO GROUPS

1. After 3 h the main filter is washed 3× (30 s, 2 × 2 min) to remove any excess of the derivatizing solution.
2. For deprotection of the β-alanine amino group the filter is incubated for 20 min in piperidine solution. The piperidine solution is removed by washing 4× (30 s, 3 × 2 min) with 15 mL of DMF followed by subsequent dehydration with 3× 15 mL EtOH (30 s, 2 × 2 min).
3. Afterwards the filter is placed between sheets of Whatman 3MM filters, additionally air dried with the help of a hair dryer (cold!), and dried further by silica gel incubation in a desiccator for 18 h. At this stage the cellulose filter should be completely derivatized with β-alanine.

3.1.1.3. INTRODUCING A SECOND β-ALANINE AS A LINEAR SPACER

For the introduction of a second β-alanine it is now advisable to select a pattern for further peptide synthesis.

1. For this the orientation of the cellulose filter in the pipetting robot has to be marked unequivocally with a soft pencil and fixed.
2. For synthesis of up to 425 different peptide sequences with the help of the robot 0.2 μL of a freshly made 0.3 *M Fmoc-β-ala-OBt solution* is spotted in a pattern of 17 × 25 spots on the cellulose membrane. After 10 min of incubation the procedure is repeated.
3. To allow for sufficient reaction time the filter is now left for about 45 min.

4. Progress of the coupling reaction can be estimated by the color change of the bromphenol blue contained in the amino acid solution observed at each spot. Immediately following the application of the activated amino acid derivative the spot appears dark blue, indicating free amino functions. The dark blue color changes to greenish yellow owing to the coupling reaction and the disappearance of free amino functions indicating the successful introduction of a second β-alanine residue.

3.1.1.4. BLOCKING OF FREE UNDERIVATIZED AMINO GROUPS

As the next amino acid should be coupled to the second β-alanine moiety the underivatized amino groups have to be blocked with acetic anhydride. The capping step after each coupling reaction ensures that in the case of incomplete coupling only shortened peptide sequences might originate and thus the generation of deletions in the amino acid sequence is prohibited.

1. For a complete capping of residual free amino groups the cellulose sheet is thoroughly rinsed (2×30 s; 1×1 h) in always freshly prepared *capping solution* (15 mL).
2. Subsequently the filter is washed in 15 mL of DMF (2×30 s; 1×2 min) followed by rinsing in EtOH (15 mL, 2×30 s; 1×2 min), drying between Whatman 3MM, and additional storage for 18 h in a desiccator.
3. The activated and with two β-alanine residues as linear spacers modified membranes are now ready for the subsequent synthesis of specific amino acid sequences in the preselected pattern (**Fig. 2**). Alternatively, at this point the filters can also be saved for future use. For this they should be dried further under high vacuum, sealed, and stored at −20 °C.

3.1.2. Synthesis of Anchored Peptide Sequences

3.1.2.1. DEPROTECTION OF SPACER ACTIVATED CELLULOSE MEMBRANES.

1. For removal of protecting groups the filter is transferred to a glass, Teflon, or polypropylene dish and incubated for 5 min with 15 mL of piperidine solution before being rinsed 4× with 15 mL of DMF (1×30 s, 3×2 min).
2. After treatment with EtOH (15 mL; 30 s, 2×2 min) the cellulose membrane is ready for further coupling steps.

3.1.2.2. PREPARATION OF ACTIVATED AMINO ACIDS

The amino acids are usually employed as Fmoc-amino acids with their particular side chain protecting groups as listed in **Subheading 2.**

1. The amount of the respective amino acid derivatives needed for 1 mL is weighed into a 2-mL reaction vial (Eppendorf tube), supplemented with the corresponding amount of 1-hydroxybenzotriazole (HOBt; to all OPfp-ester derivatives 10 mg/mL HOBt are added; no HOBt is added to the NCA derivatives; 70 mg/mL

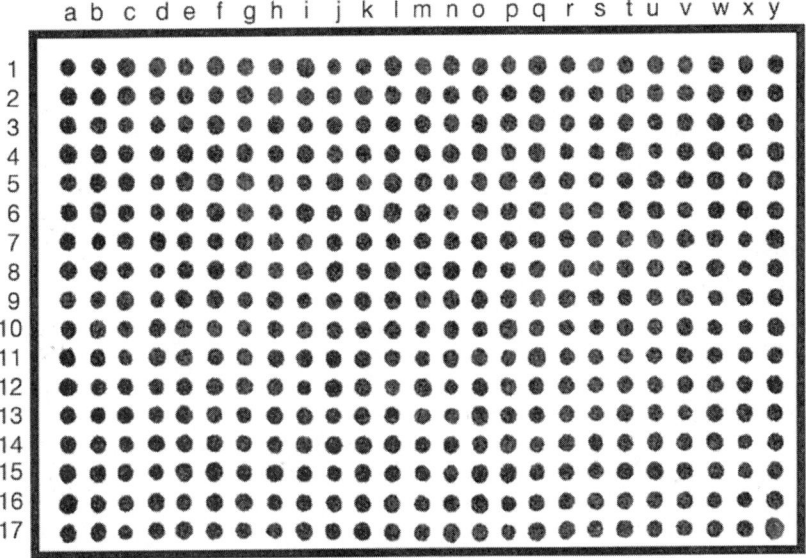

Fig. 2. Cellulose filter (9 × 13 cm) derivatized in the preselected 17 × 25 pattern with β-alanine dimers for activation.

HOBt are added to Fmoc-Ile-OH and Fmoc-Arg-(Pmc)OH, and solubilized in 1 mL of NMP (1-methyl-2-pyrrolidon).

2. Dissolved amino acid derivates are shock-frozen in 250-μL aliquots under a nitrogen atmosphere and stored at –70°C until further use.

3. Aliquots of 250 μL of the respective amino acid solutions needed on the day of synthesis are taken from storage.

4. A 7 μL aliquot of DICD is added to the Pfp-ester amino acids and free amino acids, thoroughly mixed, and subsequently centrifuged to remove any fines or amino acid crystals that might be present in the solution (to avoid potential plugging of the pipetting blunt needle). After an activation time of 45 min the amino acid solutions can be used for the synthesis.

3.1.2.3. SYNTHESIS CYCLES

The synthesis of the amino acid sequence proceeds following a repetitive cycle of the following steps:

1. The freshly deprotected filter is transferred to the pipetting robot and fixed.

2. For each amino acid coupling cycle the specific activated amino acid ester is spotted 2× in 0.2 μL aliquots. The reaction is allowed to proceed for 10 min each between the spotting cycles.

3. The filter is placed in a DMF-resistant tray.

All following steps are carried out at 20–22°C with continuous swirling of the cellulose membrane.

4. Blocking of the residual nonsaturated amino acid groups is achieved by incubation with 15 mL of capping solution (1 × 30 s; 1 × 10 min).

5. After the filters are washed 3× in 15 mL of DMF (1 × 30 s; 2 × 2 min) the Fmoc protecting groups of the newly added amino acids are removed by incubation for 10 min in piperidine solution.

6. The piperidine is removed by washing four times with 15 mL of DMF of each (1 × 30 s; 3 × 2 min).

7. The deprotected amino groups are visualized by incubation in BPB solution (1:100 diluted in DMF).

8. Rinsing 3× with EtOH removes the excess of BPB (1 × 30 s; 2 × 2 min).

9. The filter is air dried as before and returned to the pipetting robot to start the next round of amino acids in the coupling cycle. Peptide synthesis proceeds along the steps as just described.

10. In the last cycle, after the reaction with the last activated amino acid has taken place, the filter is washed directly 3× with 15 mL of DMF each (1 × 30 s; 2 × 2 min) followed by deprotection with piperidine solution, washing, and staining with BPB as described previously.

11. For the final capping step the filters are incubated two times in 15 mL of capping solution each (1× 30 s; 1 × 10 min) followed by washing with 3 × 15 mL DMF (1 × 30 s; 2 × 2 min) and a subsequent rinse in EtOH.

12. Before the activity of the synthesized peptides can be assessed the respective side chain protection groups have to be cleaved. For this 20 mL of TFE solution per filter are given in a tray, which must be sealed air tight to prevent evaporation of the dichloromethane in the TFE solution.

13. The filter is placed into the tray, which is immediately closed and incubated for 1 h with constant shaking.

14. Then the TFE solution is replaced by a fresh 20-mL TFE solution and the incubation is continued for 1 h.

15. Following the cleavage reaction the filter is rinsed 4× with dichloromethane (1 × 30 s; 3 × 2 min), followed by 3× incubation in 15 mL DMF (1 × 30 s; 2 × 2 min) to remove the TFE solution (*see* Note 6).

16. After rinsing 3× in EtOH (1 × 30 s; 2 × 2 min) the filters are now ready to be directly used for the respective experiments or, alternatively, at this stadium can be further dried as described previously, sealed under vacuum, and stored at –20°C.

3.2. ADP-Ribosylation of C-Terminally Anchored Synthetic Peptides

3.2.1. Activation of Pertussis Toxin

For ADP-ribosylation pertussis toxin (PT) has to be activated by reduction of the disulfide bridge which keeps the S1 subunit (ADP-ribosyltransferase) in its inactive form.

This is performed by incubation of PT in 50 mM dithiothreitol (DTT), 100 mM Tris-HCl at pH 8.0 for 1 h at 20–22°C.

3.2.2. ADP-Ribosylation Reaction of Control Proteins

1. For ADP-ribosylation of control proteins, such as, for example, transducin or isolated α-G$_i$ subunits, 0.1 µg of transducin are incubated in 50 µL of 100 mM Tris-HCl, pH 8.0, containing 25 mM DTT, 2 mM ATP, 8 µg/mL of activated pertussis toxin, and 10 µCi/mL of [^{32}P]NAD (approx 10^5 cpm) for 90 min at 20–22°C.
2. For the quantification of ADP-ribosylation after the addition of 50 µL of SDS-PAGE sample buffer the reaction mixture is heated to 95°C for 10 min.
3. Proteins are subsequently separated by SDS-PAGE according to Laemmli *(20)*.
4. The separated proteins are visualized by staining with Coomassie Brilliant Blue staining solution.
5. The incorporation of ^{32}P-labeled ADP-ribose into transducin or other α-G$_i$ subunits is demonstrated by exposing the SDS-PAGE for 30 min to a Bioimaging plate (e.g., Fuji BAS IIIs) which is evaluated by scanning using, for example, a Fuji BAS 1000 Bioimager. Alternatively the gel can also be exposed to an X-ray film.
6. For a faster analysis the reaction is stopped by the addition of 2 mL of binding buffer and the proteins in the mixture are sampled on a nitrocellulose filter.
7. The filter is rinsed again by suction with binding buffer (3×) to remove any residual of [^{32}]NAD and unbound protein.
8. After transfer to a scintillation vial the filter is dissolved by the addition of 1 mL of ethylene glycol monoethyl ether, 3 mL of "Instant Scint-Gel Plus" are added, and the incorporated radioactivity is measured.

3.2.3. ADP-Ribosylation of C-Terminally Anchored Synthetic Peptides

1. For the PT-mediated ADP-ribosylation of C-terminally anchored synthetic peptides the 9 × 13 cm cellulose membranes spiked with peptides are first soaked in EtOH and then incubated 3× in 15 mL of TBS buffer, pH 7.0.
2. Then the filters are incubated for 2 h in ADP-ribosylation mix containing 0.9 µg/mL of activated PT; 25 mM DTT; 100 mM Tris, pH 7.0; 2 mM ATP; and 10 µCi/mL of [^{32}]P-NAD.
3. Subsequently, the membranes are rinsed 3× for 2 min in 15 mL 5% Tween–TBS at 20–22°C, followed by incubation at 60°C in 15 mL of 5% Tween–TBS.
4. The filters are rinsed in water until no residual radioactivity can be detected in the wash.
5. For the quantitative determination of bound radioactivity the filters are exposed up to 1.5 h on an imaging plate (e.g., Fuji BAS IIIs) and scanned by a bioimager (e.g., Fuji BAS 1000).

3.3. Positional Effects on the Pertussis Toxin-Mediated ADP-Ribosylation of α-G$_{i3}$ C-Terminal Amino Acid Sequences

To demonstrate the potential of the technique described the influence of positional amino acid exchanges on the efficiency of ADP-ribosylation is assessed

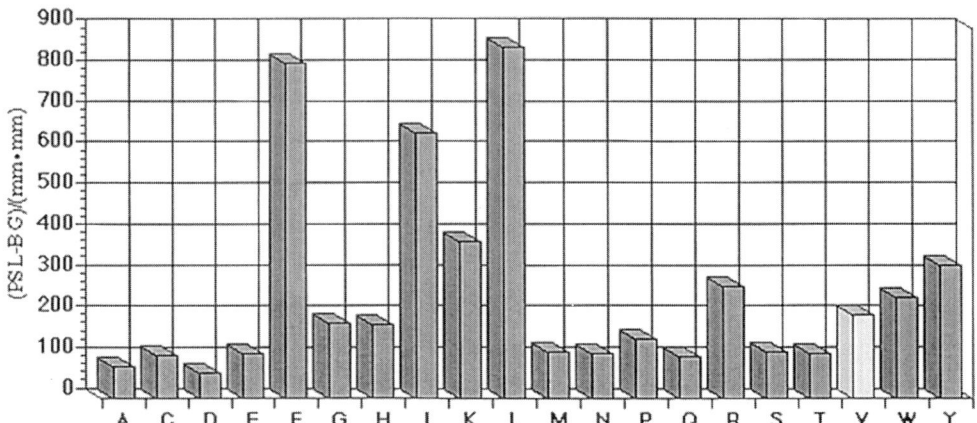

Fig. 3. Example of the positional effect of amino acid exchanges on the efficiency of pertussis toxin mediated ADP-ribosylation of the C-terminus of the α-G$_{i3}$ protein. The authentic amino acid residue at this position (Val) is *shaded lighter gray*. PSL-BG: The bioimager read-out is given in arbitrary units minus background.

employing the C-terminal amino acid sequence of the α-G$_{i3}$ protein as a template.

1. Synthetic peptides incorporating exchanges at every position with each of the natural 20 amino acids are synthesized as 16-mers as described above in **Subheading 3.1.**
2. After incubation with activated PT and [^{32}P]NAD the incorporated radioactivity is measured.

During these experiments the –4 cysteine as the sole target amino acid for PT-mediated ADP-ribosylation was confirmed. Moreover, the influence of peptide length as well as spacer composition on the efficiency of ADP-ribosylation was investigated. As shown in **Fig. 3** for the amino acid exchanges at position –16, which in the authentic sequence is occupied by a valine residue, incorporation of different amino acids might have positive as well as negative effects on ADP-ribosylation efficiencies.

The synthesis of C-terminally anchored peptides on cellulose membranes allows the elucidation and dissection of substrate requirements of enzymatic reactions as shown here for the PT-mediated ADP-ribosylation of peptide sequences modeled onto the α-G$_{i3}$ protein C-terminus. The major advantage of this technique is the parallel cost-efficient synthesis of several hundred positionally addressable synthetic peptides that can be assessed and compared under identical conditions for their activity in a given protein–protein interaction. Depending on the components involved elucidation of these protein–protein interactions might lead to epitope mapping, the identification of binding motifs, or—as indicated here—to the dissection of acceptor sites. Furthermore, this technique might also be very helpful in identifying inhibitory peptide sequences.

4. Notes

Although the synthesis of peptides on cellulose membranes can in principle also be performed manually the potential of the technique with respect to reliability, versatility, speed, ease of handling, and cost efficiency is largely dependent on having access to a more or less automated pipetting robot. Although the chemistry of the reactions worked out in principle by Frank *(18)* usually does not present a problem, the pipetting robot, however, sometimes does. Thus, in employing a pipetting robot several important parameters related to the specific chemistry should be taken care of.

1. Viscosity of the amino acid derivatives in solution: In every coupling step the amino acid esters are used in 0.3 *M* concentrations in 1-methyl-2-pyrrolidone, a concentration resulting in a highly viscous solution. Thus, the synthesis should take place at 20–22°C; otherwise the amino acid derivatives might crystallize, thereby plugging the spotting needle owing to evaporation of solvent (temperature too high) or reduced solubility (temperature too low).
2. Spotting efficiency: On contact of the blunt needle with the cellulose membrane two opposing capillary forces come into play: first, the capillary forces due to the very small inner diameter of the blunt needle necessary for the pipetting accuracy of 0.1 µL and second, the capillary forces of the cellulose membranes. These forces have to be exactly equivalent to allow for accurate and reproducible application of the reagents. The problem can be solved by replacing the normal tubing with a pressure-resistant variety and, furthermore, by replacing the Teflon tubing at the end of the blunt needle with a conical steel tip, thereby reducing the inner diameter of the blunt needle even further and resulting in an increase in capillary force.
3. Reproducible filter placement on the filter plane: Most automated pipetting robots are able to address an arbitrary point in a *xyz* frame with an accuracy of 0.1 mm. Owing to the chemical reactions during the synthesis the size of the cellulose filter membrane varies slightly because of shrinking and expansion processes. However, it turned out that the shrinking process after the basic derivatization with the two β-alanines is negligible so that the filters could be fixed and held in position for the subsequent steps with small needles.
4. Loss of reagent solution due to filter plane material: To drive every coupling step to completion it is mandatory that during the reaction time the total volume of 0.1–0.2 µL of the applied reagent is concentrated and remains at the activated spot on the filter. If the filter support is made of stainless steel, as is often the

case, the plate will be wetted by the reagent solution and so the volume and reagent concentration available for the reaction will be reduced. This problem can be overcome by replacing the stainless steel filter support with a Teflon plate. The points where the robot had to be improved are indicated by arrows in **Fig. 1**.

5. Potential disturbances due to the reagents employed: As water has to be removed from the reaction to drive the coupling steps to completion, every source of contaminating water has a devastating influence on the yield of the correct product. Thus, the quality of the solvents used and special care in handling the reagents are crucial. Contaminating primary and secondary amines in solvents such as DMF, NMP, etc. can cause side reactions and reduce coupling efficiencies dramatically. Therefore, water-free solvents and reagents of the highest available quality should be used. The cellulose filters should be thoroughly dried before use. Solvents such as DMF or NMP should be tested by the addition of bromphenol blue (color reaction to yellow or blue indicates the presence of contaminating amines).

6. Cleavage of Fmoc-Trp(Boc) protective groups: If during the synthesis amino acid derivatives with Fmoc-Trp-(Boc) protective groups have been incorporated they have to be cleaved by incubation of the filter in 15 mL of 1 M acetic acid (1×30 s; 2×5 min).

Acknowledgment

The work described was supported in part by Grant O1KI-8206 of the Bundesministerium für Bildung, Forschung, Wissenschaft und Technologie (BMBF) and by Grant SCHM 770/8-1 of the Deutsche Forschungsgemeinschaft (DFG).

References

1. Munoz, J. J. (1985) Biological activities of pertussigen (pertussis toxin), in *Pertussis Toxin* (Sekura, R. D., Moss, J., and Vaughan, M., eds.), Academic Press, London, pp. 1–18.

2. Weiss, A. (1997) Mucosal immune defenses and the response of *Bordetella pertussis*. *ASM News* **63,** 22–28.

3. Kaziro, Y., Itoh, H., Kozasa, T., Nakafuku, M., and Satoh, T. (1991) Structure and function of signal-transducing GTP-binding proteins. *Annu. Rev. Biochem.* **60,** 349–400.

4. West, R. E., Jr., Moss, J., Vaughan, M., Liu, T., and Liu, T.-Y. (1985) Pertussis toxin-catalyzed ADP-ribosylation of transducin. Cysteine 347 is the ADP-ribose acceptor site. *J. Biol. Chem.* **260,** 14,428–14,430.

5. Ui, M. (1990) Pertusis toxin as a valuable probe for G-protein involvement in signal transduction, in *ADP-Ribosylating Toxins and G-proteins: Insights into Signal Transduction* (Moss, J. and Vaughan, M., eds.), American Society for Microbiology, Washington, D.C., pp. 45–77.

6. Krueger, K. M. and Barbieri, J. T. (1995) The family of bacterial ADP-ribosylating exotoxins. *Clin. Microbiol. Rev.* **8,** 34–47.
7. Gierschick, P. (1992) ADP-ribosylation of signal-transducing guanine nucleotide-binding proteins by pertussis toxin. *Curr. Top. Microbiol. Immunol.* **175,** 69–96.
8. Dratz, E. A., Furstenau, J. E., Lambert, C. G., Thireault, D. L., Rarick, H., Schepers, T., Pakhlevaniants, S., and Hamm, H. E. (1993) NMR structure of a receptor-bound G-protein peptide. *Nature* **363,** 276–281.
9. Hamm, H. E., Deretic, D., Arendt, A., Hargrave, P. A., Koenig, B., and Hoffmann, K. P. (1988) Site of G protein binding to rhodopsin mapped with synthetic peptides from the α subunit. *Science* **241,** 832–835.
10. Price, S. R., Barber, A., and Moss, J. (1990) Structure-function relationships of guanine nucleotide-binding proteins, in *ADP-Ribosylating Toxins and G-proteins: Insights into Signal Transduction* (Moss, J. and Vaughan, M., eds.), American Society for Microbiology, Washington, D.C., pp. 397–424.
11. Osawa, S., Dhanasekaran, N., Woon, C. W., and Johnson, G. L. (1990) Gα$_i$-Gα$_s$ chimeras define the function of α chain domains in control of G protein activation and βγ subunit complex interactions. *Cell* **63,** 697–706.
12. Freissmuth, M. and Gilman, A. G. (1989) Mutations of G$_S$ α designed to alter the reactivity of the protein with bacterial toxins. Substitutions at ARG187 result in loss of GTPase activity. *J. Biol. Chem.* **264,** 21,907–21,914.
13. Neer, E. J., Pulsifer, L., and Wolf, L. G. (1988) The amino terminus of G protein α subunits is required for interaction with βγ. *J. Biol. Chem.* **263,** 8996–9000.
14. Avigan, J., Murtagh, J. J., Jr., Stevens, L. A., Angus, C. W., Moss, J., and Vaughan, M. (1992) Pertussis toxin-catalyzed ADP-ribosylation of G(o) alpha with mutations at the caboxyl terminus. *Biochemistry* **31,** 7736–7740.
15. Rothbard, J. B., Fernandez, R., Wang, L., Teng, N. H. H., and Schoolnik, G. K. (1985) Antibodies to peptides corresponding to a conserved sequence of gono-coccal pilins block bacterial adhesion. *Proc. Natl. Acad. Sci. USA* **82,** 915–919.
16. Arnon, R. and van Regenmortel, M. H. V. (1992) Structural basis of antigenic specificity and design of new vaccines. *FASEB J.* **6,** 3265–3274.
17. Graf, R., Codina, J., and Birnbaumer, L. (1992) Peptide inhibitors of ADP-ribosylation by pertussis toxin are substrates with affinities comparable to those of the trimeric GTP-binding proteins. *Mol. Pharmacol.* **42,** 760–764.
18. Frank, R. (1992) Spot-synthesis: an easy technique for the positionally address-able, parallel chemical synthesis on a membrane support. *Tetrahedron* **48,** 9217–9232.
19. Strutzberg, K., von Olleschik, L., Franz, B., Pyne, C., Schmidt, M. A., and Gerlach, G. F. (1995) Mapping of functional regions on the transferrin-binding protein (TfbA) of *Actinobacillus pleuropneumoniae*. *Infect. Immun.* **63,** 3846–3850.
20. Laemmli, U. K. (1970) Cleavage of structural proteins during the assembly of the head of the bacteriophage T 4. *Nature* **227,** 680–685.
21. von Olleschik-Elbheim, L. (1999) Dissertation, University of Münster.

13

Phage Libraries for Generation
of Anti-Botulinum scFv Antibodies

Peter Amersdorfer and James D. Marks

1. Introduction

The biotechnological generation of high affinity monoclonal antibodies (MAbs) has traditionally involved the production of hybridomas from spleen cells of immunized animals (1). This event, together with availability of increasingly sophisticated molecular biology and protein engineering techniques, opened up the field of numerous applications and benefits in not only the medical but also industrial world. Now the use of phage antibodies offers a new route for the generation of antibodies, including antibodies of human origin, which cannot be easily obtained by conventional hybridoma technology. Recent advances in the expression of antibody fragments in E. coli (2,3) and the application of the polymerase chain reaction (4) for cloning of immunoglobulin DNA (5,6) have mainly contributed to these achievements. With phage display, antibodies can be made completely in vitro, bypassing the immune system and the immunization procedures.

To display antibody fragments on the surface of phage (phage display), an antibody fragment gene is inserted into the gene encoding a phage surface protein, resulting in expression of the antibody fragment on the phage surface (**Fig. 1**). Although fusions have been made with the major phage coat protein pVIII (7,8), the most widely used vectors result in fusion with the minor coat protein pIII (9–13). There are three to five copies of pIII per phage. In phage vectors (9), each pIII exists as an antibody fusion (**Fig. 1**), whereas in phagemid vectors (10,13) there is typically only one fusion per phage owing to competition with wild-type pIII provided by helper phage. Advantages of phagemid systems, the most widely used vectors, include higher transformation efficien-

From: *Methods in Molecular Biology, vol. 145: Bacterial Toxins: Methods and Protocols*
Edited by: O. Holst © Humana Press Inc., Totowa, NJ

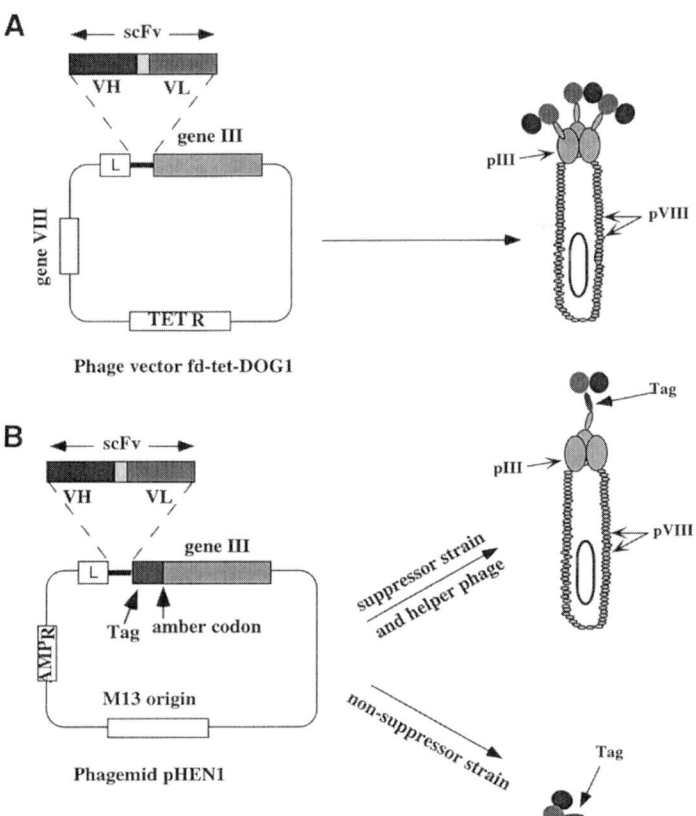

Fig. 1. Examples of different types of vectors for phage display of scFv antibody fragments. Insertion of an scFv gene in frame with bacteriophage gene III results in expression of a scFv–pIII fusion protein which is incorporated into the phage surface. The gene encoding the scFv is contained within the phage. Thus the phage serves as a vehicle that physically links genotype with phenotype. (**A**) In phage vectors, three to five copies of the scFv fragment are displayed (one on each of the pIII molecules). (**B**) In phagemid vector, fewer than three copies are displayed because pIII from the helper phage competes with the fusion protein for incorporation into the phage surface. Phagemid vectors can contain an amber codon between the antibody fragment gene and gene III. This makes it possible to easily switch between displayed and soluble antibody fragment simply by changing the host bacterial strain. L, leader exon, V_H; heavy variable region, V_L, light variable region; gene III, encoding the minor coat protein pIII, gene VIII; encoding the major coat protein pVIII, AMP$_R$, ampicillin resistance gene; TET$_R$; tetracycline resistance gene. Reproduced from **ref.** (*34*), with permission.

cies and better affinity discrimination owing to monovalent display and the absence of an avidity effect.

Because the antibody fragments on the surface of the phage are functional, phage bearing antigen binding antibody fragments can be separated from nonbinding or lower affinity phage by antigen affinity chromatography *(9)*. Mixtures of phage are allowed to bind to the affinity matrix, nonbinding or lower affinity phage are removed by washing, and bound phage are eluted by treatment with acid or alkali (**Fig. 2**). Depending on the affinity of the antibody fragment, enrichment factors of 20-fold–1,000,000-fold are obtained for a single round of affinity selection. By infecting bacteria with the eluted phage, however, more phage can be grown and subjected to another round of selection (**Fig. 2**). In this way, an enrichment of 1000-fold in one round can become 1,000,000-fold in two rounds of selection. Thus even when enrichments are low, multiple rounds of affinity selection can lead to the isolation of rare phage and the genetic material contained within which encodes the sequence of the binding antibody *(14)*. The physical linkage between genotype and phenotype provided by phage display makes it possible to test every member of an antibody fragment library for binding to antigen, even with libraries as large as 10^{10} clones *(15–17)*. For example, after multiple rounds of selection on antigen, a binding scFv that occurred with a frequency of only 1/30,000,000 clones could be recovered *(14)*. Introduction of a peptide tag, such as c-myc *(14)* or E-tag *(18)*, allows easy assaying whereas an amber stop codon (UAG) between the antibody sequence and gene III or gene VIII allows the production of phage displayed fragments when grown in supE strains of *E. coli*, or tagged, soluble fragments when grown in nonsuppressor strains where the amber codon is read through as glutamine. Practically, the translation of the amber codon as a glutamine in TG1 (suppressor strain) is only about 70% sufficient, producing antibody–gene III fusion proteins as well as soluble antibody fragments. Inclusion of a hexahistidine tag makes it possible to easily purify native scFv without the need for subcloning *(16)*.

The choice of which type of phage antibody repertoire to use depends very much on the application and the required affinity of the antibody. Ideally, antibodies to any chosen antigen are selected from universal, antigen-unbiased libraries. The first of such single human *V*-gene repertoires was made from the peripheral blood lymphocytes (PBLs) of two healthy human volunteers and contained 3×10^7 clones, from which antibodies to more than 25 different antigens were isolated *(14)*. These included antibodies to self antigens and cell surface markers *(14,19)*. The average affinity of these antibodies was in the 10^6–$10^7/M$ range, which is similar to the affinity of the naive primary immune response. Recently much larger scFv repertoires have been made by

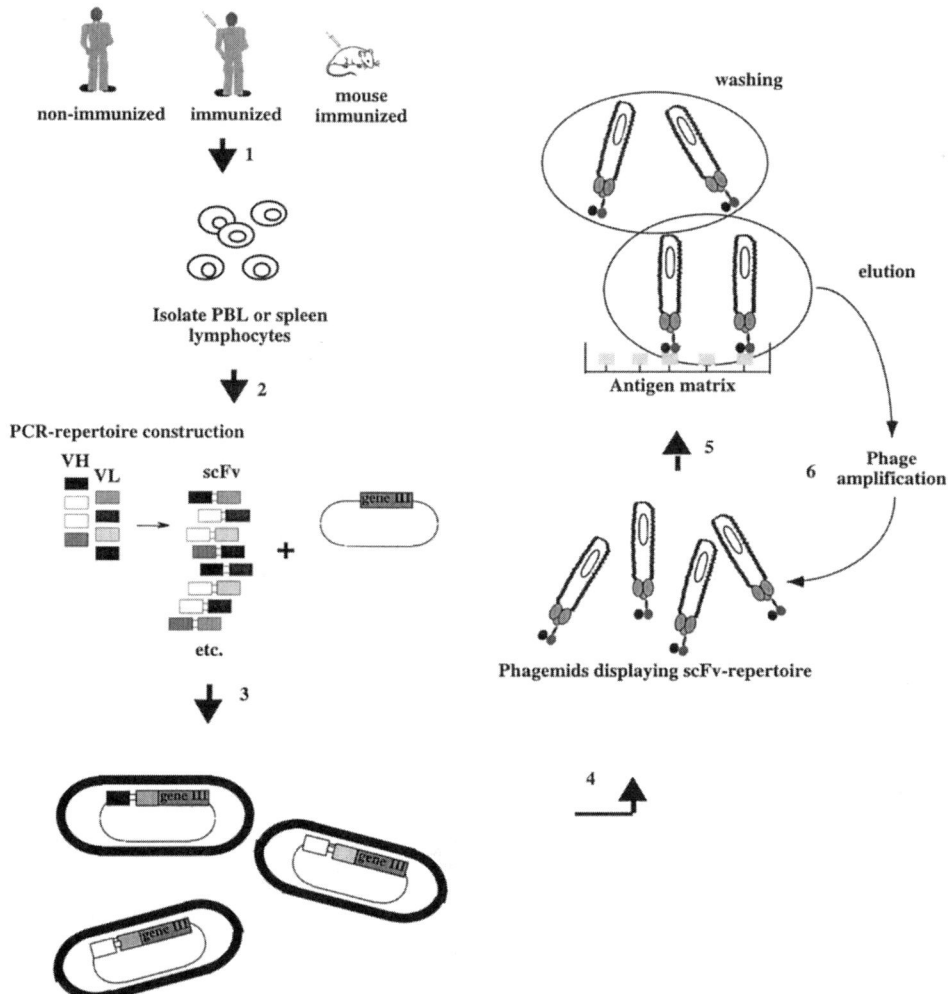

Fig. 2. Overview of the antibody phage display technique. Lymphocytes are iso-lated from mice or humans (**step 1**) and the V_H and V_L genes amplified using PCR (**step 2**). scFv gene repertoires are created by PCR assembly of V_H and V_L genes (**step 2**). The scFv repertoire is cloned into a phage display vector and the ligated DNA used to transform *E. coli* (**step 3**). Phage displaying scFv are prepared from transformed *E. coli* (**step 4**). Phage are incubated with immobilized antigen (**step 5**), non-specifically bound phage removed by washing and antigen specific phage eluted by acid or alkali (**step 5**). Eluted phage are used to infect *E. coli*, to produce more phage for the next round of selection (**step 6**). Repetition of the selection process results in the isolation of antigen specific phage antibodies present at low frequencies. Reproduced from **ref.** *(34)*, with permission.

"brute force cloning" *(16,17)*, with 10^9–10^{10} independent clones, and from these libraries higher affinity antibodies (K_a approx $10^9/M$) have been selected. Larger or more diverse phage antibody libraries should theoretically provide higher affinity antibodies against a greater number of epitopes on all antigens used for selection *(20,21)*. The alternative is to develop and use antigen-biased phage antibody repertoires, using V-genes from either immunized animals (spleen) or humans (PBLs, bone marrow), providing an enriched fraction of the antigen-specific plasma cells with high levels of mRNA *(22)*. For example, human monoclonal scFv or Fab antibody fragments have been isolated from immunized volunteers or infected patients against tetanus toxin *(13)*, botulinum neurotoxin type A *(23)*, HIV-1 gp120 *(24)*, HIV-1 gp41 *(25)*, hepatitis B surface antigen *(26)*, hepatitis C *(27)*, respiratory syncytial virus *(28,29)* and *Haemophilus influenza (30)*.

This chapter focuses on creation and manipulation of single-chain Fv phage antibody libraries using *V*-genes from spleenocytes of immunized mice.

2. Materials
2.1. Total RNA Preparation from Spleen, Tissue, and Tissue Culture Cells

1. Denaturing solution: stock solution: add 17.6 mL of 0.75 *M* sodium citrate, pH 7.0, and 26.4 mL of 10% (w/v) *N*-lauryl sarcosine to 293 mL of water. Add 250 g of guanidine thiocyanate and stir at 60°C to dissolve. Store at 20–22°C for a maximum of 3 mo; for working solution: add 0.35 mL β-mercaptoethanol to 50 mL of stock solution. Store no longer than 1 mo at 20–22°C.
2. Lysis buffer: 5 *M* guanidine monothiocyanate; 10 m*M* ethylenediaminetetraacetic acid (EDTA), 50 m*M* Tris-HCl, pH 7.5, 1 m*M* dithiothreitol (DTT).
3. RNA solubilization buffer: 0.1% sodium dodecyl sulfate (SDS), 1 m*M* EDTA, 10 m*M* Tris-HCl, pH 7.5.
4. Glass Teflon homogenizer.
5. Chloroform–isoamyl alcohol (49:1, v/v).
6. Water-saturated phenol.
7. 100% Isopropanol.
8. Diethylpyrocarbonate (DEPC), 0.2% (v/v).
9. Methacryloxypropyltrimethoxysilane (Silane).
10. DEPC treated water.
11. 2 *M* Sodium acetate, pH 4.0.

2.2. First-Strand cDNA Synthesis

1. AMV reverse transcriptase (Promega) and Vent DNA polymerase (New England Biolabs, NEB).
2. 10× First strand cDNA buffer: 500 m*M* Tris-HCl, pH 8.1, 1.4 *M* KCl, 80 m*M* MgCl$_2$.
3. RNasin (Promega).

4. 20× dNTPs (5 mM each).
5. 0.1 M DTT.
6. 70% Ethanol.
7. DEPC-treated water.
8. PCR-thermocycler.

2.3. Amplification of V$_H$ and V$_L$ genes and Linker Preparation

1. Vent DNA polymerase (NEB) and *Taq* DNA polymerase (Promega).
2. 10× Vent DNA polymerase buffer: 10 mM KCl, 10 mM (NH$_4$)$_2$SO$_4$, 20 mM Tris-HCl, pH 8.8, 0.1% Triton X-100, 20 mM MgSO$_4$.
3. 10× *Taq* DNA polymerase buffer: 500 mM KCl, 100 mM Tris-HCl, pH 9, 0.1% Triton X-100; 15 mM MgCl$_2$.
4. 20× dNTPs (5 mM each).
5. Geneclean (Bio 101).
6. PCR thermocycler.

2.4. PCR-Assembly and Library Construction

1. Water-saturated phenol.
2. *Not*I, *Sfi*I restriction enzymes (NEB).
3. T4 DNA ligase (NEB).
4. Electrocompetent TG1 bacteria (transformation efficiency >10^9/µg of DNA).
5. Diethylether.
6. SOB media: To 950 mL of deionized H$_2$O add: 20 g of Bacto-tryptone (Difco), 5 g of yeast extract (Difco), and 0.5 g of NaCl. Add 10 mL of a 250 mM KCl solution. Adjust the pH to 7.0 with 5 M NaOH and adjust the volume of the solution to 1 L with deionized H$_2$O. Sterilize by autoclaving for 20 min. Just before use add 1 mL of a sterile solution of 1 M MgCl$_2$ and 1 mL 1 M MgS0$_4$ each; otherwise it will precipitate out. SOC media is identical to SOB media except that it contains 20 mM glucose. After the SOB media has been autoclaved, allow it to cool down to 60°C or less and then add 10 mL of a sterile 20% solution of glucose.
7. Minimal media plate: 10.5 g of K$_2$HPO$_4$, 4.5 g of KH$_2$PO$_4$, 1 g of (NH$_4$)$_2$SO$_4$, 0.5 g of sodium citrate · 2H$_2$O, 15 g of agar. Adjust the volume of solution to 1 L with deionized H$_2$O and sterilize by autoclaving for 20 min. Allow the solution to cool to 60° C, then add 1 mL of 1 M MgSO$_4$ · 7 H$_2$O, 0.5 mL of 1% B1 thiamin, and 10 mL of 20% glucose as a carbon source and pour the plates.
8. 2× TY media: To 900 mL of deionized H$_2$O add: 16 g of Bacto-tryptone, 10 g of Bacto-yeast extract, and 5 g of NaCl. Adjust the pH to 7.0 with 5 M NaOH and the volume of the solution to 1 L with deionized H$_2$O. Sterilize by autoclaving for 20 min and allow the solution to cool to 60°C. Add 10 mL of 20% glucose per 100 mL of media and the appropriate antibiotics.
9. 100 µg of 2× TY/mL of ampicillin–2% glucose (2× TY–amp–Glc).
10. TYE plates: For agarose plates add 15 g of Difco agar per liter of 2× TY media. Sterilize by autoclaving for 20 min and allow the solution to cool to 60°C. Add 10 mL of 20% glucose/100 mL of media and the appropriate antibiotic.

11. 100 μg of TYE/mL of ampicillin/2% glucose agarose plates (TYE–amp–Glc).
12. Small agar plates, 100 mm × 15 mm (Fisher).
13. Large agar plates, 150 mm × 15 mm (Fisher).
14. 37°C Incubator (New Brunswick Scientific).
15. Electroporator (Gene Pulser™ Bio-Rad).
16. 16°C Water bath.
17. PCR thermocycler.
18. Phenol–chloroform.

2.5. Rescue of Phage Antibody Libraries and Phage Preparation

1. Spectrophotometer.
2. 100 μg of 2× TY/mL of ampicillin–2% glucose (2× TY–amp–Glc).
3. 100 μg of 2× TY/mL of kanamycin (2× TY–kan).
4. VCSM13 helper phage (Stratagene).
5. 20% Polyethylene glycol (PEG) 8000–2.5 M NaCl.
6. Phosphate-buffered saline (PBS) autoclaved.
7. 0.45-μm syringe filter.

2.6. Selection of Phage Antibody Libraries

1. *E. coli* TG1 grown on a minimal plate.
2. Coating buffer (PBS, pH 7.4 or 100 mM sodium bicarbonate, pH 9.6).
3. Biotinylated antigen.
4. PBS–0.1 % Tween-20.
5. PBS autoclaved.
6. 100 mM triethylamine (make fresh: 140 μL of stock/10 mL of H$_2$O, pH must be 12).
7. 1 M Tris, pH 7.0, autoclaved.
8. PBS and 5% powdered skim milk (MPBS).
9. Streptavidin-magnetic beads (Dynabeads, M–280, Dynal A.S., Oslo, Norway).
10. Avidin-magnetic beads (Controlled Pure Glan, Lincoln Park, NJ, USA).
11. Shaker–incubator (37°C and 25°C).
12. Centrifuge (adapter for microtiter plates).
13. Immunotubes (Nunc).
14. Rotator for immunotubes.
15. Magnetic rack (Dynal MPC-E, cat. no. 120.04) (Dynal AS, N–0212 Oslo, Norway).
16. Biotinylation kid (Pierce, NHS-LC-Biotin, cat. no. 21335) (Pharmacia, recombinant phage selection module).
17. 50% glycerol–5 mM MgSO$_4$, sterile filtered.
18. 2× TY–amp–Glc.

2.7. Identification of Antigen Binding Clones

1. Autoclaved toothpicks.
2. 2× TY–amp–glc and 2× TY–amp–0.1% glc.
3. 1 mM Isopropyl-β-D-thiogalactopyranoside (IPTG), sterile filtered.

4. 50% Glycerol–5 m*M* MgSO$_4$, sterile filtered.
5. Coating buffer (PBS, pH 7.4, or 100 m*M* sodium bicarbonate, pH 9.6).
6. 2% Powdered skim milk in PBS (MPBS).
7. PBS, autoclaved.
8. PBS–0.1% Tween-20.
9. 9E10 Antibody (Santa Cruz Biotechnology, CA).
10. Horseradish peroxidase conjugated anti-mouse IgG Fc monoclonal antibody (Sigma A-2554).
11. Developing solution: dissolve one 10-mg ABTS (2,2'-azino-bis-(3-ethyl-benzthiazoline–6-sulfonic acid) tablet in 10 mL of 50 m*M* citric acid and 10 mL of 50 m*M* sodium citrate.
12. 30% Hydrogen peroxide.
13. 75 m*M* NaF.
14. *BstN*1 restriction enzyme (NEB).
15. *Taq* DNA Polymerase (Promega).
16. 10× *Taq* buffer: 500 m*M* KCl; 100 m*M* Tris-HCl, pH 9.0, 1% Triton X-100, 15 m*M* MgCl$_2$.
17. 96-Well round-bottomed plates, product no. 25850 (Corning).
18. 96-Well flat-bottomed microtiter plates, product no. 3912 (Falcon).
19. 96-Well transfer device.
20. Plate sealer, product no. 3095 (Costar).
21. Shaker–incubator (37°C and 25°C) (New Brunswick Scientific).
22. Enzyme-linked immunosorbent assay (ELISA) plate reader (405 nm).
23. PCR thermocycler.
24. Multichannel-pipet.
25. Centrifuge holder for microtiter plates.
26. Nusieve agarose (3:1) (FMC Bio Products).

3. Methods

3.1 Total RNA Preparation from Spleen, Tissue, and Tissue Culture Cells (Guanidine Method)

This method is essentially described by Chomczynski et al. *(31)* (*see* **Note 1**).

1. For tissue: Add 1 mL of denaturing solution per 100 mg of tissue and homogenize with a few strokes in a glass–Teflon homogenizer. For cultured cells: After centrifuging suspension cells, discard the supernatant and add 1 mL of denaturing solution/10^7 cells and pass the lysate through a pipet 7–10×.
2. Transfer the homogenate to a 5-mL polypropylene tube. Add 0.1 mL of 2 *M* sodium acetate, pH 4.0, and mix thoroughly by inversion. Add 1 mL of water-saturated phenol, mix thoroughly, and add 0.2 mL of 49:1 (v/v) chloroform–isoamyl alcohol.
3. Mix thoroughly and incubate the suspension for 15 min at 4°C.
4. Centrifuge for 20 min at 10,000*g* at 4°C. Transfer the upper aqueous phase (containing RNA) to a new tube; DNA and proteins are left behind in the interphase and phenol layer.

5. Precipitate RNA by adding 1 mL of 100% isopropanol. Incubate the sample for 30 min at –20°C.
6. Centrifuge for 10 min at 10,000g at 4°C and discard supernatant. Dissolve the RNA pellet in 0.3 mL of denaturing solution and transfer to a 1.5-mL microcentrifuge tube.
7. Precipitate the RNA with 0.3 mL of 100% isopropanol for 30 min at –20°C.
8. Centrifuge for 10 min at 10,000g at 4°C and discard the supernatant. Dissolve RNA pellet in 2 mL of 75% ethanol, vortex-mix, and incubate 10–15 min at 20–22°C to dissolve any residual amounts of guanidine, which may be contaminating the pellet.
9. Centrifuge for 5 min at 10,000g at 4°C and discard supernatant. Air dry the RNA pellet.
10. Dissolve the RNA pellet in 100–200 µL of DEPC-treated water and store RNA at –70°C either as an aqueous solution or as an ethanol precipitate (*see* **Note 2**).

3.2. First-Strand cDNA Synthesis

1. For first-strand synthesis use either an immunoglobulin G (IgG) constant region primer (MIgG1–3) for the heavy chain, or a human κ constant-region primers (MC$_κ$ for the light chain aa listed in **Table 1A**. Prepare the following reaction mixes in a pretreated 1.5-mL Eppendorf tube (*see* **Note 1**):

10× first-strand cDNA buffer	5 µL
20× dNTPs (5 mM each)	5 µL
0.1 M DTT	5 µL
MIgG1–3 forward primer (10 pm/µL)	1 µL
or MC$_κ$ forward primer (10 pm/µL)	1 µL

2. Take an aliquot (1–4 µg) of RNA in ethanol, place into a pretreated 1.5-mL Eppendorf tube, and centrifuge 5 min in microcentrifuge. Wash the RNA pellet once with 70% ethanol, air dry in the hood, and resuspend in 28 µL of DEPC-treated water.
3. Heat at 65°C for 5 min to denature, quench on ice for 2 min, and centrifuge immediately in a microcentrifuge for 5 min. Take the supernatant and add to first-strand reaction mixture.
4. Add 2 µL (5000 U/mL) of AMV reverse transcriptase and 2 µL (40.000 U/mL) of RNasin and incubate at 42°C for 1 h.
5. Boil the reaction mixture for 5 min, quench on ice for 2 min, and centrifuge down in a microcentrifuge for 5 min. Take the supernatant and heat at 65°C for 5 min to denature, quench on ice for 2 min, and centrifuge immediately in a microcentrifuge for 5 min. Take the supernatant, containing the first-strand cDNA, and proceed with the primary PCR amplification (*see* **Note 3**).

3.3. Amplifiction of V$_H$ and V$_L$ Genes and Linker Preparation

1. Use for each *V*-gene repertoire an equimolar mixture of the designated forward/back primer sets at a final concentration of 10 pmol/µL total (V_H back–J_H forward, $V_κ$ back-$J_κ$ forward) as described by Amersdorfer et al. *(23)* and in **Table 1B**.

Table 1
Oligonucleotide Primers Used for PCR of Mouse Immunoglobin Genes

A. First strand cDNA synthesis

Mouse heavy chain constant region primers

MIgG1/2 For 5' CTG GAC AGG GAT CCA GAG TTC CA 3'
MIgG3 For   5' CTG GAC AGG GCT CCA TAG TTC CA 3'

Mouse V_κ constant region primer

MC$_\kappa$For 5' CTC ATT CCT GTT GAA GCT CTT GAC 3'

B. Primary PCRs

Mouse V_H back primers

V_H1 Back   5' GAG GTG CAG CTT CAG GAG TCA GG 3'
V_H2 Back   5' GAT GTG CAG CTT CAG GAG TCR GG 3'
V_H3 Back   5' CAG GTG CAG CTG AAG SAG TCA GG 3'
V_H4/6 Back 5' GAG GTY CAG CTG CAR CAR TCT GG 3'
V_H5/9 Back 5' CAG GTY CAR CTG CAG CAG YCT GG 3'
V_H7 Back   5' GAR GTG AAG CTG GTG GAR TCT GG 3'
V_H8 Back   5' GAG GTT CAG CTT CAG CAG TCT GG 3'
V_H10 Back 5' GAA GTG CAG CTG KTG GAG WCT GG 3'
V_H11 Back 5' CAG ATC CAG TTG CTG CAG TCT GG 3'

Mouse V_κ back primers

V_κ1 Back 5' GAC ATT GTG ATG WCA CAG TCT CC 3'
V_κ2 Back 5' GAT GTT KTG ATG ACC CAA ACT CC 3'
V_κ3 Back 5' GAT ATT GTG ATR ACB CAG GCW GC 3'
V_κ4 Back 5' GAC ATT GTG CTG ACM CAR TCT CC 3'
V_κ5 Back 5' SAA AWT GTK CTC ACC CAG TCT CC 3'
V_κ6 Back 5' GAY ATY VWG ATG ACM CAG WCT CC 3'
V_κ7 Back 5' CAA ATT GTT CTC ACC CAG TCT CC 3'
V_κ8 Back 5' TCA TTA TTG CAG GTG CTT GTG GG 3'

Mouse J_H forward primers

J_H1 For 5' TGA GGA GAC GGT GAC CGT GGT CCC 3'
J_H2 For 5' TGA GGA GAC TGT GAG AGT GGT GCC 3'
J_H3 For 5' TGC AGA GAC AGT GAC CAG AGT CCC 3'
J_H4 For 5' TGA GGA GAC GGT GAC TGA GGT TCC 3'

Mouse J_κ forward primers

J_κ1 For 5' TTT GAT TTC CAG CTT GGT GCC TCC 3'
J_κ2 For 5' TTT TAT TTC CAG CTT GGT CCC CCC 3'
J_κ3 For 5' TTT TAT TTC CAG TCT GGT CCC ATC 3'
J_κ4 For 5' TTT TAT TTC CAA CTT TGT CCC CGA 3'
J_κ5 For 5' TTT CAG CTC CAG CTT GGT CCC AGC 3'

C. Reamplification primers containing restriction sites

Mouse V_H Sfi back primers

V_H1 Sfi 5' GTC CTC GCA ACT GCG GCC CAG CCG GCC ATG GCC GAG GTG CAG CTT CAG GAG TCA GG 3'
V_H2 Sfi 5' GTC CTC GCA ACT GCG GCC CAG CCG GCC ATG GCC GAT GTG CAG CTT CAG GAG TCR GG 3'
V_H3 Sfi 5' GTC CTC GCA ACT GCG GCC CAG CCG GCC ATG GCC CAG GTG CAG CTG AAG SAG TCA GG 3'

$V_H4/6$ Sfi	5' GTC CTC GCA ACT GCG GCC CAG CCG GCC ATG GCC GAG GTY CAG CTG CAR CAR TCT GG 3'
$V_H5/9$ Sfi	5' GTC CTC GCA ACT GCG GCC CAG CCG GCC ATG GCC CAG GTY CAR CTG CAG YCT GG 3'
V_H7 Sfi	5' GTC CTC GCA ACT GCG GCC CAG CCG GCC ATG GCC GAR GTG AAG CTG GTG GAR TCT GG 3'
V_H8 Sfi	5' GTC CTC GCA ACT GCG GCC CAG CCG GCC ATG GCC GAG GTT CAG CTT CAG CAG TCT GG 3'
V_H10 Sfi	5' GTC CTC GCA ACT GCG GCC CAG CCG GCC ATG GCC GAA GTG CAG CTG CAG KTG GAG WCT GG 3'
V_H11 Sfi	5' GTC CTC GCA ACT GCG GCC CAG CCG GCC ATG GCC CAG ATC CAG TTG CTG CAG TCT GG 3'

Mouse J_κ Not forward primers

$J_\kappa 1$ Not	5' GAG TCA TTC TCG ACT TGC GGC CGC TTT GAT TTC CAG CTT GGT GCC TCC 3'
$J_\kappa 2$ Not	5' GAG TCA TTC TCG ACT TGC GGC CGC TTT TAT TTC CAG CTT GGT CCC CCC 3'
$J_\kappa 3$ Not	5' GAG TCA TTC TCG ACT TGC GGC CGC TTT TAT TTC CAG TCT GGT CCC ATC 3'
$J_\kappa 4$ Not	5' GAG TCA TTC TCG ACT TGC GGC CGC TTT TAT TTC CAA CTT TGT CCC CGA 3'
$J_\kappa 5$ Not	5' GAG TCA TTC TCG ACT TGC GGC CGC TTT CAG CTC CAG CTT GGT CCC AGC 3'

D. Reverse linker fragment PCR for scFv

Mouse J_H reverse primers

$J_H 1$	5' GGG ACC ACG GTC ACC GTC TCC TCA GGT GG 3'
$J_H 2$	5' GGC ACC ACT CTC ACA GTC TCC TCA GGT GG 3'
$J_H 3$	5' GGG ACT CTG GTC ACT GTC TCT GCA GGT GG 3'
$J_H 4$	5' GGA ACC TCA GTC ACC GTC TCC TCA GGT GG 3'

Mouse V_κ reverse primers

$V_\kappa 1$	5' GGA GAC TGT GWC ATC ACA ATG TCC GAT CCG CC 3'
$V_\kappa 2$	5' GGA GTT TGG GTC ATC AMA ACA TCC GAT CCG CC 3'
$V_\kappa 3$	5' GCW GCC TGH GTY ATC ACA ATA TCC GAT CCG CC 3'
$V_\kappa 4$	5' GGA GAY TGK GTC AGC ACA ATG TCC GAT CCG CC 3'
$V_\kappa 5$	5' GGA GAC TGG GTG AGM ACA WTT TSC GAT CCG CC 3'
$V_\kappa 6$	5' GGA GWC TGK GTC ATC WBR ATR TCC GAT CCG CC 3'
$V_\kappa 7$	5' GGA GAC TGG GTG AGA ACA ATT TGC GAT CCG CC 3'
$V_\kappa 8$	5' CCC ACA AGC ACC TGC AAT AAT GAC GAT CCG CC 3'

R = A/G, Y = C/T, S = G/C, K = G/T, W = A/T, M = A/C, V = C/G/A, B = G/C/T, and H = C/A/T.

Prepare the following reaction mixture, for example, V_H repertoire (*see* **Note 4**):

Water	31.5 µL
10× Vent buffer	5.0 µL
20× dNTPs (5 m*M* each)	2.0 µL
Acetylated BSA (10 mg/mL)	0.5 µL
V_H 1–11 back mix (10 pm/µL)	2.0 µL
J_H 1–4 forward mix (10 pm/µL)	2.0 µL
cDNA Reaction mix	5.0 µL

Back primers are an equimolar mixture of either 11 V_H back or 8 V_κ back primers at a final concentration of 10 pmol/µL total. Forward primers are an equimolar mixture of either 4 J_H or 5 J_κ primers at a final concentration of 10 pmol/µL total, respectively.

2. Overlay with paraffin oil, unless you have a thermocycler with heated lid.
3. Heat to 94°C for 5 min in a PCR thermocycler to denature the DNA. (Hotstart)
4. Add 1.0 µL (2 U) Vent DNA polymerase under the oil (*see* **Note 5**). Cycle 30× to amplify the *V*-genes at 94°C for 1 min, at 60°C for 1 min, and at 72°C for 1 min.
5. Purify the PCR fragments by electrophoresis on a 1.5% agarose gel and extract the ~340-basepair band corresponding to the V_H genes and ~325-basepair band corresponding to the V_κ gene using Geneclean.
6. To generate the linker DNA, 32 separate 50-µL PCR reactions are performed using each of the four reverse J_H primers in combination with the 8 V_κ reverse primers as described in **Table 1D**. The template for the PCR amplification is pSW2scFvD1.3 *(9)* encoding the $(Gly_4Ser)_3$ linker, which is complementary to the 3' ends of the rearranged V_H genes and the 5' end of the rearranged V_κ genes as described by Huston et al. *(32)*.

3.4. Library Construction

1. The purified V_H and V_L genes are combined in a second PCR reaction containing DNA which encodes the scFv linker (93 basepairs). Make up 25 µL of PCR reaction mixes containing (*see* **Note 6**):

Water	7.5 µL
10× Vent buffer	2.5 µL
20× dNTPs (5 m*M* each)	1.0 µL
scFv linker (80 ng)	2.0 µL
V_H repertoire (240 ng)	5.0 µL
V_k repertoire (240 ng)	5.0 µL

2. Overlay with paraffin oil, unless you have a thermocycler with a heated lid.
3. Heat to 94°C for 5 min in a PCR thermocycler.
4. Add 1.0 µL (2 U) of Vent DNA polymerase.
5. Cycle 7× without amplification at 94°C for 1 min and 72°C for 1.5 min to randomly join the fragments.
6. After 7 cycles, hold at 94°C while adding 2 µL of each flanking primer (equimolar mixture of the 11 V_H back primers at a final concentration of 10 pmol/µL

and an equimolar mixture of the 5 J_κ forward primers at a final concentration of 10 pmol/μL).

7. Cycle 30× to amplify the assembled fragments at 94°C for 1 min and 72°C for 2 min.

8. Gel-purify the assembled scFv gene repertoire (approximately 700 basepairs) on a 2% agarose gel and further purify the DNA by Geneclean. Resuspend the product in 25 μL of water.

9. Reamplify the scFv gene repertoire with primers containing the appropriate *Sfi*I and *Not*I sites as described in **Table 1C**. Make up 50 μL of PCR reaction mixes containing:

Water	37 μL
20× dNTPs (5 m*M* each)	2 μL
10× *Taq* buffer	5 μL
Back primers (10 pmol/μL)	2 μL
Forward primers (10 pmol/μL)	2 μL
scFv gene repertoire (10 ng)	1 μL

    Back primers are an equimolar mixture of 1 L V_H back *Sfi*I primers at a final concentration of 10 pmol/μL total. Forward primers are an equimolar mixture of 5 J_κ for *Not*I primers at a final concentration of 10 pmol/μL total.

10. Overlay with mineral oil, unless you have a thermocycler with a heated lid.

11. Heat to 94°C for 5 min in a PCR thermocycler.

12. Add 1.0 μL (5 U) of *Taq* DNA polymerase under the oil.

13. Cycle 30× to amplify the assembled *V*-genes at 94°C for 1 min, at 55°C for 1 min, and at 72°C for 1 min.

14. Gel-purify the assembled scFv gene repertoire (approx 700 basepairs) on a 2% agarose gel and further purify the DNA by Geneclean. Resuspend the product in 25 μL of water (*see* **Note 7**).

15. Make up 200 μL of reaction mix to digest the scFv repertoire with *Sfi*I (*see* **Note 8**):

scFv DNA (1–4 μg)	100 μL
Water	74 μL
10× NEB 2 buffer	20 μL
*Sfi*I (10 U/μL)	6 μL

16. Incubate at 50°C for 18 h

17. Phenol–chloroform extract with 100 μL of each, ethanol precipitate, wash with 70% ethanol, air dry, and resuspend in 100 μL of water.

18. Make up 200 μL of reaction mix to digest scFv repertoires with *Not*I:

scFv DNA	100 μL
Water	72 μL
Acetylated BSA (10 mg/mL)	2 μL
10× NEB 3 buffer	20 μL
Not I (10 /μL)	6 μL

19. Incubate at 37°C for 18 h.

20. Phenol–chloroform extract with 100 μL of each, ethanol precipitate, wash with 70% ethanol, air dry, and resuspend in 100 μL of water. Determine DNA concentration by analysis on an 1.5% agarose gel with markers of known size and concentration.

21. Digest 4 μg of cesium chloride purified pHEN 1 with the appropriate restriction enzymes exactly as described in **Subheading 3.4.** (*see* **Note 9**).

22. Purify the digested vector DNA on a 0.8% agarose gel, extract from the gel, and ethanol precipitate.

23. In ligation experiments, the molar ratio of insert to vector should be 2:1. Given that the ratio of sizes of assembled *scFv* (700 basepairs) to vector (4500 base–pairs) is approx 6:1, this translates into a ratio of insert to vector of 1:12 in weight terms. Make up 100 μL of ligation mixture:

10× ligation buffer	10 μL
Water	67 μL
Digested pHEN 1 (100 ng/μL)	10 μL
scFv gene repertoire (10 ng/μL)	8 μL
T4 DNA ligase (400 U/μL)	5 μL

24. Ligate at 16°C for 18 h.

25. Bring volume to 200 μL with H_2O, and extract once with phenol–chloroform and twice with ether to remove traces of phenol; ethanol precipitate and wash with 70% ethanol (*see* **Note 10**).

26. Resuspend DNA in 10 μL of water and use 2.5 μL/transformation into 50 μL electrocompetent *E. coli* TG 1.

27. Set the electroporator at 200 Ω (resistance), 25 μFD (capacitance), and 2.5 kV. After electroporation the time constant should be approx 4.5 s.

28. Grow bacteria in 1 mL of SOC at 37°C for 1 h with shaking (250 rpm) and plate serial dilutions onto small TYE–amp–Glc agar plates for determining the size of the library.

29. Centrifuge the remaining bacteria solution at 1700g for 10 min at 4°C. Resuspend the pellet in 250 μL and plate onto two small TYE–amp–Glc plates. Incubate for 18 h at 37°C.

30. Scrape bacteria from large plates by washing each plate with 3 mL of 2× TY–amp–Glc.

31. Make glycerol stocks by adding 1.4 mL of bacteria and 0.6 mL 50% glycerol in PBS (sterilized by filtration through 0.45 μm filter). Save library stock at –70°C.

3.5. Rescue of Phage Antibody Libraries and Phage Preparation

1. Calculate the number of bacteria per milliliters from your library glycerol stock (A_{600} of 1.00 corresponds to approx 10^8 cells). Usually the optical density of our bacterial stock is approx 100 or 10^{10} cells/mL. The input of cells for rescuing is dependant on the density of the original bacteria glycerol stock and should be in 10-fold excess of the number of different clones in the library, but should not exceed A_{600} of 0.05.

2. Grow with shaking (250 rpm) at 37°C to an A_{600} ~ 0.7 (corresponds to 7×10^7 bacteria/mL).

3. Transfer 10 mL (7×10^8 bacteria total) to a 50-mL Falcon tube containing the appropriate number of helper phages. To ensure rescue of all clones in the library the ratio of helper phage/bacteria should be 10:1. Therefore add 7×10^9 plaque forming units (pfus) of helper phage (VCSM13, Stratagene) to the bacterial solution. Incubate at 37°C without shaking for 30 min (*see* **Note 11**).

4. Incubate at 37°C with shaking (250 rpm) for 30 min.

5. Plate 1 μL onto TYE–kan plate to check for infectivity. Incubate at 37°C for 18 h (*see* **Note 12**).

6. Centrifuge cells at 3000*g*, 4°C, to remove glucose and resuspend in 50 mL of TYE–amp–kan. Shake at 37°C for 30 min (220 rpm).

7. Grow with shaking (250rpm) at 25°C for 18 h.

8. For preparation of phage, remove bacteria by centrifugation at 4000*g*, 10 min, 4°C. Decant the clear supernatant containing phage particles into a 500-mL centrifuge bottle and add 10 mL of 20% PEG–2.5 *M* NaCl per bottle. Mix and incubate on ice for 10 min.

9. Pellet phage by centrifuging for 15 min, 4000*g*, at 4°C. Discard the supernatant.

10. Resuspend the white phage pellet in 10 mL of PBS. To remove remaining bacteria debris, we recommend centrifuging for 10 min, 4000*g*, at 4°C.

11. Transfer the supernatant into a 15-mL Falcon tube and repeat PEG precipitation with 2 mL of 20% PEG–2.5 *M* NaCl. Incubate on ice for 15 min.

12. Centrifuge for 10 min, 4000*g*, at 4°C and resuspend the white pellet in 2.5 mL of PBS.

13. Filter supernatant through a 0.45-μm syringe filter.

14. Prior to selection, titer the phage preparation by diluting 10 μL of the eluted phage stock into 1 mL of 2× TY (10^2 dilution), vortex-mix briefly and transfer 10 μL into 1 mL of 2× TY (10^4 dilution). Vortex-mix again and transfer 10 μL into 1 mL of exponentially growing *E. coli* TG1 (A_{600} ~0.7) (10^6 dilution). Incubate at 37°C without shaking for 30 min. Transfer 10 μL into 1 mL of 2× TY (10^8 dilution). Vortex-mix again and plate 1 μL onto a small 2× TY–amp–Glc plate and incubate at 37°C for 18 h. Multiply the number of counted colonies by the factor of 10^{11}, which represents the phage preparation titer per milliliter. This protocol usually gives rise to 10^{12}–10^{13} phage (TU)/mL per 50 mL of culture (*see* **Note 13**).

3.6. Selection of Phage Antibody Libraries

1. Prior to selection, streak out *E. coli* TG1 on a minimal media plate and incubate at 37°C for 18 h (*see* **Note 14**). Toothpick one clone off the plate and incubate in 2 mL of TYE media at 25°C, shaking for 18 h. Next day, make a 1:100 dilution in 2× TY media and grow bacteria at 37°C to exponential phase. Proceed to **step 8**.

2. Block 50 μL of streptavidin–magnetic beads with 1 mL of 5% MPBS for 1 h at 37°C in a 1.5-mL Eppendorf tube. Centrifuge the beads briefly at 800*g* for 30 s and remove blocking buffer.

3. Block a 1.5-mL Eppendorf tube with 5% MPBS for 1 h at 37°C and discard the blocking buffer. Incubate biotinylated antigen (10–500 n*M*), prepared according

to the manufacturer's instruction, with 1 mL of polyclonal phage (approx 10^{12} TU) in 1% MPBS (final conc.) by rocking at 20–22°C for 30 min–1 h (*see* **Note 15**).

4. Add 100 µL of streptavidin–magnetic beads to the phage–antigen mix and incubate on a rotator at 20–22°C for 30 min. Briefly centrifuge the tube to dislodge particles in the cap at 800*g* for 30 s and then place tube in a magnetic rack for 20 s. Beads will migrate toward the magnet (*see* **Note 16**).

5. Aspirate tubes carefully, leaving the beads on the side of the Eppendorf tube. Wash beads (1 mL per wash) with PBS-Tween (0.1%) 10×, followed by 10 PBS washes. Transfer the beads after every second wash to a fresh Eppendorf tube to facilitate efficient washing. After the last washing step, transfer the beads into a new 1.5-mL Eppendorf tube, preblocked with 5% MPBS.

6. Elute phage with 500 µL of 100 m*M* triethylamine for 10 min at 20–22°C. Place tube in magnetic rack for 20 s and beads migrate toward the magnet.

7. Remove the supernatant containing eluted phage and neutralize with 1 mL of 1 *M* Tris, pH 7.4.

8. Add 0.75 mL of the phage stock to 10 mL of exponentially growing TG1 (A_{600} ~ 0.7). Store the remaining phage mix at 4°C.

9. Incubate at 37°C for 30 min without shaking.

10. Titer TG1 infection by plating 1 µL and 10 µL onto small TYE–amp–Glc plates (this is a 10^4 and 10^3 dilution, respectively) and incubate at 37°C for 18 h (*see* **Note 17**).

11. Centrifuge the remaining bacteria solution at 1700*g* for 10 min at 4°C.

12. Resuspend pellet in 250 µL of 2× TY media and plate onto two large TYE–amp–Glc plates. Incubate at 37°C for 18 h.

13. Add 3 mL of 2× TY–amp–Glc media to each plate, then scrape the bacteria from the plate with a bent glass rod.

14. Make glyercol stocks by mixing 1.4 mL of bacteria and 0.6 mL of sterile filtered 50% glycerol in PBS and save bacterial stock at –70°C.

3.7. Identification of Antigen Binding Clones

1. Toothpick 94 individual colonies per round of panning into 150 µL of 2× TY–amp–Glc in 96-well round bottom plates and grow with shaking (250 rpm) for 18 h at 37°C (= master plate). Include one positive and one negative control.

2. The next day inoculate from the master plate to a fresh 96-well round-bottom plate containing 150 µL of 2× TY–amp–0.1% Glc per well using a 96-well transfer device (use 3–4 repeated transfers or pipet 2 µL per well from the master plate) (*see* **Note 18**). This is the expression plate.

3. Grow bacteria in the newly inoculated expression plate to an A_{600} ~ 0.7 for 2–3 h at 37°C, 250 rpm.

4. To the wells of master plate add 50 µL of 50% glycerol–5 m*M* MgSO$_4$ per well.

5. Cover the surface with a plate sealer and store at –70°C.

6. To each well of the expression plate add 50 µL of 4 m*M* IPTG in 2× TY–amp (final conc. 1 m*M*) and continue shaking (250 rpm) at 25°C for 18 h.

7. Centrifuge bacterial debris at 800*g* for 10 min and transfer the supernatant containing soluble antibody fragments to a new plate.

8. For the ELISA, coat a microtiter plate with 50 μL/well of antigen at 1–10 μg/mL in PBS, pH 7.4 (*see* **Note 19**).

9. Wash wells 3× with PBS, then block with 200 μL/well of 2 % skim milk powder in PBS (MPBS) for 1 h at 20–22°C. Wash by submersing the plate into buffer and removing the air bubbles in the wells by agitation. Tap the ELISA plates on a dry paper towel, to ensure that there is no remaining buffer in the wells.

10. Wash wells 3× with PBS and add 50 μL/well of soluble antibody fragment. Incubate for 1.5 h at 20–22°C.

11. Discard the solution and wash wells 3× with PBS–0.1% Tween-20 and 3× with PBS.

12. Add 50 μL/well of 9E10 antibody at 1 μg/mL in 2% MPBS and incubate at 20–22°C for 1 h.

13. Discard primary antibody and wash wells 3× with PBS–0.1% Tween-20 and 3× with PBS.

14. Add 50 μL/well of horseradish peroxidase conjugated anti-mouse IgG Fc monoclonal antibody at 1 μg/mL in 2% MPBS. Incubate at 20–22°C for 1 h.

15. Discard secondary antibody and wash wells 3× with PBS–0.1% Tween-20 and 3× with PBS.

16. Immediately before developing, add 20 μL of 30% hydrogen peroxide to the ABTS solution and then dispense 100 μL/well. Leave at 20–22°C for 20–30 min (*see* **Note 20**).

17. Quench reaction by adding 50 μL/well of 75 m*M* sodium fluoride. Read at 405 nm (*see* **Note 21**).

18. To determine the diversity of clones between sequential rounds of selection, perform a PCR fingerprint with the restriction enzyme *BstN*I. Make up master mix, with 20 μL of PCR mix per clone containing (*see* **Note 22**):

Water	14.8 μL
10× *Taq* buffer	2. μL
dNTPs (5 m*M* each)	1. μL
LMB 3 primer (10 pmol/μL)	1. μL
fdseq 1 primer (10 pmol/μL)	1. μL
Taq DNA polymerase (5 U/μL)	0.2 μL

19. Aliquot 20 μL of PCR reaction mix into wells of a 96-well PCR-compatible microtiter plate.

20. Transfer 1 μL of a bacterial glycerol stock into the PCR tube.

21. Cycle 25× at 94°C for 1 min, then at 42°C for 1 min and at 72°C for 1 min.

22. Check PCR reactions for full-length inserts (approx 700 basepairs) on a 1.5% agarose gel.

23. Prepare a restriction enzyme mixture with *BstN*1:

Water	17.8 μL
10× NEB buffer 2	2 μL
*BstN*1 (10 U/μL)	0.2 μL

24. Add 20 µL of the above mix to each PCR-containing well under the mineral oil. Incubate the restriction digest at 60°C for 2 h.
25. Add 4 µL of 6× agarose gel electrophoresis sample dye under the oil and analyze the restriction digest on a 4% Nusieve agarose gel cast containing 0.5 µg/mL ethidium bromide (*see* **Note 23**).

4. Notes

1. As with any RNA preparative procedure, special care must be taken to ensure that solutions and glassware are free of ribonucleases. Therefore any water or salt solution should be treated with DEPC, which inactivates ribonucleases by covalent modification. We usually prefer disposable plasticware at this stage. Where glassware is used, we incubate with 0.2% DEPC solution for 1 h. In addition, Eppendorf tubes for the PCR reaction should be treated with Silane, to avoid sticking of RNA to the plastic walls. Be careful when working with DEPC, as this chemical is carcinogenic. Also, we recommend wearing gloves during experiments, to minimize contamination with RNase.
2. From a mouse spleen the yield is $5 \times 10^7 – 2 \times 10^8$ white blood cells (WBCs) of which 35–40% are B cells and lymphocytes.
3. We recommend proceeding with the primary amplification immediately upon obtaining first strand cDNA in order to guarantee good quality and yields for the following primary PCRs.
4. The Vent DNA polymerase buffer already contains 2 mM $MgSO_4$ when the buffer is diluted to its final 1× form. However, for optimal primer extensions we usually titrate the $MgSO_4$ concentration (ranging from 2 to 6 mM) for optimal results.
5. In all PCR amplification steps, the enzyme is added after the reaction mix has been heated up to 94°C, to improve amplification results by preventing extension of misannealed primers. Vent DNA polymerase is used for primary amplifications and PCR assembly because its proofreading exonuclease activity results in a lower error rate.
6. The assembly step is often the most problematic step of the whole procedure and it is crucial to use equimolar ratio of the DNA fragments. This corresponds to a mass ratio of V_H:linker/V_L of 3:1:3, given that the approximate ratio of sizes of V_H (340 basepairs) and V_L (325 basepairs) to linker (93 basepairs) is 3:1. We recommend running different dilutions of the DNA fragments on an agarose gel and comparing their intensities with a marker of known size and concentration. In the event of failure despite the above modification it is worth trying different annealing temperatures, as the temperature profile varies with different PCR blocks.
7. The cleanup step is necessary to remove *Taq* DNA polymerase and nucleotides which could fill in the generated overhangs by *Sfi*I and *Not*I digestion.
8. Overdigestion of the PCR-generated scFv repertoires is necessary owing to the poor efficiency with which PCR fragments are digested.
9. Vector DNA should be prepared using CsCl to maximize digestion efficiency and minimize background in the library.
10. Cleaning up the the ligation mixture by phenol–chloroform extraction increases transformation efficiency 10- to 100-fold. Washing with 70% ethanol removes

any traces of salts, which could lead to high-powered electrical sparks causing destruction of the electroporation cuvette.

11. The infection step must be carried out at 37°C to guarantee successful propagation of phagemid particles.

12. The plate should be nearly confluent the next day, indicating successful rescuing.

13. The shelflife of rescued phage particles is limited to approx 2–5 d owing to proteolysis of the displayed antibody fragment.

14. TG1 cells should be grown first on minimal media to select for the F-pilus; otherwise their infectivity will be lost.

15. For the first round of selection we typically use 100 nM of antigen, which is decreased 5-fold in the following rounds. The actual amount of phage and the amount of antigen will vary according to the affinities present in the library and the stage of panning. The final antigen concentration should be around or below the desired K_d of selected antibody fragments.

16. In the following rounds of selections we alternate incubation with avidin–streptavidin magnetic beads and preclear the phage mix with the appropriate beads. This will prevent the selection for streptavidin–avidin phage-binders, represented in the library.

17. Because enrichment ratios for low-affinity phage antibodies selected from immune libraries are low as 20-fold, the number of phage eluted after the first round of selection should be at least 1/100 of the original library size (10^5 eluted phage for 10^7 library). If the titer is larger than 10^5 it is likely that the washing steps have been inadequate. Repeat the first round of selection with increased washing steps. If the titer is below 10^4, you have either washed too many times, or the antigen has been poorly adsorbed to the polystyrene surface. In this case, try to use a higher antigen concentration or different buffer for coating and/or reduce the number of washes.

18. This method of expression is based on the work of DeBellis and Schwarz *(33)* and relies on the low level of glucose repressor present in the starting media being metabolized by the time the inducer (IPTG) is added.

19. We recommend checking out various coating buffers for the respective antigen to maximize the sensitivity of the ELISA.

20. Routinely check the activity of the hydrogen peroxide solution, which can go off after 4 wk storage in the refrigerator and replace if necessary.

21. If the ELISA signals are weak including the positive control, we recommend checking the expression levels of soluble antibodies, which can vary from one clone to another and range from 2 to 1000 µg/mL. Transfer 50 µL of the antibody containing supernatant onto a nitrocellulose filter and proceed with **steps 10–15**. Develop the nitrocellulose in DAB solution (dissolve one 10 mg DAB tablet [diaminobenzidine tetrahydrochloride] in 20 mL of PBS and add 20 µL of 30% hydrogen peroxide).

22. The denoted sequences for PCR screening oligonucleotides LMB 3 and fdseq 1 are as followed: LMB 3 (5'-CAG GAA ACA GCT ATG AC–3') annealing upstream from the pelB leader sequence and fdseq 1 (5'-GAA TTT TCT GTA TGA GG–3') annealing at the 5' end of gene 3.

23. The number of unique scFv binders is determined by PCR fingerprinting using primers that flank the antibody encoding gene (LMB3 and fdseq; *see* **Note 22**).

References

1. Köhler, G. and Milstein, C. (1975) Continous cultures of fused cells secreting antibody of predefined specificity. *Nature* **256,** 495–497.
2. Better, M., Chang, C. P., Robinson, R. R., and Horwitz, A. H. (1988) *Escherichia coli* secretion of an active chimeric antibody fragment. *Science* **240,** 1041–1043.
3. Skerra, A. and Pluckthun, A. (1988) Assembly of a functional immunoglobulin Fv fragment in *Escherichia coli*. *Science* **240,** 1038–1041.
4. Saiki, R. K., Gelfand, D. H., Stoffel, S., Scharf, S. J., Higuchi, R., Horn, G. T., Mullis, K. B., and Erlich, H. A. (1988) Primer-directed enzymatic amplification of DNA with a thermostable DNA polymerase. *Science* **239,** 487–491.
5. Orlandi, R., Gussow, D. H., Jones, P. T., and Winter, G. (1989) Cloning immunoglobulin variable domains for expression by the polymerase chain reaction. *Proc. Natl. Acad. Sci. USA* **86,** 3833–3837.
6. Marks, J. D., Tristrem, M., Karpas, A., and Winter, G. (1991) Oligonucleotide primers for polymerase chain reaction amplification of human immunoglobulin variable genes and design of family-specific oligonucleotide probes. *Eur. J. Immunol.* **21,** 985–991.
7. Kang, A., Barbas, C., Janda, K., and Benkovic, S. (1991) Linkage of recognition and replication functions by assembling combinatorial Fab libraries along phage surfaces. *Proc. Natl. Acad. Sci. USA* **88,** 4363–4366.
8. Huse, W., Stinchcombe, T., Glaser, S., Starr, L. M., Hellstrom, K., Hellstrom, I., and Yelton, D. (1992) Application of a filamentous phage pVIII fusion protein system suitable for efficient production, screening, and mutagenesis of F(ab) antibody fragments. *J. Immunol.* **149,** 3914–3920.
9. McCafferty, J., Griffiths, A. D., Winter, G., and Chiswell, D. J. (1990) Phage antibodies: filamentous phage displaying antibody variable domains. *Nature* **348,** 552–554.
10. Hoogenboom, H. R., Griffiths, A. D., Johnson, K. S., Chiswell, D. J., Hudson, P., and Winter, G. (1991) Multi-subunit proteins on the surface of filamentous phage: methodologies for displaying antibody (Fab) heavy and light chains. *Nucleic Acids Res.* **19,** 4133–4137.
11. Engelhardt, O., Grabher, R., Himmler, G., and Ruker, F. (1994) Two-step cloning of antibody variable domains in a phage display vector. *Bio/Techniques* **17,** 44–46.
12. Dubel, S., Breitling, F., Fuchs, P., Braunagel, M., Klewinghaus, I., and Little, M. (1993) A family of vectors for surface display and production of antibodies. *Gene* **128,** 97–101.
13. Barbas, C. F., Kang, A. S., Lerner, R. A., and Benkovic, S. J. (1991) Assembly of combinatorial antibody libraries on phage surfaces: the gene III site. *Proc. Natl. Acad. Sci. USA* **88,** 7978–7982.

14. Marks, J. D., Hoogenboom, H. R., Bonnert, T. P., McCafferty, J., Griffiths, A. D., and Winter, G. (1991) By-passing immunization: human antibodies from V-gene libraries displayed on phage. *J. Mol. Biol.* **222,** 581–597.
15. Griffiths, A. D., Williams, S. C., Hartley, O., Tomlinson, I. M., Waterhouse, P., Crosby, W. L., Kontermann, R. E., Jones, P. T., Low, N. M., Allison, T. J., Prospero, T. D., Hoogenboom, H. R., Nissim, A., Cox, J. P. L., Harrison, J. L., Zaccolo, M., Gherardi, E., and Winter, G. (1994) Isolation of high affinity human antibodies directly from large synthetic repertoires. *EMBO J.* **13,** 3245–3260.
16. Vaughan, T. J., Williams, A. J., Pritchard, K., Osbourn, J. K., Pope, A. R., Earnshaw, J. C., McCafferty, J., Hodits, R. A., Wilton, J., and Johnson, K. S. (1996) Human antibodies with sub-nanomolar affinities isolated from a large non-immunized phage display library. *Nature Biotech.* **14,** 309–314.
17. Sheets, M. D., Amersdorfer, P., Finnern, R., Sargent, P., Lindqvist, E., Schier, R., Hemingsen, G., Wong, C., Gerhart, J. C., and Marks, J. D. (1998) Efficient construction of a large non-immune phage antibody library: the production of high affinity human single-chain antibodies to protein antigens. *Proc. Natl. Acad. Sci. USA* **95,** 6157–6162.
18. Chen, F., Kuziemko, G. M., Amersdorfer, P., Wong, C., Marks, J. D., and Stevens, R. C. (1997) Antibody mapping to domains of botulinum neurotoxin serotype A in the complexed and uncomplexed forms. *Infect. Immun.* **65,** 1626–1630.
19. Griffiths, A. D., Malmqvist, M., Marks, J. D., Bye, J. M., Embleton, M. J., McCafferty, J., Baier, M., Holliger, K. P., Gorick, B. D., Hughes-Jones, N. C., Hoogenboom, H. R., and Winter, G. (1993) Human anti-self antibodies with high specificity from phage display libraries. *EMBO J.* **12,** 725–734.
20. Perelson, A. S. and Oster, G. F. (1979) Theoretical studies of clonal selection: minimal antibody repertoire size and reliability of self non-self discrimination. *J. Theor. Biol.* **81,** 645–670.
21. Perelson, A. S. (1989) Immune network theory. *Immunol. Rev.* **110,** 5–36.
22. Schibler, U., Marcu, K. B., and Perry, R. P. (1978) The synthesis and processing of the messenger RNAs specific heavy and light chain immunoglobulins in MPC-11 cells. *Cell* **15,** 1495–1509.
23. Amersdorfer, P., Wong, C., Chen, S., Smith, T., Desphande, S., Sheridan, R., Finnern, R., and Marks, J. D. (1997) Molecular characterization of murine humoral immune response to botulinum neurotoxin type A binding domain as assessed by using phage antibody libraries. *Infect. Immun.* **65,** 3743–3752.
24. Barbas, C. F., Collet, T. A., Amberg, W., Roben, P., Binley, J. M., Hoekstra, D., Cababa, D., Jones, T. M., Williamson, A., Pilkington, G. R., Haigwood, N. L., Cabezas, E., Satterthwait, A. C., Sanz, I., and Burton, D. R. (1993) Molecular profile of an antibody response to HIV-1 as probed by combinatorial libraries. *J. Mol. Biol.* **230,** 812–823.
25. Binley, J. M., Ditzel, H. J., Barbas III, C. F., Sullivan, N., Sodroski, J., Parren, P. W. H. I., and Burton, D. R. (1996) Human antibody responses to HIV type 1 glycoprotein 41 cloned in phage display libraries suggest three major epitopes are

recognized and give evidence for conserved antibody motifs in antigen binding. *AIDS Res. Hum. Retroviruses* **12,** 911–924.

26. Zebedee, S. L., Barbas, C. F., Hom, Y.-L., Cathoien, R. H., Graff, R., DeGraw, J., Pyatt, J., LaPolla, R., Burton, D. R., and Lerner, R. A. (1992) Human combinatorial antibody libraries to hepatitis B surface antigens. *Proc. Natl. Acad. Sci. USA* **89,** 3175–3179.
27. Chan, S., Bye, J., Jackson, P., and Allain, J. (1996) Human recombinant antibodies specific for hepatitis C virus core and envelope E2 peptides from an immune phage display library. *J. Gen. Virol.* **10,** 2531–2539.
28. Crowe, J., Murphy, B., Chanock, R., Williamso, R., Barbas, C., and Burton, D. (1994) Recombinant human respiratory syncytial virus (RSV) monoclonal antibody Fab is effective therapeutically when introduced directly into the lungs of RSV-infected mice. *Proc. Natl. Acad. Sci. USA* **91,** 1386–1390.
29. Barbas, C., Crowe, J., Cababa, D., Jones, T., Zebedee, S., Murphy, B., Chanock, R., and Burton, D. (1992) Human monoclonal Fab fragments derived from a combinatorial library bind to respiratory syncytial virus F glycoprotein and neutralize infectivity. *Proc. Natl. Acad. Sci. USA* **89,** 10,164–10,168.
30. Reason, D., Wagner, T., and Lucas, A. (1997) Human Fab fragments specific for the *Haemophilus influenzae* b polysaccharide isolated from a bacteriophage combinatorial library use variable region gene combinations and express an idiotype that mirrors in vivo expression. *Infect. Immun.* **65,** 261–266.
31. Chomczynski, P. and Sacchi, N. (1987) Single-step method of RNA isolation by acid guanidinium-thiocyanate-phenol-chloroform extraction. *Anal. Biochem.* **162,** 156–159.
32. Huston, J. S., Levinson, D., Mudgett, H. M., Tai, M. S., Novotny, J., Margolies, M. N., Ridge, R. J., Bruccoleri, R. E., Haber, E., Crea, R., and Oppermann, H. (1988) Protein engineering of antibody binding sites: recovery of specific activity in an antidigoxin single-chain Fv analogue produced in *Escherichia coli. Proc. Natl. Acad. Sci. USA* **85,** 5879–5883.
33. De Bellis, D. and Schwartz, I. (1990) Regulated expression of foreign genes fused to lac: control by glucose levels in growth medium. *Nucleic Acids Res.* **18,** 1311.
34. Poul, M. A. and Marks, J. D. (1997) Intracellular Antibodies, *in Intrabodies—Basic Research in Clinical Gene Therapy Applications* (Marasco, W. A., ed.), Landes Bioscience, Georgetown, TX, pp. 30,31.

14

T-Cell Cytotoxicity Assays
for Studying the Functional Interaction
Between the Superantigen Staphylococcal
Enterotoxin A and T-Cell Receptors

Alexander Rosendahl, Karin Kristensson, Kristian Riesbeck, and Mikael Dohlsten

1. Introduction

The superantigens (SAgs) staphylococcal enterotoxins (SEs) are produced by certain strains of *Staphylococcus aureus* and comprise structurally related bacterial proteins, which are among the most potent mitogens known for murine and human T lymphocytes *(1,2)*. T-cell activation induced by SEs involves binding to constant parts of major histocompatibility complex (MHC) class II molecules and subsequent interaction with T cells expressing certain T-cell receptor (TCR) Vβ chains *(3,4)*. The binding of SAgs to MHC class II molecules does not require processing and structural mutations, and biochemical experiments have demonstrated that the amino acids involved are distant from the antigen binding groove.

Activation of T cells with the SAg staphylococcal enterotoxin A (SEA) results in proliferation and production of cytokines, that is, interleukin-2 (IL-2), interferon-γ (IFN-γ), and tumor necrosis factor-α (TNF-α). Moreover, SEA directs activated human and murine T-cell lymphocytes to MHC class II+ cells and evoke substantial cytotoxic T lymphocyte (CTL) activity *(5–8)*.

In this chapter an approach to obtain human and murine SEA-reactive CTLs is reported. We outline an experimental method to record the superantigen-dependent-cell-mediated-cytotoxicity (SDCC). Finally, we describe the specific cellular and molecular phenotype linked to SEA-reactive CTLs.

From: *Methods in Molecular Biology, vol. 145: Bacterial Toxins: Methods and Protocols*
Edited by: O. Holst © Humana Press Inc., Totowa, NJ

To obtain human effector cells, peripheral blood mononuclear PBMCs are cultured in the presence of SEA and IL-2. By subsequent cultures SEA reactive cytotoxic T cells (CTLs) are recovered. Similarly, splenocytes from mice injected intravenously with SEA express CTL activity and with this model we address which phenotype is predominately responsible for the cytotoxic activity and also how to enrich this population.

The CTL activity can be analyzed in vitro in a cell assay utilizing the incorporation of ^{51}Cr to cytoplasmic protein of the relevant target cell (hereby denoted ^{51}Cr-release assay). The activated T cell attacks the ^{51}Cr incorporated SEA presenting target cells. Lysed target cells release ^{51}Cr into the media and radioactivity can be measured and correlated to CTL activity.

2. Materials

2.1. Buffers

1. PBS without Ca^{2+} and Mg^{2+}: Store at 4°C, stable up to 1 mo. Adjust to pH 7.4.
2. Flow analysis cell sorter (FACS) buffer: PBS without Ca^{2+} and Mg^{2+} supplemented with 1% bovine serum albumin (BSA). Store at 4°C, stable up to 14 d.
3. Magnetic analysis cell sorter (MACS) buffer: PBS without Ca^{2+} and Mg^{2+} supplemented with 0.5% BSA. Store at 4°C, use same day as prepared.
4. PBMC culture R medium: RPMI-1640 (store stock solution in 4°C, stable up to 5 mo), 10% fetal calf serum solution (FCS) (store stock solution in –20°C, stable up to 5 mo), 1 mM sodium pyruvate solution (store stock solution at 4°C, stable up to 4 mo), and 1 mM gentamicin solution (store stock solution at 20–22°C, stable up to 4 mo). Store at 4°C, stable up to 14 d.
5. B-cell lymphoma (BSM) culture medium: PBMC culture medium without gentamicin. Store at 4°C, stable up to 14 d.
6. Mouse effector cell medium: PBMC culture medium with 1 mM glutamine (store stock solution in 20–22°C, stable up to 5 mo) and 5×10^{-5} M β-mercaptoethanol (store stock solution at 4°C, stable up to 4 mo). Store at 4°C, stable up to 14 d.
7. GEY's solution: Stock solution 1: 1000 mL of H_2O, 35 g of NH_4Cl, 1.85 g of KCl, 1.5 g of $Na_2HPO_4 \cdot 12H_2O$, 0.119 g of KH_2PO_4, 5 g of D-glucose. Stock solution 2: 100 mL of H_2O, 0.14 g of $MgSO_4 \cdot 7H_2O$, 0.42 g of $MgCl_2 \cdot 6H_2O$, 0.34 g of $CaCl_2 \cdot 2H_2O$. Stock solution 3: 100 mL H_2O, 2.25 g of $NaHCO_3$. Store stock solutions at 4°C. Stable for 1 yr. Working solution: Mix 35 mL of H_2O, 10 mL of solution 1 and 2.5 mL of both solutions 2 and 3. Store at 4°C, stable for 1 mo.

2.2. Reagents

1. Mitomycin C (MMC): Carcinogenic. Dissolve in distilled water to 0.5 mg/mL. Store in the dark at 4°C, stable for 1 wk. Use extreme caution when working with the substance.

2. rhIL-2: Store stock solution at −70°C. Dilute working solution in R medium. Store at 4°C, use the same day as prepared.
3. Monoclonal antibodies (MAbs; anti-CD4, -CD8, -CD11a, -CD32/-CD16) conjugated for FACS analysis (PharMingen, San Diego, CA): Dilute in FACS buffer to working dilution. Store at 4°C, stable up to 2 d. (*See* **Note 1**.)
4. Monoclonal antibodies (MAbs; anti-CD54, -CD18, -CD11a) described to block intercellular adhesion with respective ligand (PharMingen, San Diego, CA): Dilute mouse effector cell medium. Store at 4°C, use the same day as diluted.
5. ^{51}Cr: Store at 4°C. Half-life of 27 d. Use within 30 d for best results.
6. rSEA: Store stock solution in −70°C. Dilute working solution in R medium. Store at 4°C, use the same day as prepared. The substance is extremely toxic and symptoms are acute. Use with extreme caution.
7. MACS magnetically labeled antibodies (Miltenyi Biotec, CA): Store at 4°C, stable for 6 mo.
8. Ficoll-Isopaque (F-I): Store at 4°C. Avoid direct contact with light.
9. Sodium dodecyl sulfate (SDS): Working solution (1% SDS) diluted in distilled H_2O.
10. Cell dissociation solution (Sigma).

2.3. Cells

1. Primary peripheral blood lymphocytes in buffy coats: Keep at 20–22°C until used.
2. Raji: Keep at 37°C until used.
3. A20: Keep at 37°C until used.
4. DR transfected B16 melanoma: Keep at 37°C until used.
5. C57Bl/6 splenocytes: Keep on ice. Use within 4–8 h after the single-cell suspension is made.
6. Perforin knockout (KO) splenocytes: Keep on ice. Use within 4–8 h after the single-cell suspension is made.
7. BSM: Keep at 37°C until used.
8. EA.hy926 (human umbilical vein endothelial cell [HUVEC]-derived endothelial cell line): Keep at 37°C until used.

2.4. Utilities and Equipment

All equipment is stored at 20–22°C.

1. MACS columns: Miltenyi Biotec, CA.
2. MACS board: Miltenyi Biotec, CA.
3. γ-Counter: Capable to read 96-well microtiter plates.
4. Lymphocyte separation tube: Greiner.
5. 6-Well plates: NUNC, Roskilde, Denmark.
6. U- and V-shaped 96-well microtiter plates: NUNC, Roskilde, Denmark.
7. Harvest 96-well plates (Packard).

3. Methods

3.1. Induction of CTL Activity in Human PBMCs In Vitro

3.1.1. Separating Human PBMCs

This separation system provides an easy procedure to separate granulocytes and erythrocytes from mononuclear cells (MNC) in the upper part.

1. Equilibrate Ficoll-Isopaque (F-I) to 20–22°C.
2. Pipet 3 mL or 15 mL equilibrated F-I into 15-mL or 50-mL lymphocyte separation tubes, respectively.
3. Centrifuge at 250g for 20 s at 20–22°C. Check that F-I now is below the filter disc.
4. Pour freshly defibrinated or anticoagulated blood into the tube; 3–6 mL for the 15-mL tube and 15–30 mL for the 50-mL tube.
5. Centrifuge at 300g for 20 min.
6. The content in the tube is at this stage (top to bottom): plasma–MNC–F-I–Filter DISC–F-I–erythrocytes + granulocytes. Gently transfer the MNC with a thin pipet to a new 50-mL tube.
7. Wash twice with 40 mL phosphate-buffered saline (PBS) buffer at 250g for 10 min.
8. Resuspend cells in PBMC medium, count, and adjust the cell concentration to 24×10^6/mL.

3.1.2. Establishment of Effector Cell Lines

Human SEA reactive effector cell lines are easily established and are ready to use approx 3 wk after initiation of cultures and can be kept for up to 3 mo. Human PBMCs are cultured with SEA alone for 3–4 d. New antigen presenting cells (BSMs) are then cocultured with the PBMCs as the MHC class II expressing cells that existed in the PBMCs have been killed by the activated T cells.

1. Transfer 1 mL of the freshly isolated PBMC (24×10^6 cells/mL) to one well in a six-well plate (NUNC). Add 7 mL of PBMC-medium.
2. Add rSEA to the culture at final concentration of 12.5 pg/mL.
3. Culture for 3–4 d at 37°C. *See* **Note 2**.
4. If the medium needs to be changed gently remove 4 mL and replace with 4 mL of fresh medium containing 40 U/mL of rIL-2. If the culture is strongly confluent split 1/3 in fresh PBMC medium.
5. Culture BSM in BSM medium.
6. Dissolve Mitomycin C (MMC) in distilled water (0.5 mg/mL).
7. 100 μL of 10^7 BSM/mL is incubated with 100 μg of MMC for 45 min at 37°C.
8. Wash 3× at 250g for 10 min in BSM medium.
9. Incubate the BSM cells in 1 mL BSM medium containing rSEA (100 ng/mL) for 30 min at 20–22°C.

A

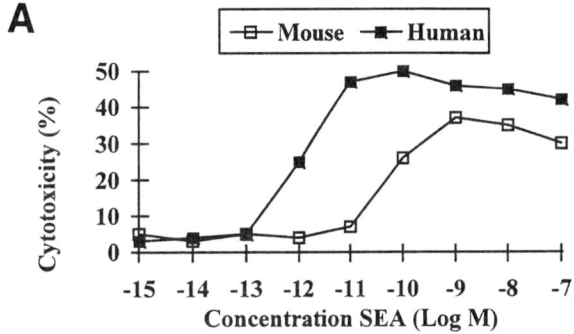

B

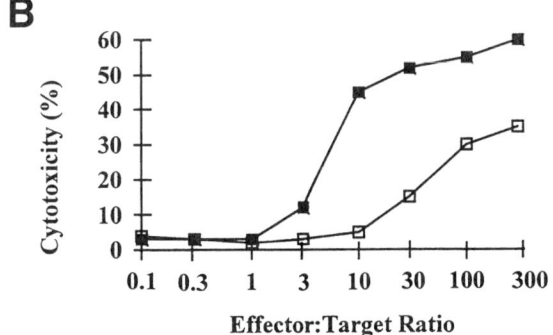

Fig. 1. SEA-dependent cell-mediated cytotoxicity by a human SEA-reactive T-cell line or *in vivo* activated splenocytes from C57Bl/6 mice. Cytotoxicity was measured in a standard 4-h ^{51}Cr-release assay against SEA-coated Raji cells at **(A)** various concentrations of SEA or **(B)** SEA-coated Raji cells (1 µg/mL) at various E/T ratios. The human cell line was used 4 wk after the culture was initiated and the spleens were removed 48 h after an injection of 10 µg of SEA. Mean values from triplicate wells are shown. Standard deviation in the assay was routinely <15%. Data are presented from one out of two similar experiments.

10. Wash 3× at 250*g* for 10 min in BSM medium.
11. Gently remove 4 mL of the medium from the PBMC cultures.
12. Add 3 mL of fresh PBMC medium.
13. Add 1 mL of the treated BSM (2×10^6/mL).
14. Repeat **step 4** once per week and **steps 5–13** 3 d later.

3.2. Detection of CTL Activity In Vitro by ^{51}Cr-Release Assay

This method is rapid and provides reliable reproducible results. *See* **Fig. 1A**. *See* **Note 3**.

3.2.1. Titration of Antigen

To determine at which concentration a superantigen induces CTL activity in vitro, a dose–response experiment should be conducted.

1. Centrifuge the target cells twice at 250g for 5 min at 20–22°C (in PBMC or mouse effector cell medium, depending on which effector cell is used).
2. Count the cells and adjust the cell number (50 × 10^3 cells/mL). Calculate the total volume required in the assay (later 50 µL of the target cells will be added per well).
3. Centrifuge target cells at 250g for 5 min at 20–22°C in PBMC or mouse effector cell medium, depending on which effector cell is used.
4. Resuspend the pellet in a total volume of 80 µL in PBMC or mouse effector cell medium, depending on which effector cell is used and add ^{51}Cr according to:
    Volume ^{51}Cr = 0.08 × [specific activity (mCi/mL)/decay factor].
5. Incubate for 60 min at 37°C. Keep the tube in a lead container to minimize radiation.
6. Add 10 mL of PBMC or mouse effector cell medium, depending on which effector cell is used and wash twice for 5 min at 250g at 20–22°C.
7. Add 5 mL of PBMC or mouse effector cell medium, depending on which effector cell is used, and incubate 45 min.
8. Wash twice at 250g for 5 min at 20–22°C in PBMC or mouse effector cell medium, depending on which effector cell is used.
9. Resuspend the pellet in volume calculated under **step 2** to 50 × 10^3 cells/mL.
10. Wash effector cells twice at 250g for 5 min at 20–22°C in PBMC or mouse effector cell medium, depending on which effector cell is used.
11. Count and adjust effector cell concentration to 0.75 × 10^6 cells/mL (mouse) and 0.25 × 10^6 cells/mL (PBMC).
12. Add 100 µL of the effector cell suspension to each well in the U-shaped microtiter plate, except the A-line which remains empty.
13. Add 50 µL of the superantigen to the cells in a dose-dependent manner as triplicates.
14. Add 50 µL of the ^{51}Cr incorporated target cells to each well including the A-line.
15. Add 150 µL R medium (background release) and 150 µL of 1% SDS (total release) to five wells in the A-line, respectively.
16. Incubate for 4 h at 37°C.
17. Gently remove 40 µL of the supernatant without suspending the cells and transfer to absorbent plastic plates (harvest plates). Incubate the culture plate for additional 16 h.
18. Incubate the collected supernatant at 20–22°C for 16 h or at 56°C for 2 h.
19. Read plate in a γ-counter according to standard settings.
20. Calculate the specific cytotoxic effect according to the formula:
    Cytotoxicity % = 100 × (cpm experimental release – cpm background release)/(cpm total release – cpm background release).
21. Repeat **steps 17–20** after 20 h of culture.

Table 1
Vβ-Specificity of Some Bacterial Superantigens in Mice[a]

Superantigen	Vβ-specificity	Suggested strains
SEA	1, 3, 10, 11, 17	C57Bl/6, NOD, SJL/J
SEB	7, 8.1–8.3	BALB/c, C3H/He
SEC1	3, 8.2, 8.3, 11	AKR/J, BALB/c, C57Bl/6
SEC2	3, 8.2, 10, 17	AKR/J, BALB/c, C57Bl/6
SEC3	7, 8.2	BALB/c, C3H/He
SED	3, 11, 17	C57Bl/6, NOD, SJL/J
SEE	11, 15, 17	C57Bl/6, NOD, SJL/J
TSST–1	15, 16	AKR/J, BALB/c, C3H/He

[a]Data for this table were taken from **refs. (9–11)**.

3.2.2. Effector/Target Ratio

To determine how efficient CTL is induced by a superantigen, it is recommended to analyze cytotoxicity on a per cell basis. To do this, the ratio of effector/target cells (E/T ratio) is titrated from 300:1 to 1:1 (**Fig. 1B**).

1. Follow **Subheading 3.2.1.**, steps **1–11**.
2. Add 450 μL of the effector cells in a 6-well plate.
3. Transfer 100 μL of the effector cells to three wells in the U-shaped microtiter plate. Remember to leave the A-line empty.
4. Add 300 μL of PBMC or mouse effector cell medium, depending on which effector cell is used, to the remaining 150 μL of effector cells in the six-well plate. Repeat **steps 3–4** four times.
5. Add 50 μL of the superantigen at a concentration determined in **Subheading 3.2.1.**, **step 13**, to be most efficient, to the effector cells in the microtiter plate.
6. Follow **Subheading 3.2.1.**, step **14–21**.

3.3. Activation of CTL Activity In Vivo

Mouse mammary tumor viruses (MTVs) are retroviruses that produce a viral SAg (vSAg) in the 3' long terminal repeats of their DNA. Like bacterial SAg, the vSAg activates T cells by binding to MHC class II molecules and then engaging a T cell expressing a specific TCR Vβ chain on the T cells (*9–11*). The T-cell subsets that respond to the vSAg have been shown to be deleted during thymic maturation (*12*). Thus, to induce SAg-derived CTL activity in vivo mice carrying a high frequency of SAg reactive TCR Vβ chains should be used. Hence mice carrying Mls[2a] and IE should **not** be used when SEA is the bacterial superantigen used because TCR Vβ3, Vβ11, and Vβ17 chains have been deleted in those mice. Instead, the C57Bl/6 strain (Mls[1b–2b]) is one of the most preferred (*see* **Note 5a**). **Table 1** summarizes which TCR Vβ chains can be used by certain bacterial SAg. In addition, certain mouse strains that could be used with a specific SAg are suggested.

3.3.1. Kinetics of CTL Induction In Vivo

The in vivo responses to SAg are very rapid. Substantial cytokine production with high levels of IL-2, TNF-α, and IFN-γ can be detected at mRNA levels after 0.5–1 h and at the protein level already after 1.5–4 h *(7)*. Thus, to record the maximal CTL response after stimulation with a SAg in vivo a kinetic analysis has to be conducted. Splenocytes should be prepared from treated animals between 12 and 120 h after the injection and the CTL activity determined. This is particularly important when a new mouse strain is used. We routinely detect maximal cytotoxicity 48–72 h after a single SEA intravenous injection (**Fig. 2A**, *see* **Note 4**).

1. Dissolve rSEA in PBS supplemented with 1% normal mouse serum (NMS) to a concentration of 50 µg/mL.
2. Inject 0.2 mL of the substance per animal intravenously.
3. Dissect the spleen at various time points after the injection and make a single-cell suspension by carefully mincing the spleen through a nylon mech with the back of a sterile syringe.
4. Follow **Subheading 3.2.2.**

3.3.2. Characterization of Dose–Response Relationship of SAg-Induced CTL Activity In Vivo

Induction of CTL activity in vivo depends on the strength of the signals to the T cell. Numerous antigens induce CTL activity, but the concentration required varies markedly. Thus, to induce maximal CTL activity, in any given mouse strain, after stimulation with a SAg a dose–response study needs to be conducted (**Fig. 2B**).

1. Dissolve rSEA in PBS supplemented with 1% NMS to concentrations varying from 0.005–500 µg/mL.
2. Inject 0.2 mL of the substance per animal intravenously.
3. Dissect the spleen, at the time point determined under **Subheading 3.3.1.** after the injection, and make a single-cell suspension by carefully mincing the spleen through a nylon mech with the back of a sterile syringe.
4. Follow **Subheading 3.2.2.**

3.4. Characterization of Cytotoxic Effector Cells

SEA interacts with TCR Vβ chains on both CD4+ and CD8+ T cells. To pinpoint which cells are the main cytotoxic effector cells, a number of functional assays can be conducted.

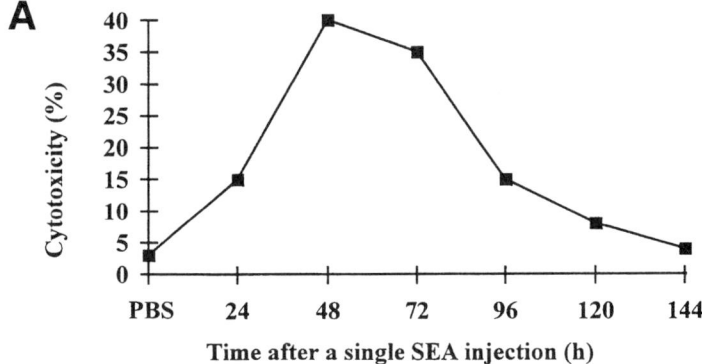

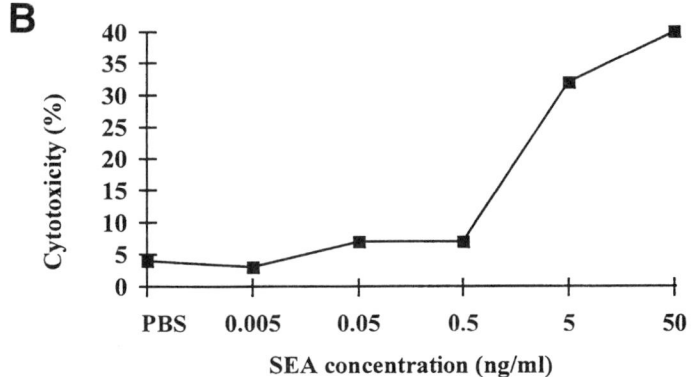

Fig. 2. SEA-dependent cell-mediated cytotoxicity by *in vivo* activated splenocytes from C57Bl/6 mice. Cytotoxicity was measured in a standard 4-h ⁵¹Cr-release assay against SEA-coated (1 μg/mL) A20 cells at (**A**) various time points after an injection of 30 μg of SEA or (**B**) after injections of SEA at increasing concentrations. Mean values from triplicate wells are shown. Standard deviation in the assay was routinely <15%. Data are representative for two similar experiments.

3.4.1. CTL Activity Is Confined to the CD8⁺ T Cells. Enrichment of CD4⁺ and CD8⁺ T Cells by MAb-Coated Beads and Magnetic Separation

This is a rapid and reliable method to positively enrich CD4⁺ or CD8⁺ T cells. The CD4⁺ or CD8⁺ T cells are magnetically captured and a >95% pure population is obtained (**Fig. 3**).

1. Centrifuge the single-cell suspension of SEA activated splenocytes at 250g for 5 min at 4°C in mouse effector cell medium.

A

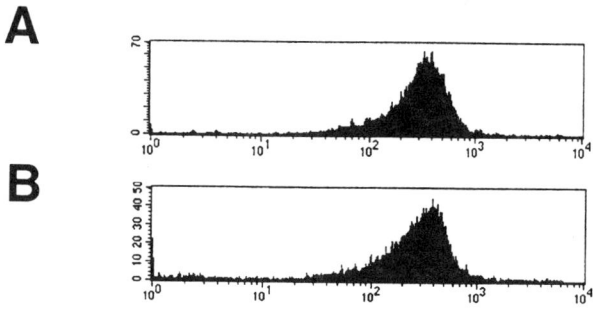

B

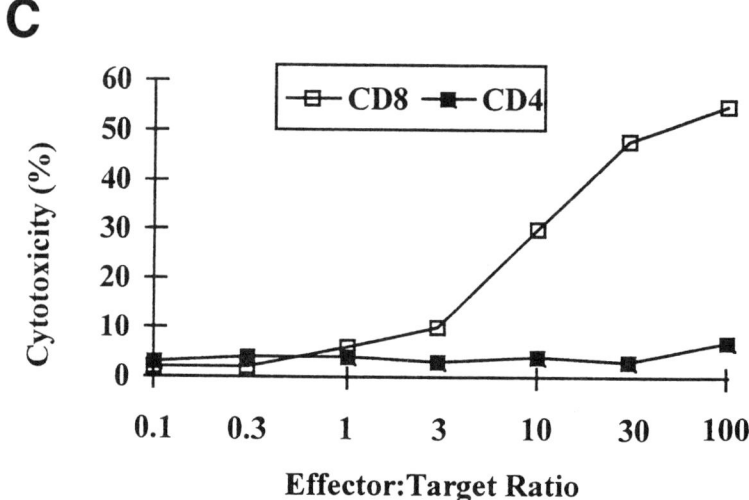

C

Fig. 3. SEA-dependent cell-mediated cytotoxicity by *in vivo* activated splenocytes from C57Bl/6 mice is mediated by CD8+ T cells. Spleens were removed 48 h after an injection of 10 μg of SEA. MACS sorting of (**A**) CD4+ or (**B**) CD8+ T cells were performed and >95% pure subpopulations were obtained. Cytotoxicity was measured in a standard 4-h ^{51}Cr-release assay against SEA-coated (1 μg/mL) A20 cells at various E/T ratios. Mean values from triplicate wells are shown. SD in the assay was routinely <15%. Data are representative for two similar experiments.

2. Add 1 mL of GEY's solution for 2 min at 20–22°C . Shake gently after 1 min.
3. Add 5 mL of MACS buffer.
4. Pass the cells through a nylon filter.
5. Spin twice at 250*g* for 5 min at 4°C in MACS buffer.

6. Add 10 μL of magnetically labeled CD4 or CD8 MACS antibody and 90 μL of MACS buffer per 10^7 cells.
7. Incubate at 4°C for 20 min.
8. Wash twice at 250g for 5 min at 4°C in MACS-buffer.
9. Resuspend the pellet in 0.5 mL MACS buffer.
10. Place the MACS column in the magnetic field and let 0.5 mL MACS-buffer flow through the column. Avoid letting the column dry out.
11. Add the cell suspension to the column and collect the eluate. Place the eluate on the column once more. The flow-through is now discharged.
12. Wash twice with 0.5 mL of MACS buffer.
13. Remove the column from the magnetic field and place on a test tube.
14. Add 1 mL of MACS buffer and press the fluid through rapidly with the back of the syringe.
15. Centrifuge the cells twice at 250g for 5 min at 4°C in mouse effector cell medium. Resuspend in mouse effector cell medium.
16. Perform a ^{51}Cr-release assay as described under **Subheading 3.2.2.**

3.4.2. High Surface Expression of CD11a Is Important for CTL Activity. Cell Sorting by Flow Cytometry

It may be important to determine if purified CD8$^+$ T cells or potentially CD4$^+$ T cells from treated animals have different capacities to perform CTL activity. By sorting for surface expression of various molecules on the activated T cell, cytotoxic function can be discriminated to a CD8$^+$ T cell subpopulation expressing high surface density of the α-chain (CD11a) of the LFA-1 heterodimer (CD11a/CD18) (**Fig. 4A–C,** *see* **Note 5b**). LFA-1 belongs to the β$_2$-integrins and represents an important T cell adhesion molecule crucial for, for example, T-cell activation and stable CTL–target interactions *(13,14)*.

1. Follow **steps 1–15** of **Subheading 3.4.1.,** but stain and sort for CD8$^+$ T cells only. Resuspend in FACS buffer.
2. Transfer 20×10^6 cells/200 μL per FACS tube.
3. Add 40 μL of 25 μg/mL of purified anti-CD32/CD16 MAb to block unspecific binding. Incubate 5 min at 20–22°C.
4. Add 50 μL of a mixture of the FACS-labeled MAb antibodies CD11a-PE/TCR Vβ3-flourescein isothiocyanate (FITC) diluted in FACS-buffer. *See* **Note 1**.
5. Incubate at 4°C for 30 min in the dark.
6. Add 1 mL of FACS buffer and centrifuge the cells twice at 250g for 5 min at 4°C.
7. If biotinylated antibodies are used under **step 4** then add 50 μL of steptavidin-Tricolor or other desired fluorescent and repeat **steps 5–6**.
8. Take up the cells in 1 mL of mouse effector cell medium.
9. Sort the cells on a FACStar into Vβ3$^+$/CD11a$^{(high)}$ and Vβ3$^+$/CD11a$^{(low)}$ populations.
10. Perform a modified ^{51}Cr assay as described in **Subheading 3.2.2.** Use V-shaped microtiter plates. Only use 1000 target cells/well and an E/T ratio of 10:1 down to 0.1:1.

A

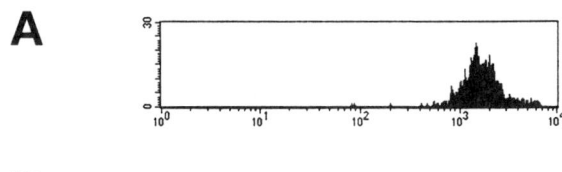

B

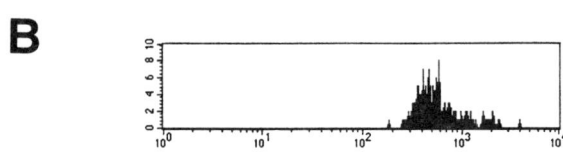

C

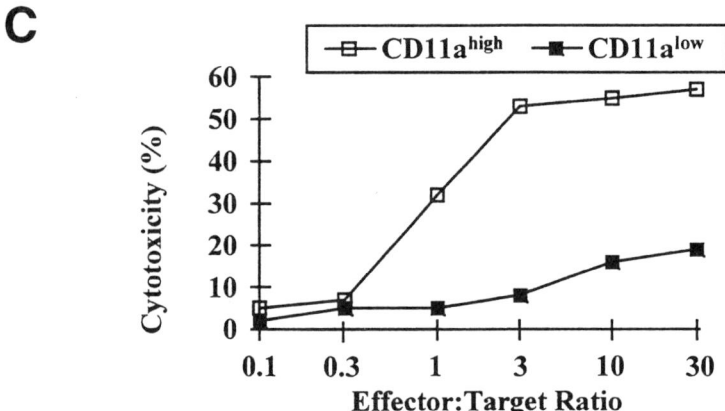

D

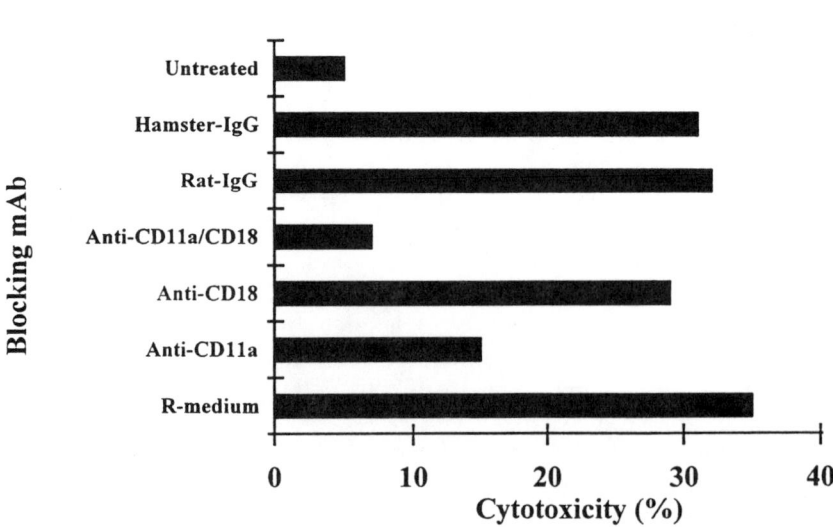

3.4.3. Functional Interaction of CD11a with ICAM-1 Is Required for Optimal CTL Activity. Antibody Blocking Experiments

To determine if certain surface molecules on the T cell or the target cell have direct functional importance for the CTL activity these molecules are blocked by addition of MAbs in vitro (**Fig. 4D**).

1. Follow **steps 1–15** under **Subheading 3.4.1.** and resuspend in mouse effector cell medium.
2. Perform a modified ^{51}Cr-release assay as described in **Subheading 3.2.2.** with the alterations stated below.
3. Add blocking antibodies toward the CD11a (80 µg/mL), CD18 (100 µg/mL), CD54 (100 µg/mL), isotype Ig (100 µg/mL), or medium alone (PBMC- or mouse effector cell medium, depending on which effector cell is used) to three wells, respectively.
4. Use an E/T ratio of 100:1 (mouse) or 30:1 (human).

3.4.4. Interaction Between ICAM-1 and LFA-1 Strongly Augments CTL Activity Against Inflammatory Endothelial Cells

Several different types of target cells may be sensitive to SAg-directed CTL. This may include normal MHC class II⁺ leukocytes or several MHC class II⁺ malignant tumor cells. During inflammatory responses several normal cells may be activated to express MHC class II and ICAM-1 and serve as CTL targets. Endothelial cells are activated at inflammatory sites and begin to express high surface levels of MHC class II and important adhesion molecules such as intercellular adhesion molecule-1 (ICAM-1) (*15,16*). These cells represent potential target cells in vivo, as they can present SAg to SEA reactive CTL and express important adhesion molecules such as ICAM-1 in high surface density. To mimic an inflammatory phenotype and target a T-cell attack toward MHC class II-inducible cells, the HUVEC-derived endothelial cell line EA.hy926 is activated with either TNF-α, IFN-γ or the two cytokines combined.

Fig. 4. (*previous page*) SEA-dependent cell-mediated cytotoxicity by *in vivo* activated splenocytes from C57Bl/6 mice is dependent on high surface expression of CD11a on CD8⁺ T cells. Spleens were removed 48 h after an injection of 10 µg of SEA. MACS sorting of CD8⁺ T cells were performed and >95% pure subpopulations were obtained. Flow cytometric sorting of TCR Vβ3⁺ (**A**) CD11a(high) or (**B**) CD11a(low) expressing cells were performed. CTL activity was examined in a standard 4-h ^{51}Cr-release assay against SEA-coated (1 µg/mL) A20 cells at (**C**) various E/T ratios or (**D**) in the presence of 25 µg/mL of mAbs against various surface molecules as indicated. Mean value from triplicate wells is shown. Standard deviation in the assay was routinely <15%. Data are presented from one out of two similar experiments.

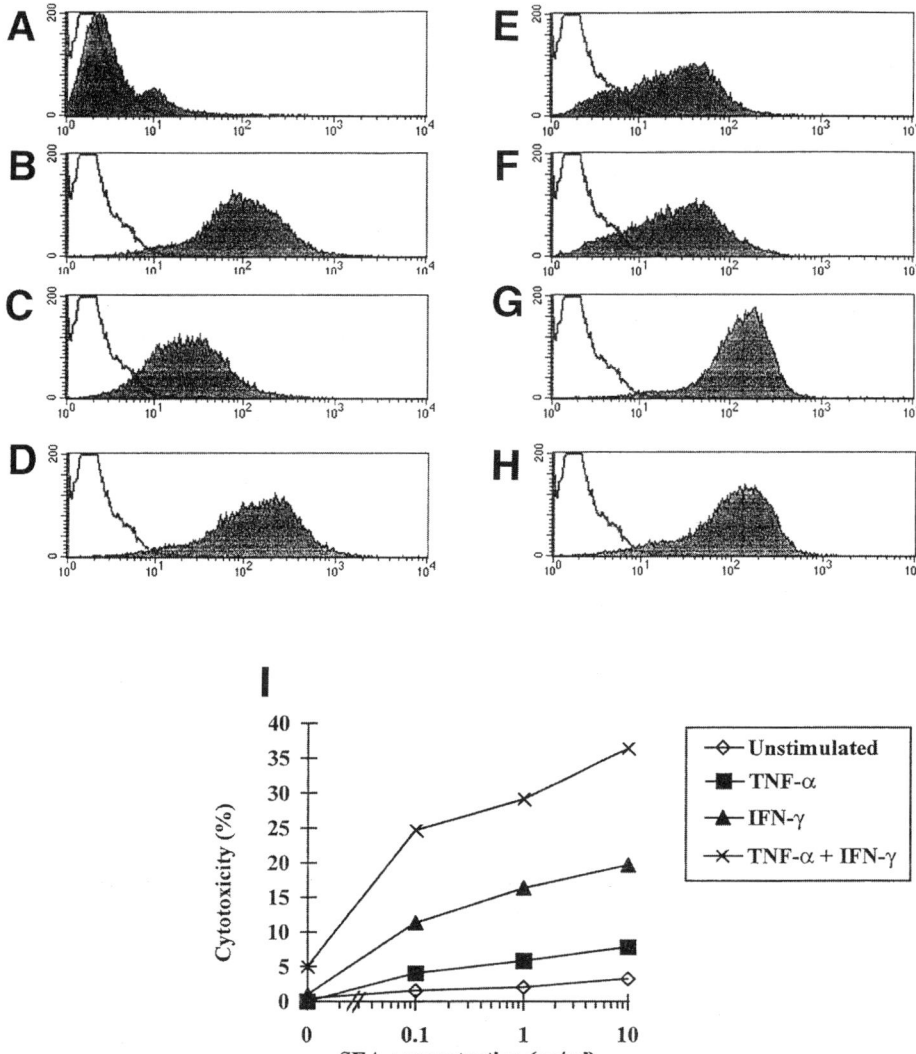

Fig. 5. SEA-dependent cell-mediated cytotoxicity by *in vitro* activated PBM is dependent on high surface expression of CD54 and MHC class II on EA.hy926 cells. Flow cytometry profiles of are shown for ICAM-1 (CD54; **A–D**) and MHC class II (**E–H**). Unstimulated cells (**A,E**) were compared to cultures activated with TNF-α (**B,F**), IFN-γ (**C,F**), or TNF-α and IFN-γ in combination (**D,H**). EA.hy926 cells were treated with TNF-α (250 U/mL) and/ or IFN-γ (100 U/mL) for 18 and 48 h, respectively. (*I*) Cytotoxicity was measured in a standard 4 h ^{51}Cr-release assay against EA.hy926 cells coated with various SEA concentrations at an E/T ratio of 40:1. Mean values from triplicate wells are shown. Standard deviation in the assay was routinely less than 15%. Data are presented from one out of two similar experiments.

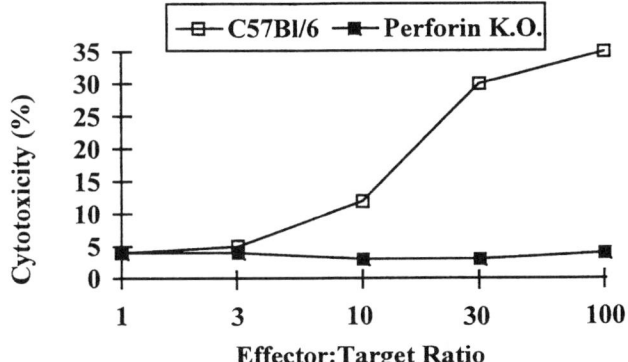

Fig. 6. CTL activity in perforin KO and normal C57Bl/6 mice. SEA (10 μg) was injected intravenously to perforin KO or C57Bl/6 mice. After 48 h spleens were removed and CTL activity was examined in a standard 4-h ^{51}Cr-release assay against SEA-coated (1 μg/mL) A20 cells at various E/T ratios. Mean values from triplicates wells are shown. Standard deviation in the assay was routinely <15%. Two experiments were performed with similar results.

This will result in high expression of ICAM-1 (TNF-α treatment) and MHC class II (IFN-γ) (**Fig. 5**).

1. Activate EA.hy926 cells (5 mL of 0.5×10^6 cells/mL in a T-25 culture flask) with recombinant TNF-α (final concentration 250 U/mL), IFN-γ (100 U/mL), or the cytokines combined.
2. Harvest cells after 18 h (TNF-α-treated cells) or 48 h (IFN-γ); wash the monolayer twice with 5 mL of PBS, add 1 mL of cell dissociation solution (Sigma), and incubate at 37°C. After 5 min, check that all cells are detached from the plastic.
3. Take up the cells in 5 mL of R medium and centrifuge at 250*g* for 5 min at 20–22°C.
4. Count the cells and wash once more (as described in **step 3**).
5. Perform a ^{51}Cr assay as described under **Subheading 3.2.1.**

3.4.5. CD8-Mediated CTL Activity Is Mediated by Release of Perforin

Elimination of target cells by T cells can be conducted by several mechanisms involving perforin release, cytokine secretion, and when FasL expressing T cells interacts with Fas expressing target cells *(17–20)*. To determine whether the SEA-induced CTL activity depends on perforin release, perforin KO mice can be used (**Fig. 6**).

1. Follow **Subheading 3.3.** and also include perforin KO mice. In addition, gld or lpr mice can be used to characterize the importance of Fas–FasL interaction.

2. Follow **Subheading 3.2.2.** and perform a ^{51}Cr release assay with various E/T ratios.

4. Notes

1. The required concentration varies and must be titrated the first time used.
2. Highly activated T cells proliferate rapidly and consume all growth factors in the medium and releases waste products. This results in a color change. It is time to change the medium when the medium turns yellow.
3. If the background CTL activity is too high the target cell may be in poor condition. Always check the status of the target cell under a microscope before use. Moreover, certain cell types may not be suitable for ^{51}Cr incorporation experiments owing to leakage.
4. SAgs are very potent and may induce acute toxicity in the animals. Therefore, close observation during the first 6 h after the injection has to be conducted.
5. If no or marginal CTL activity is detected:
   a. Characterize by FACS the TCR Vβ profile in the mouse strain used. Some major TCR Vβ families may have been deleted.
   b. Investigate whether the CD8$^+$ T cells express high CD11a on the surface. If no up-regulation can be detected it may be due to suboptimal dosage or unsuccessful immunization with the SAg.

Acknowledgments

We thank Ms. I. Andersson for skillful technical assistance with in vitro cell cultures, Ms. K. Behm for excellent assistance with in vivo experiments, Ms. L. Karlsson for excellent assistance with the endothelial cytotoxicity assays, and Ms. A. Åberg for skillful technical assistance with the flow cytometer.

References

1. Marrack, P. and Kappler, J. (1990) The staphylococcal enterotoxins and their relatives [published erratum appears in Science 1990 1;248(4959):1066]. *Science* **248,** 705–711.
2. Betley, M. J., Borst, D. W., and Regassa, L. B. (1992) Staphylococcal enterotoxins, toxic shock syndrome toxin and streptococcal pyrogenic exotoxins: a comparative study of their molecular biology. *Chem. Immunol.* **55,** 1–35.
3. Fraser, J. D. (1989) High-affinity binding of staphylococcal enterotoxins A and B to HLA-DR. *Nature* **339,** 221–223.
4. Fischer, H., Dohlsten, M., Lindvall, M., Sjogren, H. O., and Carlsson, R. (1989) Binding of staphylococcal enterotoxin A to HLA-DR on B cell lines. *J. Immunol.* **142,** 3151–3157.
5. Fischer, H., Dohlsten, M., Andersson, U., Hedlund, G., Ericsson, P., Hansson, J., and Sjogren, H. O. (1990) Production of TNF-alpha and TNF-beta by staphylococcal enterotoxin A activated human T cells. *J. Immunol.* **144,** 4663–4669.
6. Carlsson, R. and Sjogren, H. O. (1985) Kinetics of IL-2 and interferon-gamma production, expression of IL-2 receptors, and cell proliferation in human

mononuclear cells exposed to staphylococcal enterotoxin A. *Cell Immunol.* **96,** 175–183.

7. Dohlsten, M., Bjorklund, M., Sundstedt, A., Hedlund, G., Samson, D., and Kalland, T. (1993) Immunopharmacology of the superantigen staphylococcal enterotoxin A in T-cell receptor V beta 3 transgenic mice. *Immunology* **79,** 520–527.

8. Rosendahl, A., Hansson, J., Antonsson, P., Sekaly, R. P., Kalland, T., and Dohlsten, M. (1997) A mutation of F47 to A in staphylococcus enterotoxin A activates the T-cell receptor V beta repertoire in vivo. *Infect. Immun.* **65,** 5118–5124.

9. Janeway, C. A., Jr., Yagi, J., Conrad, P. J., Katz, M. E., Jones, B., Vroegop, S., and Buxser, S. (1989) T-cell responses to Mls and to bacterial proteins that mimic its behavior. *Immunol. Rev.* **107,** 61–88.

10. Scherer, M. T., Ignatowicz, L., Winslow, G. M., Kappler, J. W., and Marrack, P. (1993) Superantigens: bacterial and viral proteins that manipulate the immune system. *Annu. Rev. Cell Biol.* **9,** 101–128.

11. Kotzin, B. L., Leung, D. Y., Kappler, J., and Marrack, P. (1993) Superantigens and their potential role in human disease. *Adv. Immunol.* **54,** 99–166.

12. Kappler, J. W., Staerz, U., White, J., and Marrack, P. C. (1988) Self-tolerance eliminates T cells specific for Mls-modified products of the major histocompatibility complex. *Nature* **332,** 35–40.

13. Dustin, M. L. and Springer, T. A. (1989) T-cell receptor cross-linking transiently stimulates adhesiveness through LFA–1. *Nature* **341,** 619–624.

14. Sundstedt, A., Hoiden, I., Hansson, J., Hedlund, G., Kalland, T., and Dohlsten, M. (1995) Superantigen-induced anergy in cytotoxic CD8+ T cells. *J. Immunol.* **154,** 6306–6313.

15. Scholz, M., Cinatl, J., Gross, V., Vogel, J. U., Blaheta, R. A., Freisleben, H. J., Markus, B. H., and Doerr, B. H. (1996) Impact of oxidative stress on human cytomegalovirus replication and on cytokine-mediated stimulation of endothelial cells. *Transplantation* **61,** 1763–1770.

16. Viac, J., Schmitt, D., and Claudy, A. (1994) Adhesion molecules and inflammatory dermatoses. *Allerg. Immunol. Paris* **26,** 274–277.

17. Kagi, D., Ledermann, B., Burki, K., Seiler, P., Odermatt, B., Olsen, K. J., Podack, E. R., Zinkernagel, R. M., and Hengartner, H. (1994) Cytotoxicity mediated by T cells and natural killer cells is greatly impaired in perforin-deficient mice. *Nature* **369,** 31–37.

18. Podack, E. R. and Kupfer, A. (1991) T-cell effector functions: mechanisms for delivery of cytotoxicity and help. *Annu. Rev. Cell Biol.* **7,** 479–504.

19. Podack, E. R., Young, J. D., and Cohn, Z. A. (1985) Isolation and biochemical and functional characterization of perforin 1 from cytolytic T-cell granules. *Proc. Natl. Acad. Sci. USA* **82,** 8629–8633.

20. Rosendahl, A., Kristensson, K., Hansson, J., Riesbeck, K., Kalland, T., and Dohlsten, M. (1998) Perforin and IFN-γ are involved in the anti-tumor effects of antibody targeted superantigens. *J. Immunol.* **160,** 5309–5313.

15

In Vitro Physiological Studies on Clostridial Neurotoxins

Biological Models and Procedures
for Extracellular and Intracellular Application of Toxins

Bernard Poulain, Marie-France Bader, and Jordi Molgó

1. Introduction

Botulinum (BoNT, serotypes A–G, *see also* Chapter 2) and tetanus (TeNT) neurotoxins are known under the generic term of clostridial neurotoxins. These dichainal proteins comprise a light (M_r ~50) and a heavy (M_r ~100) chain that are disulfide linked. In mammals, these proteins are the causative agents of two severe neuroparalytic diseases, botulism and tetanus. Botulism manifests as a flaccid muscle paralysis caused by a near irreversible and selective inhibition of acetylcholine release at the skeletal neuromuscular junction. Tetanus is characterized by a spastic neuromuscular paralysis that results from motoneuron disinhibition following the specific blockage of inhibitory glycinergic or γ-aminobutyric acid-ergic (GABAergic) synapses by TeNT in the central nervous system (CNS). The cellular action of BoNT and TeNT can be depicted according to several steps. After binding to specific acceptors located at the nerve ending membrane, TeNT and BoNT are endocytosed. Subsequently, their active moiety (the light chain) is translocated from the endocytic compartment into the cytosol. Here, it cleaves one among three synaptic proteins (viz. the vesicle-associated membrane protein [VAMP]/synaptobrevin, syntaxin, and synaptosomal-associated protein of M_r 25 kDa [SNAP-25] which are also known under the collective term of SNAREs) involved in docking and fusion of synaptic vesicles at the active zone (i.e., release site) (for reviews *see [1,2]*).

The biological preparations that have been used to study the blocking action of clostridial neurotoxins come from species that belong to the various branches

From: *Methods in Molecular Biology, vol. 145: Bacterial Toxins: Methods and Protocols*
Edited by: O. Holst © Humana Press Inc., Totowa, NJ

of the animal phylogenic tree. The vertebrates that are currently exploited include mammals (e.g., mouse, rat), birds (chick), amphibians (frog), and fishes (goldfish and *Torpedo*). Other useful models comprise molluscs (*Aplysia* and squid), arthropods (*Drosophila*), and annelids (leech). However, not all species are equally sensitive to the various serotypes of clostridial neurotoxins. As detailed below, the resistance to clostridial toxins might be attributed to differences in *(1)* the membrane binding, *(2)* intracellular trafficking of the toxins, or *(3)* in the susceptibility of the synaptic target to proteolytic attack.

The neurotoxin receptors are not yet identified; however, they appear to be well conserved in nerve cells during evolution. For example, cholinergic neurons in *Aplysia* exhibit a sensitivity to extracellularly applied BoNT or TeNT that is very close to that of the neuromuscular junction. In fact, the neurotoxin receptors seem to be expressed by all neuronal cells but not by secretory cells (e.g., chromaffin cells, islet cells from the pancreas). This explains why nerve–muscle and neuroneuronal preparations have been widely exploited before the advent of intracellular application procedures that allow the experimenter to bypass the membrane limiting steps. In addition, there are differences in the intracellular trafficking of the neurotoxins. An example is provided by comparing the action of TeNT at cholinergic peripheral and central nerve endings: TeNT exhibits a very low efficiency in blocking acetylcholine release at the rat motor endplate *(3)* whereas it is very potent at cholinergic synapses in the rat striatum *(4)*. Because it is well established that TeNT is avidly taken up by the motor nerve terminal (this is the way by which TeNT gains access to the CNS by axonal retrograde transport), the difference observed at the neuromuscular junction and striatal synapses indicates that the intraneuronal routing of a given toxin may differ even in cells that are believed to be related (in this case cholinergic neurons).

In neurons or neuroendocrine cells, an important source of variation in the susceptibility of exocytotic mechanisms to be inhibited by the various clostridial neurotoxins is the presence of mutations in the amino acid sequence of their synaptic targets (VAMP/synaptobrevin, SNAP-25, or syntaxin). This form of resistance to the action of the toxins is now well documented. For example, in the rat and the chick, there is a point mutation (Gln to Val substitution) at the cleavage site for TeNT and BoNT/B in the VAMP/synaptobrevin isoform 1 but not in isoform 2 *(5,6)* and, indeed, rat VAMP/synaptobrevin 1 but not isoform 2 is resistant to either TeNT or BoNT/B *(5)*. Apparently, this confers to the rat and chick a lower susceptibility to tetanus compared to mice or humans, in which both VAMP/synaptobrevin 1 and 2 exhibit the same sequence (Gln-Phe) at the cleavage site. Note that the SNAP-25 of the leech, *Hirudo officinalis*, cannot be attacked by BoNT/A, also owing to a mutation at its cleavage site *(7)*. In nonneuronal cells, an additional level of toxin resistance that is encountered when examining the toxin action on exocytosis is the implication of proteins that are functional homologues of VAMP/synapto-brevin, SNAP-25, and syntaxin 1 but are not targets for the clostridial neuro-toxins. For instance, SNAP-23 cannot be cleaved by BoNT/A *(8)* and the TeNT-insensitive VAMP homologue Ti-VAMP is not attacked by TeNT *(9)*.

The aim of this chapter is to comment on the merits and limitations of several popular animals models used in studying the potent action of clostridial neurotoxins. Special emphasis is given to preparations in which cell-to-cell connectivity is maintained intact (i.e., mainly the neuromuscular junction). We do not detail the many preparations in which the cell integrity is broken (such as synaptosomes). In addition, we briefly review the advantages of several nonneuronal secretory cells widely used in the study of the effects of clostridial neurotoxins. We also describe several of the methods used to apply the neuro-toxins or their derivatives, extracellularly or intracellularly.

2. Materials

2.1. Chemicals

2.1.1. Clostridium Toxins and Derivatives
(See *Chapter 2 of This Volume for Purification of Botulinum Toxin*)

1. Botulinum toxin (BoTx, seven serotypes A–G) corresponds to a complex comprising the botulinum neurotoxin (BoNT) and several nonneurotoxic proteins (*see* **Note 1**). Available as crystalline powder.
2. Botulinum neurotoxin (BoNT, seven serotypes A–G).
3. Tetanus toxin or neurotoxin (TeNT).
4. Reduced BoNT or TeNT. Depending on the cells used, reduction of toxin may be needed before intracellular application (*see* **Note 1**).
5. TeNT-L chain, BoNTs-L chain, highly purified from reduced TeNT or BoNTs or recombinant.
6. TeNT-H chain, BoNTs-H chains, highly purified.

2.1.2. Injection Procedure

1. Paraffin oil.
2. Fluorinert™ Liquid FC-77 (3M, Minnesota).
3. A vital dye. It may be fast green FCF (Sigma), Texas red dextran, or fluorescein isothiocyanate (FITC) dextran (Molecular Probes).

2.1.3. Labeling of Catecholamine Stores

1. [^{3}H]Noradrenaline.
2. Dulbecco's modified Eagle's medium (DMEM) containing 125 n*M* noradrenaline but no amino acids.
3. Locke's solution.
4. Ca^{2+}-free Locke's solution.

2.1.4. Permeabilization Procedure

1. Streptolysin-O (SLO) (Institut Pasteur or Sigma).

2.2. Buffers

2.2.1. Neurotransmitter Release from Excised Preparations

1. Mammalian saline: 136.8 mM NaCl, 5 mM KCl, 2 mM CaCl$_2$, 1 mM MgCl$_2$, 24 mM NaHCO$_3$, 1 mM NaH$_2$PO$_4$, 11 mM glucose, pH 7.4, gassed with 5% CO$_2$ + 95% O$_2$.
2. Chick saline: 144 mM NaCl, 3 mM KCl, 3 mM CaCl$_2$, 5 mM NaHCO$_3$, 10 mM N-hydroxyethlpiperazine-N'-ethanesulfonic acid (HEPES), 12 mM glucose, pH 7.2.
3. Amphibian saline: 115 mM NaCl, 2.5 mM KCl, 1.8 mM CaCl$_2$, 2.15 mM Na$_2$HPO$_4$, 0.85 mM NaH$_2$PO$_4$, pH 7.25.
4. Marine fish saline (elasmobranch saline): 280 mM NaCl, 7 mM KCl, 4.4 mM CaCl$_2$, 1.3 mM MgCl$_2$, 5 mM Na$_2$HCO$_3$, 20 mM HEPES, 5.5 mM glucose, 300 mM urea, pH 7.2, gassed with 5% CO$_2$ + 95% O$_2$.
5. Fresh Water Fish Saline (goldfish saline): 132 mM NaCl, 3.1 mM KCl, 3 mM CaCl$_2$, 1 mM MgCl$_2$, 2.15 mM Na$_2$HPO$_4$, 0.85 mM NaH$_2$PO$_4$, 5.5 mM glucose, pH 7.2.
6. *Drosophila* saline: 128 mM NaCl, 2 mM KCl, 2 mM CaCl$_2$, 4 mM MgCl$_2$, 5 mM HEPES, pH 7.2.
7. *Aplysia* saline: 460 mM NaCl, 10 mM KCl, 11 mM CaCl$_2$, 25 mM MgCl$_2$, 28 mM MgSO$_4$, 10 mM Tris-HCl, pH 7.4. To diminish spontaneous neuron activity, the concentration of divalent cations may be increased (*10*). However, the [Ca^{2+}]/[Mg^{2+}] ratio should be kept constant to avoid change in neurotransmitter release.
8. Squid saline: 466 mM NaCl, 10 mM KCl, 11 mM CaCl$_2$, 54 mM MgCl$_2$, 3 mM Na$_2$HCO$_3$, 50 mM HEPES; pH 7.2.

2.2.2. Catecholamine Secretion from Chromaffin Cells

1. Locke's solution: 140 mM NaCl, 4.7 mM KCl, 2.5 mM CaCl$_2$, 1.2 mM KH$_2$PO$_4$, 1.2 mM MgSO$_4$, 11 mM glucose, 0.56 mM ascorbic acid, 15 mM HEPES, pH 7.5.
2. Ca^{2+}-free Locke's solution: 140 mM NaCl, 4.7 mM KCl, 1 mM EGTA, 1.2 mM KH$_2$PO$_4$, 1.2 mM MgSO$_4$, 11 mM glucose, 0.56 mM ascorbic acid, 15 mM HEPES, pH 7.5.

2.2.3. Permeabilization Medium

1. 150 mM Glutamate potassium salt, 10 mM 1,4-piperazine *bis*-ethanesulfonic acid (PIPES), 5 mM nitrilotriacetic acid, 0.5 mM EGTA, 5 mM MgATP, 4.5 mM magnesium acetate, 0.2% bovine serum albumin, pH 7.2, adjusted with 1 M KOH.

2.2.4. Poisoned Frog Maintenance Medium

1. 0.125 mL of Ringer + 2 g of glucose per liter of fresh water.

2.3. Preparation of Toxins Solutions

2.3.1. Stock Solutions

Crystalline BoTx, purified BoNT, or TeNT may be dissolved in sodium acetate or sodium phosphate buffer (70 mM; pH 6.5) Alternatively, BoNT and TeNT may be dissolved in 140 mM Tris-HCl, pH 7.9. Gelatine is usually added (0.2%, w/v) to diminish nonspecific inactivation of toxins.

Other buffers may be used. However, when the purpose of experiments is to apply toxins or their chains using the permeabilization procedure, direct addition of toxin sample to the permeabilization medium can be made provided the stock solution has been prepared in Ca^{2+}-free medium. This avoids Ca^{2+} entry in permeabilized cells and elicitation of secretion upon toxin application.

Note that treatments aimed to prevent degradation of the toxins may interfere with their biological activity. For example:

1. Protease inhibitors used to prevent toxin degradation may alter the proteolytic activity of BoNT or TeNT.
2. Metabolic inhibitors used to prevent bacterial growth (e.g., sodium azide) may exert toxic action in animal or on excised preparations if they are not removed before toxin application.

When solutions of purified H- or L-chains are prepared, addition of DTT (1 mM) is recommended to prevent dimerization of chains. However, dithiothreitol (DTT) may affect the biological activity of the L-chain or the secretion process (*see* **Note 1**).

Prior to use, biological activity of BoTx , BoNT, or TeNT could be determined using the mouse bioassay (*see* **Note 2**) for determination of MLD$_{50}$/mg of protein. Typical values for BoNTs and TeNT range between 10^5 and 10^8 MLD$_{50}$/mg of protein, depending on the purity and serotype of toxin.

2.3.2. Toxin Storage

When dissolved, BoTx, TeNT, BoNTs, and their purified chains can be stored for few days at 4°C or for years at –20°C or –80°C. In our hands, the TeNT-L chain could be kept biologically active for years when stored at 4°C in sodium phosphate buffer.

2.3.3. Toxin Solution for Extracellular Application

Ideally, neurotoxins or fragments from stock solution are diluted (or dialyzed) into the appropriate physiological buffers used for studying neurotransmitter release or hormone secretion (*see* **Subheading 2.2.**). When samples are added directly to the tissue bath, special care should be taken when a stock solution of toxin contains phosphate or high K^+ concentration. Indeed, (1) in the presence of Ca^{2+}, there is formation of insoluble $Ca_2(PO_4)_2$ precipitate that may induce changes in extracellular Ca^{2+} concentration, thus modifying Ca^{2+}-dependent neurotransmitter release; and (2) extracellular K^+ concentration determines transmembrane polarization and cell excitability.

2.3.4. Toxin Solution for Intracellular Injection

In principle, any buffer compatible with cell life may be used. When stock solution of toxins or their purified chains have been prepared in phosphate buffer, it is better to dialyze the samples against phosphate-free buffer and low (~1 mM) DTT before use. Indeed, (1) before impalement, when the injection micropipet is plunged into the extracellular bath (i.e., Ca^{2+}-containing medium) an insoluble calcium phosphate precipitate forms that can block the tip of the micropipet, and (2) it is better to avoid media containing high concentration of DTT owing to possible interference with the exocytotic machinery (*see* **Note 1**).

For example, the buffers used for intracellular injection in *Aplysia* neurons or squid giant nerve terminals are given:

1. Injection in *Aplysia*: 100 mM NaCl, 20–50 mM Tris-HCl, pH 7.8.
2. Injection in squid nerve terminal: 100 mM KCl, 250 mM K isothionate, 100 mM taurine, 50 mM HEPES, pH 7.7.

2.3.5. Application to Permeabilized Cells

Sample should be dissolved (or dialyzed) in permeabilization medium (*see* **Subheading 2.2.3.**) prior to their application. If another buffer is used, it should be Ca^{2+} free (*see* **Subheading 2.3.1.**).

2.3.6. Streptolysin-O Solution

Streptolysin-O (SLO) is dissolved in water containing 1 mM (DTT) to give a stock solution that may be stored for several days at 4°C. The hemolytic activity of the SLO stock solution is assayed against 2.5% rabbit erythrocytes. The SLO dilution hemolyzing 50% of the erythrocytes is taken as the number of hemolytic units per milliliter of the undiluted SLO stock solution.

2.4. Biological Models

2.4.1. The Skeletal Neuromuscular Junction (NMJ) Preparations

Motor nerve endings are the natural sites of action for not only BoNT but also TeNT. Indeed, TeNT is primarily taken up at the motor terminal before being delivered to the CNS. Moreover, at low concentrations, TeNT causes flaccid paralysis of the goldfish owing to the blockage of acetylcholine release at the NMJ, a situation found in mammals at higher doses (as in the flaccid paralysis that occurs in cephalic tetanus). Nerve–muscle preparations excised from killed rats, mice, frogs, birds, and fishes provide easy models for characterizing the blocking action of both BoNT and TeNT on neurotransmission. Several of these preparations are favorable for characterizing the toxin action via twitch muscle contraction measurements (*see* **Note 3**). All of them also allow the study of the neurotoxin mechanisms in its most elementary (i.e., quantal) aspects by the way of electrophysiological approaches. A particular advantage for using the NMJ is the fact that, in general, there is a 1:1 relationship between the pre- and postsynaptic elements. Hence, analysis of the postsynaptic responses recorded electrophysiologically refer only to the release events coming from the afferent presynaptic nerve terminal. This situation contrasts strongly with the vast majority of neuro–neuronal synapses where the postsynaptic cell receives inputs from hundred to thousand of different presynaptic neurons (a notable exception is the giant synapse of the squid). A limitation of the NMJ is the relative small size of its nerve terminal (<1 μm in diameter) which precludes intracellular recording or injection.

2.4.1.1. RODENT (MOUSE, RAT, GUINEA PIG)

1. Main neuromuscular preparations: the most popular preparations used for characterizing in vitro the action of BoNT and TeNT at the NMJ are the phrenic nerve-hemidiaphragm (rat, mouse, guinea pig), extensor digitorium longus, soleus muscles, triangularis sterni, levator auris longus, and the plantar nerves–lumbrical muscles (for further details *see* **Note 4**).
2. Advantages: Several NMJ preparations isolated from rat and mouse allow an easy characterization of the clostridial neurotoxins action via nerve-evoked twitch muscle contraction measurements (*see* **Note 3**) or conventional electrophysiological recording of spontaneous and evoked postsynaptic activities. In addition, several thin muscles permit the morphological characterization of the trophic events that are associated with the functional recovery from BoNT poisoning (*see* **Subheading 3.1.2.1.** and **Note 4**).
3. Recording conditions: Recording can be performed either at 22°C or at higher temperatures (30–37°C) using a temperature-regulated organ bath. A good oxygenation of the medium (mammalian saline) is needed (*see also* **Note 3** on the twitch muscle bioassay and comment on temperature and poisoning in **Subheading 3.1.1.**).

2.4.1.2. FROG

1. Species used: *Rana esculenta, R. temporaria, R. pipiens, R. nigromulata.*
2. Main neuromuscular preparations used: These are the nerve pectoralis propius–cutaneous pectoris and sciatic nerve–sartorius muscle.
3. Advantages: The frog NMJ is one of the best characterized preparations, with a nerve terminal extension that ranges between 350 and 1100 μm. It presents a typical longitudinal organization of the release sites (i.e., active zones) spaced at regular intervals of about 1 μm. The frog NMJ is very favorable for electrophysi-

ological recordings of neurotransmission and spontaneous and nerve-evoked quantal transmitter release (for a review *see [11]* and references therein). In addition, electrophysiological techniques can be coupled to nerve terminal imaging *in situ* using confocal laser scanning microscopy or with morphological studies. Owing to its well-defined architecture, frog neuromuscular junctions are very suitable for optical monitoring of synaptic vesicle endocytosis, recycling, and exocytosis resolved by the microfluorometric imaging of lipophilic dyes. These approaches mainly involve styryl dyes (e.g., FM1-43, RH414) that partition reversibly into the outer leaflet of the surface membrane and are trapped in endocytic vesicles or released during exocytosis (reviewed in *[12]*). The frog NMJs are also very suitable for immunodetection of the impressive array of presynaptic proteins organized in bands believed to play a role in regulated exocytosis *(13)*.

4. Limitations: We observed that BoNT/B and TeNT are ineffective in producing flaccid muscle paralysis; nevertheless, TeNT produces a strong spastic paralysis of central origin (tetanus).

5. Recording conditions: Because frog is a cold-blooded species, electrophysiological recordings can be made in a wide temperature range (4–22°C), but *see* comment in **Subheading 3.1.2.3.** No oxygenation of the medium (frog saline) is required for maintaining the release properties for hours, unless temperature is raised over 22°C.

2.4.1.3. CHICK

The ciliary ganglion–iris muscle preparation has also been used for assessing the dose-dependent effects of BoNT/A and /E or the interaction between the isolated light and heavy chains of BoNT/A by measuring the contractile force in response to nerve stimulation *(14)* (*see* **Note 3**). Another very suitable preparation for nerve-evoked twitch muscle contraction recording during several hours is the biventer cervicis muscle *(15,16)*.

2.4.1.4. GOLDFISH

The effects of TeNT have been particularly studied in nerve–muscle preparations isolated from the fins of goldfish (*Carassius auratus* L.). The abductor surperficialis or abductor ventralis nerve–muscle preparations of the goldfish are suitable for an in vitro assay of the toxin effects. For a detailed description of the isolation of NMJ and the recording conditions (electrophysiological or by muscle contraction measurement) *see (17)*. A striking characteristic of the goldfish is the inability of TeNT to produce spastic paralysis as it does in frog and mammals; in fact it produces only flaccid muscle paralysis. Note that the glycinergic inhibitory neuronal system acting on the Mauthner cells is highly insensitive to TeNT *(17)*; this is in marked contrast with the preferential action of TeNT on glycinergic and GABAergic mammalian spinal neurons.

2.4.2. Non-Skeletal Neuromuscular or Related Preparations

2.4.2.1. MAMMALIAN AUTONOMIC NEUROMUSCULAR JUNCTION

Botulism presents several autonomic alterations that can predominate over neuromuscular symptoms in the early stage of the disease. Examination of the action of BoNT on autonomic neurotransmission can be performed by evaluating the contractile response of several smooth muscle preparations. The longitudinal smooth muscle of the ileum allows examination of cholinergic autonomic transmission. The anococcygeus muscle is very convenient for investigating sympathetic noradrenergic transmission. Nonadrenergic, noncholinergic nerves have a widespread distribution within the gastrointestinal and urogenital tracts. Examination of clostridial neurotoxins on nonadrenergic, noncholinergic peripheral nervous system is possible by using the taenia coli muscle or the detrusor strips of the urinary bladder. For a comparative study *see (18)*.

2.4.2.2. TORPEDO ELECTRIC ORGAN

The electromotor system of *Torpedo marmorata* and other "electric fishes" is ontogenetically homologous to the neuromuscular junction. It has been studied either electrophysiologically or using a biochemical assay for acetylcholine release. TeNT and BoNT/A blocks the electrical discharge of the electric organ prisms by impairing the release of acetylcholine from nerve endings *(4,19)*. The electromotor system preparation of *Torpedo* is also suitable for characterizing the different steps of the mechanism of action of clostridial neurotoxins. For example, after its injection in the electric organ, [^{125}I]TeNT specifically binds to the neuronal plasma membrane, is internalized into nerve terminals, and is retrogradely transported to the electric lobe *(20)*.

2.4.2.3. DROSOPHILA LARVA NMJ

A glutamatergic motor nerve–muscle preparation of *D. melanogaster* larvae is a model widely exploited to examine neurosecretion using genetic approaches. For example, transgenic expression of the light chain of TeNT in the nervous system of *Drosophila* allows the consequence of the functional deletion of VAMP/synaptobrevin on neuroexocytosis to be investigated *(21)*. Furthermore, another interest for using *Drosophila* is the existence of mutants with very large sized motor nerve endings that allow direct recording of the synaptic buttons using patch-clamp recording techniques *(22)*.

2.4.3. Neuro–Neuronal Preparations from Invertebrates

2.4.3.1. APLYSIA

The sea mollusc, *Aplysia* (e.g., *A. californica, A. punctata, A. depilans*) contains several nerve ganglia with easily identifiable neurons. In *A. californica*, identified cholinergic and noncholinergic synapses have been exploited for the study of the structure–function relationship and mode of action of the BoNT and TeNT *(5,23–26)*. Indeed, *Aplysia* neurons combine a high sensitivity to TeNT or BoNT and afford an easy access to the intracellular medium by way of microinjection procedures. Neurotransmitter release can be easily quantified by measuring the amplitude of postsynaptic responses evoked either by a presynaptic action potential or a long depolarization of the presynaptic neuron, using conventional electrophysiological (current- or voltage-clamp) techniques (for further details *see [10,23]*). Among the advantages encountered in this preparation is the possibility of injecting neurotoxins, their corresponding fragments, or mRNA encoding toxin fragments inside the presynaptic cell bodies from which they can reach (in 5–15 min) the nearby nerve endings (300–500 µm away). The cell body behaves as a reservoir that allows the maintenance of a stable concentration of toxins or derivatives for hours.

Note that in *Aplysia*, certain neurotoxin serotypes (BoNT/F and BoNT/C) are ineffective (B. Poulain, *unpublished observation*). For an unknown reason, for all the BoNT serotypes tested, the presence of the heavy chain, applied to the extracellular space from which it can enter the cytosol or directly injected inside the neuron, is required for the BoNT light chain to exert its blocking activity *(23–27)*. This is not the case for TeNT *(26,27)*. Usually, the temperature for recording is ideally near 20–22°C. No oxygenation of the physiological medium is required but continuous superfusion allows longlasting experiments (more than 20 h of continuous recordings).

2.4.3.2. SQUID

The stellate ganglion dissected from the mantle of the squid (*Loligo pealii*) has been used to determine the effects of TeNT and its light chain on neurotransmitter release, as well as on voltage-dependent presynaptic calcium current, using electrophysiological techniques *(28–30)*. The main interest of the preparation resides in the extraordinarily large size of the nerve terminal. It branches in several fingerlike extensions (e.g., 250 µm long on 70 µm diameter). This unique characteristic allows injection of neurotoxins or their constituent chains directly at the vicinity of the release sites. However, the other digits and axon represents a very large diffusional sink that leads to a rapid decrease in the local concentration of injected material.

2.4.3.3. LEECH

Cultured Retzius neuron cells from *Hirudo officinalis* are a valuable alternative to *Aplysia* synapses. Using immunoblotting techniques, this in vitro model provides the unique opportunity to characterize the cleavage of the neurotoxins' target in a single neuron *(7)*.

2.4.4. Neuro-Neuronal Preparations from Vertebrates

Owing to the inherent intricacy and heterogeneity of neuronal connections, tremendous difficulties are encountered when examining the action of TeNT or BoNT in the CNS. Very often, it is difficult to distinguish the direct effects of the toxins on a given synapse from those attributable to changes in the activity of neuronal networks (for review *see [31,32]*). In general, the isolated preparations (brain slices, dissociated cells, synaptosomes) that have been used to characterize the action of TeNT or BoNT are heterogeneous. In addition their short life time requires the use of high doses of neurotoxins to obtain rapid and significant effects. Primary cultures in which neurons establish synapses in vitro appeared soon as valuable preparations, notably because they permit long-term studies. Because cultured neurons provide an easy access to the cell somata, this allows conventional electrophysiological recordings of neuro-transmission *(31,32)* and also investigation of the membrane events induced by the clostridial neurotoxins (viz. the creation of membranes pores *[33]*). Note, however, that cultured neurons remain as heterogeneous as the brain structures from which the cells are isolated. This problem has been partially solved by studying the action of the toxins on homogeneous populations of neurons making autapses (i.e., making synapses on themselves) *(34)*. An elegant alternative consists of studying the toxins on hippocampal slices maintained in organotypic culture. This preparation offers the double advantage of presenting the very well defined neuronal connection organization of the hippocampus (glutamatergic terminals can be studied independently of the GABAergic ones) while still affording easy access to the nerve cell somata as in conventional neuron cultures (reviewed in *[35]*).

A serious limitation of most neuronal preparations is that there is no direct access to the cytosolic compartment of the nerve endings. In theory, this is possible via injection of toxins or their derivatives into the cell body. However, diffusion/axonal transport of large peptides (the light chain of the neurotoxins is ~50 kDa) into the presynaptic arborization can take several hours, a time scale that is very often incompatible with the life time of the preparations. Few preparations provide an easy access to the nerve ending of central neurons. The giant calyx-type nerve terminal of the chick ciliary ganglion synapse is large enough to allow the use of patch-clamp techniques directly at the release sites.

Analysis of the Ca^{2+}-current at this synapse isolated from chick pretreated *in ovo* with BoNT/C revealed that the cleavage of syntaxin results in an alteration of G-protein regulation of presynaptic calcium channels *(36)*. Mammals offer very few large synapses. In the brainstem (in the rat: medial nucleus of the trapezoid body), a giant synapse called the calyx of Held presents all the characteristics required for allowing direct recording of neuroexocytotic events at the release sites as well as affording access to the cytosol (by dialysis from a patch clamp pipet) *(37)*.

2.4.5. Nonneuronal Secretory Cells

2.4.5.1. CHROMAFFIN AND PC12 CELLS

Endocrine and neuroendocrine cells release hormones and neuropeptides by Ca^{2+}-dependent fusion of secretory granules with the plasma membrane. This exocytotic process shares many molecular similarities with neurotransmitter release. In general, nonneuronal secretory cells are nearly insensitive to externally applied clostridial neurotoxins owing to the inability of these cell types to bind and internalize the toxins. Nevertheless, this limitation has not precluded their use in studying the clostridial neurotoxins mechanisms because purified toxins or their constitutive chains can be easily applied intracellularly using membrane permeabilization (*see* **Subheadings 3.2.3.** and **3.2.4.**; for related procedures *see* **Note 5**). It should be emphasized that cultured secretory cells offer the interesting possibility to correlate the toxic effect on secretion with the extent of substrate proteolytic cleavage generally assessed biochemically by immunodetection on cell extracts (e.g., *see [38]*).

The most studied nonneuronal secretory cells are the adrenal medullary chromaffin cells (which derive embryonically from the neural crest) and their tumor cell derivatives, PC12 cells. These cells express a large number of neuronal-specific proteins and have been widely used to study the recruitment, docking, and fusion of a class of regulated secretory vesicles that are very similar to the large dense-core vesicles present in certain neurons.

Key steps in the actions of clostridial neurotoxins have been achieved using chromaffin or PC-12 cells. For example, in blocking Ca^{2+}-dependent exocytosis by intracellular application of TeNT or BoNT/A into chromaffin cells, Penner et al. *(39)* were first to demonstrate that the targets of clostridial neurotoxins are intracellular ubiquitous components essential for regulated exocytosis. The use of digitonin- or SLO-permeabilized chromaffin cells demonstrated that the light chain of the toxin is responsible for the intracellular blockade of exocytosis *(40,41)*.

Furthermore, calcium-regulated exocytosis can be dissected into distinct ATP-dependent and ATP-independent phases in permeabilized cell models

(42,43). These sequential stages exhibit a different sensitivity to clostridial neurotoxins, consistent with the idea that these steps require a distinct protein machinery *(44)*.

2.4.5.2. OTHER SECRETORY CELLS

Many other secretory cell types have been used to probe the effects of clostridial toxins on the intracellular processes underlying exocytosis, including pancreatic insulin-secreting cells *(8)*, and, more recently, intestinal endocrine cell lines *(45)*. Note that amylase secretion from pancreatic acinar exocrine cells is apparently resistant to BoNT/A and TeNT *(46)*, perhaps implicating resistant SNAREs in the amylase secretory mechanism.

2.4.5.3. MONITORING EXOCYTOSIS ON SECRETORY CELLS WITH HIGH RESOLUTION

Despite the absence of a postsynaptic target cell that can be used to probe secretion of hormone with high resolution, the analysis of elementary exocytotic events is nevertheless possible. Fusion of the vesicle membrane at the release site increases the plasma membrane area. Electrically speaking, the lipid bilayer behaves as a capacitor whose capacitance is directly proportional to its surface area. Hence, the secretory/recycling activity can be monitored as time-resolved capacitance changes of the plasma membrane under whole cell patch-clamp configuration (for an early example in the study of BoNT/A and TeNT, *see [39]*). This approach allows the detection of single fusion events attributable to exocytosis of large granules (200 nm up to 1 μm diameter) (reviewed in *[12]*). Transmitter release can be detected with a carbon fiber using electrochemical detection methods. However, this approach is limited to the few transmitters that can be readily oxidized or reduced (catecholamines, serotonin, NO). Several optical monitoring techniques can be used. For example, intragranule markers (dopamine β-hydroxylase immunoreactivity) that incorporate into the plasma membrane during exocytosis proved effective in monitoring secretion *(47)*. Note, however, that owing to the complexity of the granule-membrane recycling pathway (via the post-Golgi apparatus), the use of styryl dyes (FM1-43) is probably restricted to imaging endocytosis.

2.4.6. Other Cell Models

During foodborne or infant botulism, BoNT must cross the intestinal epithelium before disseminating into the body and reaching their final target, the motor nerve endings. Maksymowych and Simpson *(48)* recently introduced the first model that permits the study of this crucial step for the disease. Using T-84 and Caco-2 human colon carcinoma cells they demonstrated specific binding and transcytosis of dichain BoNT/A and /B in intestinal cells.

3. Methods

3.1. Extracellular Application of Clostridial Neurotoxins

3.1.1. In Vitro Extracellular Application

Acute actions of the neurotoxins may be studied on preparations to which the BoNT or TeNT, their constituent chains, or toxin fragments are applied in vitro by adding the toxin sample directly to the appropriate extracellular physiological medium (*see* **Subheading 2.2.**). Note that gelatin (0.01%, w/v, final) may be added to physiological solution to minimize inactivation of toxins.

Despite the requirement of oxygenation for maintaining stable neurotransmission at mammalian excised NMJ preparations, it is safer for the experimenter to stop the extracellular superfusion during the duration needed for binding and internalization of neurotoxins (e.g., 10–20 min). This procedure diminishes the formation of toxin-containing spray when drops of saline are falling into the tissue bath.

Great care should be taken with regard to the temperature at which the toxins are applied to the preparation because their internalization and intracellular action is strongly temperature dependent *(3,24,25,49)*. This should be kept in mind because when neurotransmitter release is examined at NMJ, it is tempting to reduce the temperature to keep a stable response over period of several hours in vitro.

3.1.2. In Vivo Application, Prior to In Vitro Experiments

The in vivo–*in vitro* approach allows the study of BoNT or TeNT toxic action in a time scale that is incompatible with the life span of an isolated preparation. This procedure has been also widely used for characterizing the trophic relations between nerve and muscle and the synaptic remodeling that occurs during the functional recovery of neurotransmission at BoNT- or TeNT-poisoned neuromuscular junctions. A variant of the in vivo–*in vitro* approach has been developed for the chick: the toxin is injected *in ovo* 16–20 h before isolation of the ciliary ganglion and subsequent electrophysiological investigation *(36)*.

The general procedure consists of injecting in vivo sublethal doses of the toxin (dissolved in physiological solution containing 0.2% w/v gelatin) into a given muscle group several hours or days prior to examining neurotransmission on nerve–muscle preparations dissected from poisoned animals. Ideally, the toxin should be injected in the region of the NMJ to maximize the local concentration of toxin and to reduce the systemic toxic effects. The amount of toxin to be injected depends on its purity or association with nontoxic components (BoTx vs BoNT; *see also* **Note 1**) or on the site of injection (*see* **Sub-**

headings 3.1.2.2.–3.1.2.4.). Variations on the procedure depend on the animal species used (*see* **Subheadings 3.1.2.2.–3.1.2.4.**).

3.1.2.1. MURINE

Sublethal doses of BoNT or BoTx have to be used for in vivo poisoning. To obtain a complete block of neurotransmission in a given muscle in vivo, it is necessary to inject BoTx or BoNT in the immediate vicinity of the muscle *(50)*. Owing to the restricted diffusion of the toxin around the injection site, a "sublethal dose" can correspond to several LD_{50} as evaluated by the intraperitoneal injection procedure *(51)*. When using the auricular nerve–levator auris longus muscle of the mouse, it is sufficient to administer BoNT or BoTx subcutaneously from where it can reach almost immediately the nerve endings, inducing a complete blockade of neurotransmission without generalized action. For other nerve–muscle preparations (e.g., extensor digitorum longus muscle) BoNT or BoTx is injected directly into the muscle mass. However, this procedure causes inflammatory reactivity, which accelerates the functional recovery by sprouting formation.

The in vivo poisoning procedure permits also study of the TeNT-induced inhibition of acetylcholine release at the NMJ. Sublethal doses of TeNT are injected into hindleg muscles of the mouse *(52)* or the soleus muscle of the rat *(53)*. Under these conditions, TeNT undergo motor axon retrograde ascent from the nerve terminals close to the site of injection. Because it generally affects only a spinal chord segment by transcytosis, this induces a local tetanus within 20 h that may persist for weeks. In the periphery, the injected muscle exhibits a delayed (150 h) flaccid muscle paralysis attributable to the direct inhibition of nerve-evoked acetylcholine release by TeNT. Interestingly, TeNT appears to have much less effect on motor nerve terminals innervating fast rather than slow contracting muscle.

3.1.2.2. FROG

BoTx can be injected either subcutaneously or in the lymphatic sacs to obtain a generalized muscle paralysis, or by intramuscular injection for localized effects. The BoNT doses that can be used range from 5 ng to several micrograms. Indeed, the cutaneous respiration allows the frogs to support very high doses of clostridial neurotoxins without dying (>100,000 mice LD_{50}). Note that when BoNT is applied in vivo, the onset of flaccid paralysis is slower at lower temperature. For example, at 14°C, a delay of about 5 d can be observed with a dose of BoTX-A of 0.4 µg.

Because BoNT-induced neuroparalysis persists for several months without recovery, animals need special care. They should be kept at low temperature

(4–14°C), partially covered by water supplemented with glucose and NaCl (i.e., poisoned frog maintenance medium in **Subheading 2.2.4.**). The skin should be daily cleaned (water flow or brushing). Moreover, the urinary bladder needs to be emptied (daily or twice a day) to avoid urine retention.

3.1.2.3. FISH

TeNT or BoNT dissolved in "fish saline" solution containing 0.2% gelatin as protective colloid is injected in vivo at the base of the tail or into the belly of the pectoral fin muscles. Investigations on the NMJ or the Mauthner cells are performed in vitro on preparations isolated from fishes that display erratic and abnormal swimming.

3.2. Intracellular Application of Clostridial Neurotoxins

3.2.1. Intracellular Injection: Principle

Analysis of the intracellular actions of various toxins is hampered by the plasma membrane. The intracellular injection procedure was initially developed with the aim of applying clostridial neurotoxins to neurons or chromaffin cells in such a way that the membrane-limiting steps (binding, endocytosis) are bypassed. This proved to be a very powerful approach allowing examination of the toxins action on cell types devoid of toxin receptors and also to study the effect of nonpermeant derivatives of toxins *(5,24–26,39)*. This technique also applies for the injection of mRNA encoding toxin chains or fragments *(27)* or antibodies. Intracellular application into intact cells may be performed either by an air pressure injection procedure (via an intracellular micropipet, *see* **Subheading 3.2.2.**) or via the dialysis of the contents of a patch pipet into the cell *(39)*. Note that in this latter case, key factors can diffuse from the inside of the cell to the interior of the pipet (the volume of the pipet is ~100× that of the cytosol).

3.2.2. Air Pressure Injection into Aplysia Neurons

Although this **Subheading** refers mainly to the procedure followed in *Aplysia* experiments, it can be applied to other neurons.

1. Injection micropipets (0.5–1.5 MOhm) are pulled from glass tubing without a capillary. Then, the injection pipet may be "coated" with paraffin oil. This procedure greatly facilitates the removal of the micropipet after the intracellular injection. The injection micropipet may contain a silver wire (50 μm in diameter, plunging into the solution to be injected). This allows the electrophysiological monitoring of the impalement.
2. The general procedure for filling of the micropipet consists in filling the tip of the injection micropipet by suction. However, we suggest first filling the injection

micropipet with an inert nonmiscible compound (e.g., Fluorinert™) to prevent desiccation of the sample (a few picoliters of total volume) contained in the tip of the injection micropipet.

To allow the visual or epifluorescence monitoring of the injection, we suggest to mix (prior to filling the micropipet) the toxin sample with a vital dye. Useful dyes are (1) fast green FCF for eye-monitoring *(23–27)*, (2) Texas red-dextran *(29)* or (3) FITC-dextran *(28)* for epifluorescence monitoring.

3. The samples are injected using commercial air (or any inert gas) pressure injectors. In our hands, it appears important to keep a short resting period (~5 min) between the impalement of the cell and the injection, with another rest period of several minutes (~20 min) before removal of the pipet. Note that when injection micropipets with large diameter are used, the content of the pipet can dialyze into the cells. For large cells, the injected volume cannot exceed 5% of that of the cell body without seriously damaging the plasma membrane. Smaller cells seem to recover more easily. When the injected neurons look damaged (appearance of puffs, depolarization of the resting membrane potential >–45 mV, loss of membrane resistance, changes in the kinetics of action potential) we have often observed spontaneous changes in the secretion properties that are unrelated to the treatment.

3.2.3. Permeabilization Using Detergents or Pore-Forming Toxins: Principle

Nonionic detergents such as digitonin have been widely used to gain access to the intracellular environment by permeabilization of the plasma membrane. Digitonin interacts with membrane cholesterol and creates lesions (0.3–2 μm) in the plasma membrane that are large enough to introduce toxins into the cytoplasm. However, digitonin treatment induces a significant leakage of cytosolic proteins and as a consequence, digitonin-permeabilized cells progressively lose their capacity to secrete in response to micromolar Ca^{2+} *(54)*. Nevertheless, digitonin has been employed successfully for the introduction of BoNT and TeNT into cultured chromaffin cells *(38,40,42,44)*. A popular alternative is to render secretory cells leaky with bacterial pore-forming toxins. Note, however, that the 2-nm pores formed by the *Staphylococcus aureus* α-toxin render the cells freely permeable to ions and small metabolites but do not permit the introduction of proteins and toxins into the cytoplasm. The streptococcal cytotoxin SLO generates pores of 30 nm size in the cell plasma membrane that are sufficient to permit fluxes of large proteins *(41,42,46)*. These lesions preserve the subplasmalemmal cytoskeleton *(55)* and despite the substantial leakage of cytosolic proteins, SLO-permeabilized secretory cells remain responsive to calcium for long periods after permeabilization *(56)*. In our hands, maximal catecholamine secretion observed in SLO-treated chromaffin cells is always greater than that obtained in digitonin-permeabilized cells. Other procedures are briefly summarized in **Note 5**.

3.2.4. SLO-Permeabilization of Chromaffin Cells

1. Chromaffin cells are prepared from bovine adrenal medullas by collagenase digestion and purification on self-generating Percoll gradients. The preparation and maintenance of adrenal medullary chromaffin cells in vitro have been reviewed *(57)*. Secretion experiments are usually performed with 24-well culture dishes containing 250,000 cells/well (130,000 cells/cm²), 3–6 d after cell preparation.

2. Perform labeling of catecholamine stores by incubating cells with [³H]noradrenaline in DMEM containing 125 nM noradrenaline but no amino acids. The amount of radioactivity taken up by the cells is in the range of 15–20%. Cells are then washed several times with Locke's solution and 2× with calcium-free Locke's solution containing 1 mM EGTA to remove calcium from the incubation medium before permeabilization.

3. Cells are permeabilized at 25°C by incubation with 20–50 hemolytic units/mL SLO in calcium-free permeabilizing medium. Cells become sufficiently permeable to introduce proteins into the cytosol after 2–4 min of incubation with SLO. Incubation volume is 200 µL in wells of a 24-well plate.

4. For clostridial neurotoxin application, the SLO solution is replaced with calcium-free permeabilizing medium containing the clostridial neurotoxins at the concentrations to be tested for 10–20 min. As discussed in **Note 1**, dichainal toxins should be reduced before application.

5. Secretion. Catecholamine secretion is subsequently triggered for 10 min with fresh permeabilizing medium (i.e., SLO- and BoNT- or TeNT-free) containing buffered calcium *(58)*. Radioactivity is determined in the supernatant and in the cells after precipitation with 10% trichloroacetic acid (v/v). [³H]Noradrenaline secretion is generally expressed as the percentage of total radioactivity present in cells before stimulation.

   Note that cells are irreversibly permeabilized and continue to release cytosolic components even after the removal of the SLO-containing medium *(56)*. Thus, the secretory response declines with time after permeabilization. Typically, 25–35% of the catecholamines are secreted in 10 min after 2 min of permeabilization and drop to 15–25% if a 10-min incubation period is introduced between permeabilization and calcium-evoked stimulation. The extent of secretion from SLO-permeabilized cells is half-maximal between 1 and 2 µM free Ca²⁺ and becomes maximal at 20 µM Ca²⁺ *(59)*.

6. As variations of the protocol, cells can be permeabilized and stimulated in the absence of MgATP to investigate the effects of clostridial toxins on the ATP-independent stages of the exocytotic machinery *(43)*. However, secretion that occurs in the absence of MgATP is labile and usually disappears within 5 min after permeabilization.

4. Notes

1. Structure of the neurotoxins and their biological activity: BoNT (but not TeNT) are secreted by bacteria of the *Clostridium* genus together with several other non-

toxic proteins, several of which exhibit hemagglutinin activity. This secreted molecular complex can reach ~900 kDa in the case of BoNT/A and corresponds to the so-called botulinum toxin or BoTx. The nontoxic proteins do not seem to participate in the inhibition of neurotransmitter release inhibition *(60)*; they are proposed to protect the neurotoxin from proteases in the digestive tract.

The various clostridial neurotoxins are synthesized as a single chain protoxin of ~150 kDa. In most of the *C. botulinum* strains, this protoxin is subsequently cleaved into the bacteria to give rise to the dichain neurotoxin that is secreted. Several functional studies provide evidence indicating that, to be intracellularly active, the light chain needs to be free in the cytosol. This means that when purified as a single chain (100% of BoNT/E and 80% BoNT/B), the toxin has to be proteolytically processed or "nicked" to be active *(24,61)*.

In the cytosol, the disulfide bridge that links the L- and H-chains has to be reduced. In most intact cells, the reduction appears to occur spontaneously and the nonreduced dichain neurotoxin appears as potent as the reduced form, no matter which of the latter is dialyzed into the cell *(39)*, pressure-injected *(23,26)* or applied intracellularly by lipofection *(62)*. For an unknown reason, TeNT and BoNT must be reduced in order to block exocytosis when microinjected at the squid giant synapse *(28,29)*. When intracellular application is performed using a procedure in which the plasma membrane integrity is broken (permeabilization, electroporation, cell cracking), the toxins need to be reduced before application *(63)*, suggesting perhaps that the cytosolic reductive system is lost after permeabilization. Note that, at concentrations >5 mM, the reducing agent DTT can have an effect on the transmitter release machinery. In addition, DTT interferes with the proteolytic activity of the light chain of TeNT *(64)*. Hence, when possible, it is better to use the purified or recombinant light chain of BoNT or TeNT to avoid these problems.

Note that when recombinant L-chains of TeNT or BoNT are prepared, the presence of fusion protein or His-tag in the N-terminal position may affect biological activity of L-chains. In this view, deletion of the first 10 N-terminus amino acids of TeNT or BoNT/A L-chain has been reported to abolish blocking activity in neurons *(27)*.

2. Botulinum toxin assay procedure: Routine assessment of the biological activity of BoTx or BoNTs is based on lethality assay performed in mice *(65)*. In brief, BoTx or BoNT samples are serially diluted in sodium phosphate buffer containing 0.2% (w/v) gelatin. Diluted toxin is then injected intraperitoneally (0.5 mL/mouse) into groups of at least four mice. Death is monitored over a period of 4 d. The MLD$_{50}$/mL is estimated from the toxin dilution that killed half the mice in the group over a 4-d period.

3. Twitch tension recording: This is the most sensitive in vitro assay for quantifying alterations of neuromuscular transmission by clostridial neurotoxins. For example, a 90% paralysis can be induced in ~300 min with only 0.01 nM of purified BoNT/A *(49)*. The basis of this approach is to evoke muscle contraction by stimulating the motor nerve and to transduce the contraction into a signal that

can be recorded. This assay was originally described by Bulbring in 1946 *(66)*. The obvious simplicity of this bioassay makes it easy to set up. For typical examples recorded under various conditions in the rat, mouse, or guinea pig phrenic nerve–hemidiaphragm preparations *see (3,16,49,62,67)*. For the neuromuscular preparation of the goldfish, *see (17)*; for the chick, *see (14,16)*; for smooth muscle contraction, *see (18)*.

Basically, the dissected nerve–muscle preparations are mounted into a tissue bath containing Krebs solution continuously oxygenated (95% O_2 + 5% CO_2). The temperature is usually kept at 37°C. A stimulatory external electrode (e.g., platinum ring) is placed around the nerve trunk for stimulation. Twitches are elicited by stimulating the motor nerve (typically at a frequency of 0.1 Hz with pulses of 0.1 ms duration). Intensity of stimulation needs to be supramaximal to recruit all nerve fibers and to elicit maximal contraction. Isometric contraction is recorded using a force/displacement transducer attached to the muscle. The resting tension applied to the muscle depends on the force of the muscle (typically 15 mN for a mouse hemidiaphragm). Neuromuscular preparations have to stabilize for at least 20 min and should be used only if reproducible muscle contraction amplitudes in response to nerve stimulation are obtained for at least 10 min prior to incubation with the clostridial toxins. Incubation with a given toxin is usually carried out at 27°C or 37°C (*see also* comments in **Subheading 3.1.1**) for at least 30 min without perfusion, to ascertain that the binding step has been accomplished. Inhibition of evoked neurotransmitter release is detected as a failure to initiate muscle contraction. However, muscle paralysis does not mean that evoked neurotransmitter exocytosis is abolished; it indicates only that acetylcholine release is depressed so that the amplitude of postsynaptic responses is below the threshold for initiating a muscle action potential.

Special care should be taken in (1) keeping supramaximal the nerve stimulation; otherwise a modification of the muscle tension may indicate a change in the number of nerve fibers that are recruited at each stimulus and (2) oxygenating the physiological medium. Indeed, anoxia of the motor nerve of the neuromuscular preparation can cause block of nerve conduction and can produce a muscle paralysis irrelevant to the studied toxin action.

4. The NMJ preparations from rat and mouse: The phrenic nerve–hemidiaphragm muscle *(66,67)* and plantar nerve–lumbrical muscle *(68)* are suitable neuromuscular preparations for evaluating synaptic transmission and for recording nerve elicited skeletal muscle contractions by recording the muscle twitch tension.

The extensor digitorium longus and soleus muscles of either rat or mouse are suitable nerve–muscle preparations for electrophysiological recordings of synaptic quantal events.

The intercostal nerves–triangularis sterni muscle of the mouse *(69)* is a preparation that has been extensively used for recording extracellular currents from visible nerve terminals and, in particular, the presynaptic currents occurring along the unmyelinated nerve terminal branches. Individual NMJ are easily observed at high magnification (×400) by Nomarski interference contrast optics. This allows intracellular or extracellular electrophysiological recordings from well-defined microelectrode positions. The insertion of a microelectrode underneath the perineurial sheath surrounding small nerve bundles permits the recording of local circuit currents flowing between nerve terminals and their parent axons. This technique demonstrated for the first time that presynaptic calcium currents remain unaffected during the action of BoNT/A or TeNT at doses that block neuromuscular transmission *(60,70)*.

The mouse auricular nerve-levator auris longus muscle *(71)* is very convenient for studying the short- and long term effects of the local application of BoNT in living mice. The fact that this muscle lies just under the skin makes it particularly suitable for subcutaneous injection of BoNT in vivo (*see* **Subheading 3.1.2.1.**). The toxin has ready access to the muscle and exerts its effects within 18–24 h when used at sublethal concentrations. The animals are able to survive for several months. The processes of pre- and postsynaptic remodeling of the NMJ can be followed and evaluated, at given times after toxin injection, with morphological and/or electrophysiological techniques. The levator auris longus muscle is sufficiently thin to be stained as a whole mounted preparation. This allows morphological observations of the extent of terminal sprouts and the localization of synaptic proteins and nicotinic acetylcholine receptors. The analysis of the functional properties of nerve terminal sprouts elicited by an in vivo injection of BoNT/A prior to in vitro studies revealed an active impulse propagation of the action potential over most of their length. Furthermore, Ca^{2+} influx and Ca^{2+}-dependent K^+ currents in the sprout membrane were found to be similar to those described in unpoisoned endings. The terminal sprouts induced by BoNT/A have the molecular machinery for acetylcholine release and play a role in the recovery of neuromuscular transmission after BoNT/A poisoning *(72)*.

The plantar nerves–lumbrical muscles of the hindpaw of the mouse is a preparation that is particularly useful for assaying the effects of BoNT on the mammalian NMJ. Indeed, each mouse provides four to eight preparations. The muscles are thin enough to easily localize the neuromuscular junctions with Nomarski optics for intracellular recordings. Moreover, the small number of junctions and their arrangement permit application of low concentrations of BoNT to produce a total paralysis in a convenient period of time *(68)*.

5. Other methods to bypass the plasma membrane: Several methods have been developed with the aim of giving access to the intracellular space. They are briefly summarized.

   a. Electroporation or electropermeabilization: High-voltage capacitor discharge creates transient membrane pores of approx 1 μm at the sites facing the electrodes *(73,74)*. As demonstrated functionally *(74)* or by electron microscopy using both [^{125}I]- and gold-labeled TeNT *(75)* macromolecules diffuse through these transient openings. In contrast to toxin- or detergent-permeabilized cells, electroporated cells remain viable because the pores created in the plasma membrane reseal. However, electroporation is suitable only for cells in suspension. Note that following the electroporation protocol, cells transiently lose their secretory properties *(74)*. This permits studies on cells several days after intracellular application of the toxins.

b. Cell cracking: Cultured cells are passed through the narrow clearance (2.5 μm) of a ball homogenizer *(76)*. This results in a transient mechanical disruption of the plasma membrane (i.e., "cracking"). Immediately after permeabilization, cells are incubated at 0°C to deplete their soluble intracellular components. These cell ghosts retain the subcortical membrane cytoskeleton and associated intracellular organelles. Cell cracking renders also the secretory apparatus directly accessible to macromolecules. Using this technique on PC-12 cells, the possible membranous or cytoskeletal-site of action of BoNT was investigated *(77)*.

c. Lipofection. The motor nerve terminal has a very small diameter (<1 μm) that makes it inaccessible to micropipet injection. Moreover, the permeabilization techniques are not applicable to NMJ preparations. This problem was elegantly solved by DePaiva and Dolly *(62)* by delivering purified BoNT chains to the motor nerve endings via liposomes. Interestingly, the fusion of liposomes with the plasma membrane efficiently delivers encapsulated material but the extralipids incorporated to the plasmalemma (phosphatidylcholine, phosphatidylserine) do not seem to affect dramatically the excitatory properties of the nerve endings. Indeed, neuromuscular transmission could be monitored for hours using the twitch contraction assay.

d. Transfection and targeted expression of clostridial neurotoxins: The gene encoding TeNT light chain has been expressed in *Drosophila* and its expression in embryonic neurons removed detectable VAMP/synaptobrevin and specifically eliminated nerve-evoked, but not spontaneous quantal neurotransmitter release as revealed by electrophysiological recordings *(21)*. Successful transfection of the light chain of TeNT into the ACTH secretory cell line (AtT-20) has also been reported *(78)*.

e. Preincubation of cells with gangliosides to increase internalization: Incubation of chromaffin cells with a mixture of gangliosides (GM1, GD1a, GD1b, GT1b) makes these cells sensitive to extracellularly applied BoNT/A and TeNT *(79)*. Most probably, these polysialogangliosides provide membrane binding sites that allow the subsequent nonspecific internalization of the toxins. This technique could be implemented in cells devoided of specific toxins receptors.

f. Extracellular pH drop: As observed for many bacterial toxins, the translocation of BoNT or TeNT light chain from the endocytic uptake compartment to the cytosol is pH dependent. Acidification of the extracellular environment facilitates the uptake of TeNT into NG108-15 and NBr-10A neurohybridoma cells *(80)*. This suggests that transient acidification of the extracellular milieu could be useful to induce nonspecific internalization of clostridial neurotoxins.

g. Carrier-mediated internalization: Several bacterial toxins exhibit the ability to pass the plasma membrane in a wide variety of cell types. This makes it possible to construct chimeric proteins comprising the light chain of TeNT or

BoNT fused to a fragment of bacterial toxin that can act as a nonspecific carrier. This possibility has been recently illustrated by the report that the C2II component of the binary actin-ADP-ribosylating *Clostridium botulinum* C2 toxin causes the cellular uptake of C3 toxin fused to the N-terminal 225 amino acid residues of C2I *(81)*.

Acknowledgments

This work was supported by grants from A. F. M. (Association Française contre les Myopathies) to B. P., from A. R. C. (Grant N° 9101 from Association pour la Recherche contre le Cancer) to M.-F. B., and from D. S. P (Grant N°98–151 to J. M. and 99-34-038 to B. P. from Direction des Systèmes des forces et de la Prospective).

References

1. Niemann, H., Blasi, J., and Jahn, R. (1994) Clostridial neurotoxins: new tools for dissecting exocytosis. *Trends Cell Biol.* **4,** 179–185.
2. Montecucco, C. and Schiavo, G. (1995) Structure and function of tetanus and botulinum neurotoxins. *Q. Rev. Biophys.* **28,** 423–472.
3. Schmitt, A., Dreyer, F., and John, C. (1981) At least three sequential steps are involved in the tetanus toxin-induced block of neuromuscular transmission. *Naunyn Schmiedebergs Arch. Pharmacol.* **317,** 326–330.
4. Rabasseda, X., Blasi, J., Marsal, J., Dunant, Y., Casanova, A., and Bizzini, B. (1988) Tetanus and botulinum toxins block the release of acetylcholine from slices of rat striatum and from the isolated electric organ of *Torpedo* at different concentrations. *Toxicon* **26,** 329–336.
5. Schiavo, G., Benfenati, F., Poulain, B., Rossetto, O., Polverino de Laureto, P., DasGupta, B. R., and Montecucco, C. (1992) Tetanus and botulinum-B neurotoxins block neurotransmitter release by proteolytic cleavage of synaptobrevin. *Nature* **359,** 832–835.
6. Patarnello, T., Bargelloni, L., Rossetto, O., Schiavo, G., and Montecucco, C. (1993) Neurotransmission and secretion. *Nature* **364,** 581–582.
7. Bruns, D., Engers, S., Yang, C., Ossig, R., Jeromin, A., and Jahn, R. (1997) Inhibition of transmitter release correlates with the proteolytic activity of tetanus toxin and botulinus toxin A in individual cultured synapses of Hirudo medicinalis. *J. Neurosci.* **17,** 1898–1910.
8. Sadoul, K., Berger, A., Niemann, H., Weller, U., Roche, P. A., Klip, A., Trimble, W. S., Regazzi, R., Catsicas, S., and Halban, P. A. (1997) SNAP-23 is not cleaved by botulinum neurotoxin E and can replace SNAP-25 in the process of insulin secretion. *J. Biol. Chem.* **272,** 33,023–33,027.
9. Galli, T., Zahraoui, A., Vaidyanathan, V. V., Raposo, G., Tian´y, J. M., Karin´y, M., Niemann, H., and Louvard, D. (1998) A novel tetanus neurotoxin insensitive vesicle associated membrane protein (TI-VAMP) in SNARE complexes of the apical plasma membrane of epithelial cells. *Mol. Biol. Cell* **9,** 1437–1448.

10. Doussau, F., Clabecq, A., Henry, J. P., Darchen, F., and Poulain, B. (1998) Calcium-dependent regulation of rab3 in short-term plasticity. *J. Neurosci.* **18,** 3147–3157.

11. Van der Kloot, W. and Molgó, J. (1994) Quantal acetylcholine release at the vertebrate neuromuscular junction. *Physiol. Rev.* **74,** 899–991.

12. Angleson, J. K. and Betz, W. J. (1997) Monitoring secretion in real time: capacitance, amperometry and fluorescence compared. *Trends Neurosci.* **20,** 281–287.

13. Raciborska, D. A., Trimble, W. S., and Charlton, M. P. (1998) Presynaptic protein interactions *in vivo*: evidence from botulinum A, C, D and E action at frog neuromuscular junction. *Eur. J. Neurosci.* **10,** 2617–2628.

14. Lomneth, R., Suszkiw, J. B., and DasGupta, B. R. (1990) Response of the chick ciliary ganglion–iris neuromuscular preparation to botulinum neurotoxin. *Neurosci. Lett.* **113,** 211–216.

15. Ginsborg, B. L. and Warriner, J. (1960) The isolated chick biventer cervitis nerve-muscle preparation. *Br. J. Pharmacol.* **15,** 410–411.

16. Simpson, L. L. (1982) The interaction between aminoquinolines and presynaptically acting neurotoxins. *J. Pharmacol. Exp. Ther.* **222,** 43–48.

17. Diamond, J. and Mellanby, J. (1971) The effect of tetanus toxin in the goldfish. *J. Physiol. (Lond.)* **215,** 727–741.

18. MacKenzie, I., Burnstock, G., and Dolly, J. O. (1982) The effects of purified botulinum neurotoxin type A on cholinergic, adrenergic and non-adrenergic, atropine-resistant autonomic neuromuscular transmission. *Neuroscience* **7,** 997–1006.

19. Dunant, Y., Esquerda, J. E., Loctin, F., Marsal, J., and Muller, D. (1987) Botulinum toxin inhibits quantal acetylcholine release and energy metabolism in the *Torpedo* electric organ. *J. Physiol. (Lond.)* **385,** 677–692.

20. Herreros, J., Blasi, J., Arribas, M., and Marsal, J. (1995) Tetanus toxin mechanism of action in *Torpedo* electromotor system: a study on different steps in the intoxication process. *Neuroscience* **65,** 305–311.

21. Sweeney, S. T., Broadie, K., Keane, J., Niemann, H., and O'Kane, C. J. (1995) Targeted expression of tetanus toxin light chain in *Drosophila* specifically eliminates synaptic transmission and causes behavioral defects. *Neuron* **14,** 341–351.

22. Martinez-Padron, M. and Ferrus, A. (1997) Presynaptic recordings from *Drosophila*: correlation of macroscopic and single-channel K+ currents. *J. Neurosci.* **17,** 3412–3424.

23. Poulain, B., Tauc, L., Maisey, E. A., Wadsworth, J. D., Mohan, P. M., and Dolly, J. O. (1988) Neurotransmitter release is blocked intracellularly by botulinum neurotoxin, and this requires uptake of both toxin polypeptides by a process mediated by the larger chain. *Proc. Natl. Acad. Sci. USA* **85,** 4090–4094.

24. Poulain, B., Wadsworth, J. D., Shone C. C., Mochida, S., Lande, S., Melling, J., Dolly, J. O., and Tauc, L. (1989) Multiple domains of botulinum neurotoxin contribute to its inhibition of transmitter release in *Aplysia* neurons. *J. Biol. Chem.* **264,** 21,928–21,933.

25. Poulain, B., De Paiva, A., Deloye, F., Doussau, F., Tauc, L., Weller, U., and Dolly, J. O. (1996) Differences in the multiple step process of inhibition of neu-

rotransmitter release induced by tetanus toxin and botulinum neurotoxins type A and B at *Aplysia* synapses. *Neuroscience* **70,** 567–576.

26. Mochida, S., Poulain, B., Weller, U., Habermann, E., and Tauc, L. (1989) Light chain of tetanus toxin intracellularly inhibits acetylcholine release at neuro-neuronal synapses, and its internalization is mediated by heavy chain. *FEBS Lett.* **253,** 47–51.

27. Kurazono, H., Mochida, S., Binz, T., Eisel, U., Quanz, M., Grebenstein, O., Poulain, B., Tauc, L., and Niemann, H. (1992) Minimal essential domains specifying toxicity of the light chains of tetanus toxin and botulinum neurotoxin type A. *J. Biol. Chem.* **267,** 14,721–14,729.

28. Hunt, J. M., Bommert, K., Charlton, M. P., Kistner, A., Habermann, E., Augustine, G. J., and Betz, H. (1994) A post-docking role for synaptobrevin in synaptic vesicle fusion. *Neuron* **12,** 1269–1279.

29. Llinas, R., Sugimori, M., Chu, D., Morita, M., Blasi, J., Herreros, J., Jahn, R., and Marsal, J. (1994) Transmission at the squid giant synapse was blocked by tetanus toxin by affecting synaptobrevin, a vesicle-bound protein. *J. Physiol. (Lond.)* **477,** 129–133.

30. Marsal, J., Ruiz-Montasell, B., Blasi, J., Moreira, J. E., Contreras, D., Sugimori, M., and Llinas, R. (1997) Block of transmitter release by botulinum C1 action on syntaxin at the squid giant synapse. *Proc. Natl. Acad. Sci. USA* **94,** 14,871–14,876.

31. Habermann, E. and Dreyer, F. (1986) Clostridial neurotoxins: handling and action at the cellular and molecular level. *Curr. Top. Microbiol. Immunol.* **129,** 93–179.

32. Wellhöner, H. H. (1992) Tetanus and botulinum neurotoxins, in *Selective Neurotoxicity, Handbook of Experimental Pharmacology,* Vol. 102 (Herken, H. and Hucho, F., eds.), Springer-Verlag, Berlin, pp. 357–417.

33. Beise, J., Hahnen, J., Ansersen-Beckh, B., and Dreyer, F. (1994) Pore formation by tetanus toxin, its chain and fragments in neuronal membranes and evaluation of the underlying motifs in the structure of the toxin molecule. *Naunyn Schmiedebergs Arch. Pharmacol.* **349,** 66–73.

34. Owe-Larsson, B., Kristensson, K., Hill, R. H., and Brodin, L. (1997) Distinct effects of clostridial toxins on activity-dependent modulation of autaptic responses in cultured hippocampal neurons. *Eur. J. Neurosci.* **9,** 1773–1777.

35. Gähwiler, B. H., Capogna, M., Debanne, D., McKinney, R. A., and Thompson, S. M. (1997) Organotypic slice cultures: a technique has come of age. *Trends Neurosci.* **20,** 471–477.

36. Stanley, E. F. and Mirotznik, R. R. (1997) Cleavage of syntaxin prevents G-protein regulation of presynaptic calcium channels. *Nature* **385,** 340–343.

37. Borst, J.G. and Sakmann, B. (1996) Calcium influx and transmitter release in a fast CNS synapse. *Nature* **383,** 431–434.

38. Lawrence, G. W., Foran, P., Mohammed, N., DasGupta, B. R., and Dolly, J. O. (1997) Importance of two adjacent C-terminal sequences of SNAP–25 in exocytosis from intact and permeabilized chromaffin cells revealed by inhibition with botulinum neurotoxin A and E. *Biochemistry* **36,** 3061–3067.

39. Penner, R., Neher, E., and Dreyer, F. (1986) Intracellularly injected tetanus toxin inhibits exocytosis in bovine adrenal chromaffin cells. *Nature* **324,** 76–78.

40. Bittner, M. A., DasGupta, B. R., and Holz, R. W. (1989) Isolated light chains of botulinum neurotoxins inhibit exocytosis. Studies in digitonin-permeabilized chromaffin cells. *J. Biol. Chem.* **264,** 10,354–10,360.

41. Ahnert-Hilger, G., Weller, U., Dauzenroth, M. E., Habermann, E., and Gratzl, M. (1989) The tetanus toxin light chain inhibits exocytosis. *FEBS Lett.* **242,** 245–248.

42. Bittner, M. A., and Holz, R. W. (1992) Kinetic analysis of secretion from permeabilized adrenal chromaffin cells reveals distinct components. *J. Biol. Chem.* **267,** 16,219–16,225.

43. Vitale, N., Gensse, M., Chasserot-Golaz, S., Aunis, D., and Bader, M. F. (1996) Trimeric G proteins control regulated exocytosis in bovine chromaffin cells: sequential involvement of Go associated with secretory granules and Gi$_3$ bound to the plasma membrane. *Eur. J. Neurosci.* **8,** 1275–1285.

44. Lawrence, G. W., Weller, U., and Dolly, J. O. (1994) Botulinum A and the light chain of tetanus toxin inhibit distinct stages of MgATP-dependent catecholamine exocytosis from permeabilized chromaffin cells. *Eur. J. Biochem.* **222,** 325–333.

45. Nemoz-Gaillard, E., Bosshard, A., Regazzi, R., Bernard, C., Cuber, J. C., Takahashi, M., Catsicas, S., Chayvialle, J. A., and Abello, J. (1998) Expression of SNARE proteins in enteroendocrine cell lines and functional role of tetanus toxin-sensitive proteins in cholecystokinin release. *FEBS Lett.* **425,** 66–70.

46. Stecher, B., Ahnert-Hilger, G., Weller, U., Kemmer, T. P., and Gratzl, M. (1992) Amylase release from streptolysin O-permeabilized pancreatic acinar cells. Effects of calcium, guanosine 5'-[γ-thio]triphosphate, cyclic AMP, tetanus toxin and botulinum A toxin. *Biochem. J.* **283,** 899–904.

47. Chasserot-Golaz, S., Vitale, N., Sagot, I., Delouche, B., Dirrig, S., Pradel, L. A., Henry, J. P., Aunis, D., and Bader, M. F. (1996) Annexin II in exocytosis: catecholamine secretion requires the translocation of p36 to the subplasmalemmal region in chromaffin cells. *J. Cell Biol.* **133,** 1217–1236.

48. Maksymowych, A. B. and Simpson, L. L. (1998) Binding and transcytosis of botulinum neurotoxin by polarized human colon carcinoma cells. *J. Biol. Chem.* **273,** 21,950–21,957.

49. Simpson, L. L. (1980) Kinetic studies on the interaction between botulinum toxin type A and the cholinergic neuromuscular junction. *J. Pharmacol. Exp. Ther.* **212,** 16–21.

50. Cull-Candy, S. G., Lundh, H., and Thesleff, S. (1976) Effects of botulinum toxin on neuromuscular transmission in the rat. *J. Physiol. (Lond.)* **260,** 177–203.

51. Molgó, J., Siegel, L. S., Tabti, N., and Thesleff, S. (1989) A study of synchronization of quantal transmitter release from mammalian motor endings by the use of botulinal toxins type A and D. *J. Physiol. (Lond.)* **411,** 195–205.

52. Duchen, L. W. and Tonge, D. A. (1973) The effects of tetanus toxin on neuromuscular transmission and on the morphology of motor end-plates in slow and fast skeletal muscle of the mouse. *J. Physiol. (Lond.)* **228,** 157–172.

53. Bevan, S. and Wendon, L. M. (1984) A study of the action of tetanus toxin at rat soleus neuromuscular junctions. *J. Physiol. (Lond.)* **348,** 1–17.

54. Sarafian, T., Aunis, D., and Bader, M. F. (1987) Loss of proteins from digitonin-permeabilized adrenal chromaffin cells essential for exocytosis. *J. Biol. Chem.* **262,** 16,671–16,676.

55. Sontag, J. M., Aunis, D., and Bader, M. F. (1988) Peripheral actin filaments control calcium-mediated catecholamine release from streptolysin-O-permeabilized chromaffin cells. *Eur. J. Cell Biol.* **46,** 316–326.

56. Sarafian, T., Pradel; L. A., Henry, J. P., Aunis, D., and Bader, M. F. (1991) The participation of annexin II (calpactin I) in calcium-evoked exocytosis requires protein kinase C. *J. Cell Biol.* **114,** 1135–1147.

57. Livett, B. G. (1984) Adrenal medullary chromaffin cells in vitro. *Physiol. Rev.* **64,** 1103–1161.

58. Fohr, K. J., Warchol, W., and Gratzl, M. (1993) Calculation and control of free divalent cations in solutions used for membrane fusion studies. *Methods Enzymol.* **221,** 149–157.

59. Vitale, N., Mukai, H., Rouot, B., Thierse, D., Aunis, D., and Bader, M. F. (1993) Exocytosis in chromaffin cells. Possible involvement of the heterotrimeric GTP-binding protein G(o). *J. Biol. Chem.* **268,** 14,715–14,723.

60. Mólgo, J., Comella, J. X., Angaut-Petit, D., Pécot-Dechavassine, M., Tabti, N., Faille, L., Mallart, A., and Thesleff, S. (1990) Presynaptic actions of botulinal neurotoxins at vertebrate neuromuscular junctions. *J. Physiol. (Paris)* **84,** 152–166.

61. Simpson, L. L. and DasGupta B. R. (1983) Botulinum neurotoxin type E: studies on mechanism of action and on structure-activity relationships. *J. Pharmacol. Exp. Ther.* **224,** 135–140.

62. DePaiva, A. and Dolly, J. O. (1990) Light chain of botulinum neurotoxin is active in mammalian motor nerve terminals when delivered via liposomes. *FEBS Lett.* **277,** 171–174.

63. Stecher, B., Gratzl, M., and Ahnert-Hilger, G. (1989) Reductive chain separation of botulinum A toxin—a prerequisite to its inhibitory action on exocytosis in chromaffin cells. *FEBS Lett.* **248,** 23–27.

64. Cornille, F., Goudreau, N., Ficheux, D., Niemann, H., and Roques, B. P. (1994) Solid-phase synthesis, conformational analysis and in vitro cleavage of synthetic human synaptobrevin II 1-93 by tetanus toxin L chain. *Eur. J. Biochem.* **222,** 173–181.

65. Lamana, C. and Carr, C. J. (1967) The botulinal, tetanal, and enterostaphylococcal toxins: a review. *Clin. Pharmacol. Ther.* **8,** 286–332.

66. Bulbring, E. (1997) Observations on the isolated phrenic nerve diaphragm preparation of the rat, 1946. *Br. J. Pharmacol.* **120(4 Suppl.),** 3–26.

67. Simpson, L. L. and Tapp, J. T. (1967) Actions of calcium and magnesium on the rate of onset of botulinum toxin paralysis of the rat diaphragm. *Int. J. Neuropharmacol.* **6,** 485–492.

68. Clark, A. W., Bandyopadhyay, S., and DasGupta, B. R. (1987) The plantar nerves–lumbrical muscles: a useful nerve–muscle preparation for assaying the effects of botulinum neurotoxin. *J. Neurosci. Methods* **19,** 285–295.

69. McArdle, J. J., Angaut-Petit, D., Mallart, A., Bournaud, R., Faille, L., and Brigant, J. L. (1981) Advantages of the triangularis sterni muscle of the mouse for investigations of synaptic phenomena. *J. Neurosci. Methods* **4,** 109–115.

70. Dreyer, F., Mallart, A., and Brigant, J. L. (1983) Botulinum A toxin and tetanus toxin do not affect presynaptic membrane currents in mammalian motor nerve endings. *Brain Res.* **270,** 373–375.

71. Angaut-Petit, D., Molgó, J., Connold, A. L., and Faille, L. (1987) The levator auris longus muscle of the mouse: a convenient preparation for studies of short- and long-term presynaptic effects of drugs or toxins. *Neurosci. Lett.* **82,** 83–88.

72. Angaut-Petit, D., Molgó, J., Comella, J. X., Faille, L., and Tabti, N. (1990) Terminal sprouting in mouse neuromuscular junctions poisoned with botulinum type A toxin: morphological and electrophysiological features. *Neuroscience* **37,** 799–808.

73. Lambert, H., Pankov, R., Gauthier, J., and Hancock, R. (1990) Electroporation-mediated uptake of proteins into mammalian cells. *Biochem. Cell Biol.* **68,** 729–734.

74. Bartels, F. and Bigalke, H. (1992) Restoration of exocytosis occurs after inactivation of intracellular tetanus toxin. *Infect. Immun.* **60,** 302–307.

75. Erdal, E., Bartels, F., Binscheck, T., Erdmann, G., Frevert, J., Kistner, A., Weller, U., Wever, J., and Bigalke, H. (1995) Processing of tetanus and botulinum A neurotoxins in isolated chromaffin cells. *Naunyn Schmiedebergs Arch. Pharmacol.* **351,** 67–78.

76. Martin, T. F. and Walent, J. H. (1989) A new method for cell permeabilization reveals a cytosolic protein requirement for Ca^{2+}-activated secretion in GH3 pituitary cells. *J. Biol. Chem.* **264,** 10,299–10,308.

77. Lomneth, R., Martin, T. F., and DasGupta, B. R. (1991) Botulinum neurotoxin light chain inhibits noradrenaline secretion in PC12 cells at an intracellular membranous or cytoskeletal site. *J. Neurochem.* **57,** 1413–1421.

78. Aguado, F., Gombau, L., Majo, G., Marsal, J., Blanco, J., and Blasi, J. (1997) Regulated secretion is impaired in AtT-20 endocrine cells stably transfected with botulinum neurotoxin type A light chain. *J. Biol. Chem.* **272,** 26,005–26,008.

79. Marxen, P., Fuhrmann, U., and Bigalke, H. (1989) Gangliosides mediate inhibitory effects of tetanus and botulinum A neurotoxins on exocytosis in chromaffin cells. *Toxicon* **27,** 849–859.

80. Kalz, H. J. and Wellhöner, H. H. (1996) Acidification of the cytosol inhibits the uptake of tetanus toxin in NG108–15 and NBr–10A neurohybridoma cells. *Naunyn Schmiedebergs Arch. Pharmacol.* **353,** 606–609.

81. Barth, H., Hofmann, F., Olenik, C., Just, I., and Aktories, K. (1998) The N-terminal part of the enzyme component (C2I) of the binary *Clostridium botulinum* C2 toxin interacts with the binding component C2II and functions as a carrier system for a Rho ADP-ribosylating C3-like fusion toxin. *Infect. Immun.* **66,** 1364–1369.

16

The Biology of Endotoxin

Volker T. El-Samalouti, Lutz Hamann, Hans-Dieter Flad, and Artur J. Ulmer

1. Introduction

Approximately 100 yr ago Richard Pfeiffer, a co-worker of Robert Koch in Berlin, discovered that cholera bacteria produced, in addition to heat-labile exotoxin, another toxin *(1)*. In contrast to the secreted exotoxins this new, heat-stable toxin was found to be a constituent of the bacterial cell, and therefore Pfeiffer termed it endotoxin. Today we know that endotoxin (lipopolysaccharide, LPS) is the main outer membrane component of Gram-negative bacteria and plays a key role during severe Gram-negative infection, trauma, and shock *(2,4)*. Despite its destructive effects, the presence of low amounts of LPS, which gain access to body fluids and organs by infection and translocation from the gut, are rather beneficial for the host, causing immunostimulation leading to enhanced resistance to infections and malignancy *(5)*. This picture changes completely when larger amounts of LPS are present in the bloodstream, as observed during severe Gram-negative bacterial infections (notably after application of antibiotics) or possibly caused by translocation of enterobacteria from the gut. Released LPS causes various pathophysiological reactions including fever, leukopenia, tachycardia, tachypnea, hypotension, disseminated intravascular coagulation, and multiorgan failure. This may culminate in septic shock which is associated with a mortality rate of 20–50% and causes approx 100,000 deaths annually only in the United States *(6)*.

The harmful as well as the beneficial host responses to LPS are not induced directly, but are rather mediated by immune modulator molecules such as tumor necrosis factor α (TNF-α), members of the interleukin family (IL-1, IL-6, IL-8, IL-12), interferon α, reduced oxygen species, and lipids. These mediators are released mainly by monocytes/macrophages, but also other cells such

From: *Methods in Molecular Biology, vol. 145: Bacterial Toxins: Methods and Protocols*
Edited by: O. Holst © Humana Press Inc., Totowa, NJ

as vascular cells, polymorphonuclear cells, and T cells participate in the response to LPS *(7–12)*. In addition, polyclonal stimulation of B lymphocytes, at least in mice, was observed.

In the past, much effort has been directed to identifying the chemical structure and the structural principles of endotoxin to elucidate the bioactive principle behind this molecule. In addition, much progress was achieved in the understanding of the mechanisms of its biological action. This chapter summarizes our knowledge about the biological mechanism of LPS action.

2. Structural Requirements for the Bioactivity of LPS

Although there is a great compositional variation among endotoxins derived from different bacterial serotypes, they all share a common structural principle. Endotoxins are amphiphilic molecules consisting of a hydrophilic polysaccharide part and a covalently bound hydrophobic lipid component, termed lipid A. The polysaccharide part can be divided into two subdomains, the core region and the O-specific chain (**Fig. 1**), composed of a sequence of repeating units of identical polysaccharides *(13)*.

The O-specific chain consists of up to 50 repeating units, each composed of two to eight sugar monomers *(13–15)*. There is a great variability in the monosaccharides building the O-specific chain, and therefore this part of the endotoxin molecule is unique for a given endotoxin structure and characteristic for its bacterial origin (i.e., the serotype) *(16)*. Nevertheless a large number of pathogenic Gram-negative bacterial strains express endotoxin without an O-specific chain. The core portion is structurally less variable and it can formally be subdivided into the O-chain-proximal outer core and the lipid A-proximal inner core *(17)*. The outer core portion exhibits variabilities in sugar composition and linkage *(18)*. Common elements are D-glucose (Glc), D-galactose (Gal), and 2-amino-2-desoxy-D-glucose (GlcN). The inner core region of LPS of various Gram-negative bacteria, e.g., *S. enterica* and *E. coli* and many other genera, have a most conserved structure containing mostly 3-deoxy-D-*manno*-oct-2-ulosonic acid (Kdo) and heptose (Hep) residues, which are, in general, phosphorylated *(19)*. This conserved common structure allowed the development of monoclonal antibodies, which recognize an epitope within this region of the endotoxin molecule, and therefore cross-react with all serotypes of *S. enterica*, *E. coli* and other genera *(20,21)*.

The lipid A component is in many cases composed of a phosphorylated β-(1 → 6)-linked D-GlcN disaccharide which carries up to seven acyl residues. It was found to be highly conserved among bioactive endotoxins *(22)*. Nevertheless there are variations in the length, position, and number of the fatty acids. Lipid A can be separated from the polysaccharide part by mild acid treatment and has thereby been shown to constitute the endotoxic principle of LPS, as the

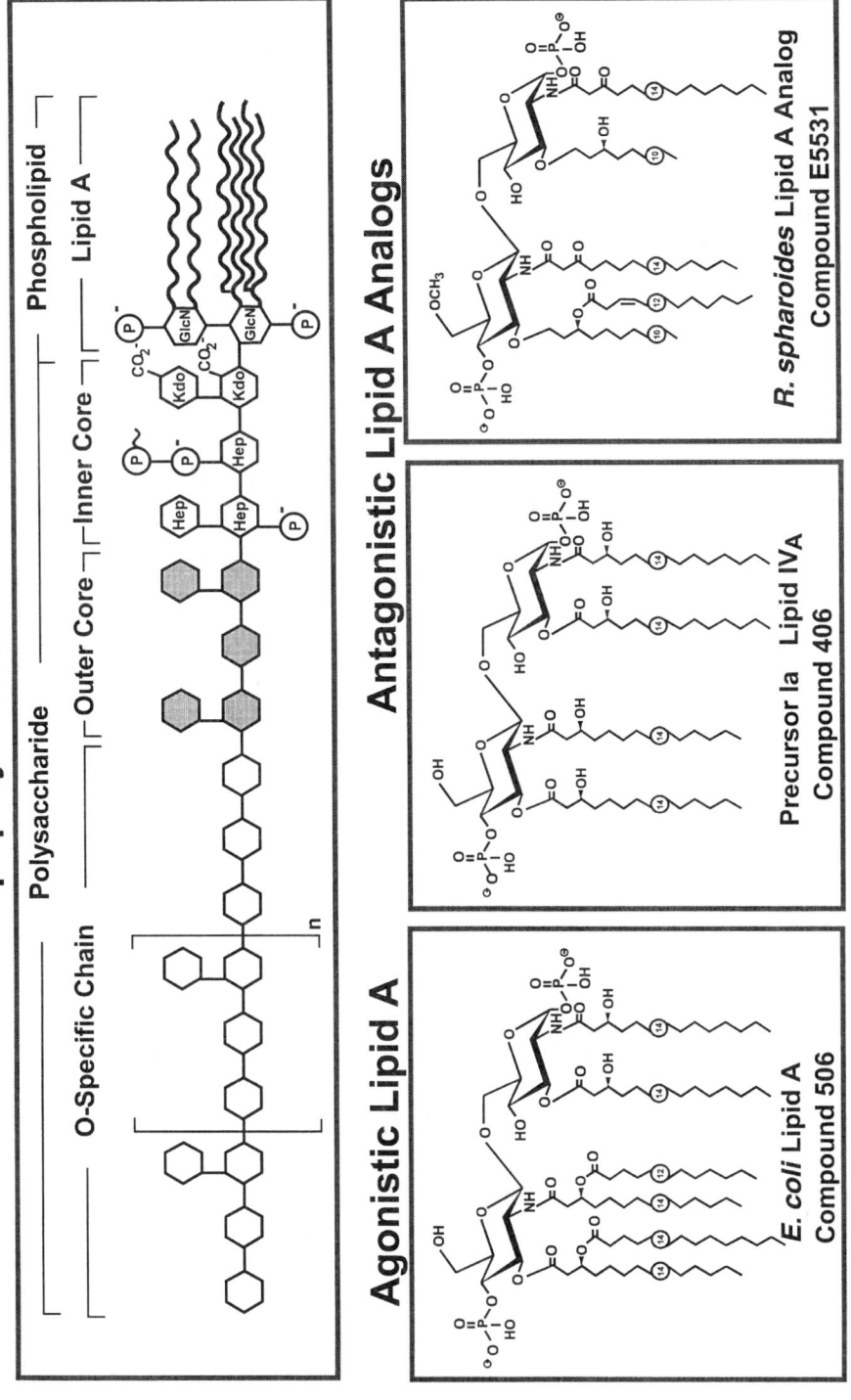

Fig. 1. Chemical structure of endotoxin, lipid A, and lipid A analogs.

biological effects of LPS are reproduced by free lipid A *(23)*. This finding was later confirmed by using the complete chemically synthesized lipid A and its corresponding lipid A partial structures, such as *E. coli* lipid A (named compound 506 or LA-15-PP) (**Fig. 1**) *(24,25)*. Furthermore, these compounds were the basis for the investigation of the structure–activity relationship of LPS *(26–29)*.

The minimal requirement for lipid A bioactivity, referred to the cytokine inducing capacity, is a molecule having two *gluco*-configurated hexosamine residues, two phosphoryl groups, and six fatty acids as present in *E. coli* lipid A or compound 506 *(30)*. Lipid A partial structures deficient in one of these elements are less active or even inactive regarding the inducing of monokines in human monocytes. The significant influence of the two phosphoryl groups on the bioactivity of synthetic lipid A partial structures has been shown by stimulation with the 1-dephospho derivative (compound 504) and the 4'-dephospho derivative (compound 505), which were less active than the bisphosphorylated compound 506.

Also, the presence as well as the position of the six fatty acids are important for complete bioactivity of LPS. The lack of the two secondary fatty acids, as in the tetraacylated lipid A precursor Ia (compound 406, also known as lipid IVa or LA-14-PP) (**Fig. 1**), makes this molecule completely inactive in inducing IL-1, IL-6, and TNF-α release in human monocytes *(28,31–33)*. In contrast, additional acylation, as in the highly acylated heptaacyl lipid A (*S. enterica* sv. Minnesota lipid A, compound 516), leads to less bioactivity, and even the change of the location of the secondary acyl residues has a negative effect on stimulatory activity of the compound. The compound LA-22-PP, which consists of a symmetrical distribution of the two secondary fatty acids, is a very weak monocyte activator. These findings are further supported by the endotoxical inactivity of the natural lipid A species of *Rhodobacter sphaeroides* (**Fig. 1**) and *R. capsulatus*, which also differ in their acylation pattern from *E. coli* lipid A *(34,35)*. Both of these natural substances and the synthetic lipid A precursor Ia showed unaffected binding activity and therefore they are used as LPS antagonists for competition of the natural LPS during inflammation. A potent synthetic LPS antagonist is the compound E5531 (**Fig. 1**), which was synthesized based on the structure of the *R. sphaeroides* lipid A *(36)*.

Interestingly, chemically synthesized monosaccharide analogs related to the nonreducing part of lipid A (GLA27 and GLA60) are capable in the stimulation of monocytes, whereas the synthetic hexosamine monosaccharide precursor lipid X was found to be completely inactive. However, in comparison to *E. coli* lipid A (compound 506), the effective concentration of GLA27 is about 30 times higher, and no pyrogenic activity or Shwartzman reactivity has been found *(37,38)*. Therefore, a lipid A partial structure having a disaccharide with two hexosamine residues is necessary for exerting optimal endotoxic activity in vitro and in vivo.

In contrast to human monocytes, different structure requirements for the induction of monokine production by LPS are found in murine macrophages and are assumed to be necessary for other species. For example, compound 406 possesses endotoxic activity in mice in vitro and in vivo *(39–41)*, but is inactive in humans *(28,31–33)* and in various other primates *(unpublished data)*. At present the reason for these different structural requirements of lipid A is unclear but is found to be unrelated to the expression of murine or human CD14, respectively *(41)*.

In addition to the chemical structure of LPS or lipid A, the physical structure of LPS also has an effect on its bioactivity. LPS or lipid A, as amphiphilic substances, form aggregates in aqueous solution with a distinct three-dimensional structure. The bioactive lipid A was shown to form nonlamellar structures, which are either cubic (*S. enterica* sv. Minnesota) or hexagonal (*R. gelatinosus*) *(42)*, whereas the inactive lipid A from *R. capsulatus* forms lamellar structures *(43)*. However, these conformational differences between the compounds may be of significance only at higher LPS/lipid A concentrations, as it has been found that endotoxicity is expressed by monomeric LPS molecules *(44)*.

From the large body of data concerning the structure–activity relationship one can conclude that the "natural" form of the synthetic compound 506, derived from *E. coli* LPS, represents the best configuration for optimal mono-cyte activating capacity. All chemically different substructures of compound 506, if differing in the phosphorylation or in the acylation pattern of the hexosamine disaccharide, are less or even not active in inducing monokines.

3. LPS Responsive Cells

The majority of LPS responsive cells are components of the cellular immune system, and most of these cells are activated by minute amounts of LPS. Each individual cell type reacts in a typical way, but in general these reactions are the production of mediators, phagocytosis, proliferation, and/or differentiation (**Fig. 2**).

3.1. Monocytes/Macrophages

The most prominent LPS-sensitive cell population consists of cells of the monocyte/macrophage lineage. These cells produce a large variety of bioactive protein mediators in response to LPS including interleukin 1 (IL-1), IL-6, IL-8, and in particular TNF-α. A large number of host cells respond to these cytokines and thereby initiate the typical acute phase response (e.g., leukocytosis, moderate fever, attraction of defense cells to the infectious focus and/or

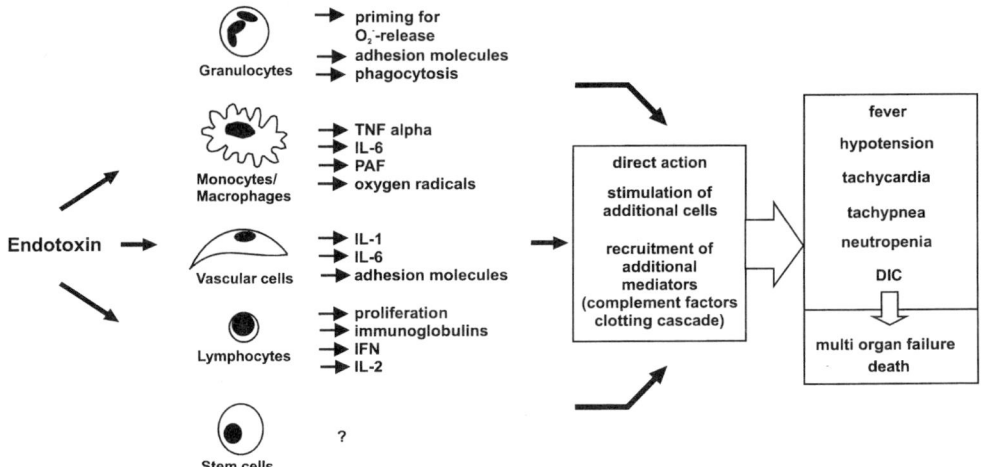

Fig. 2. Schematic representation of the mode of action of endotoxin in the pathogenesis of septic shock. DIC, Disseminated intravasal coagulation.

activation of microbicidal mechanisms) that helps to eliminate the invading
microorganisms. Massive release of cytokines, however, becomes hazardous
for the organism by causing shock, cell damage, and multiorgan failure. Thus
it can be considered that the overproduction of those protein mediators leads to
the manifestation of septicemia, multiorgan failure, and lethal septic shock syndrome *(3,16)*.

In addition to cytokines, macrophages produce reduced oxygen species
(superoxide anion, hydrogen peroxide, hydroxyl radicals, and nitric oxide),
bioactive metabolites of arachidonic acid (prostaglandins, thromboxane, and
leukotrienes) and of linoleic acid (*S*-13-hydroxylinoleic acid) as well as platelet-activating factor on exposure to endotoxin. All these nonprotein mediators
have also been shown to be involved in the pathophysiology of septicemia.

The importance of monocytes/macrophages during septicemia was demonstrated on LPS-resistant C3H/HeJ mice which become LPS-sensitive after
application of macrophages from an LPS-sensitive mice strain *(45)*.

3.2. Polymorphonuclear Leukocytes

The function of polymorphonuclear leukocytes (PMN) is the uptake of
invading bacteria and bacterial fragments and therefore PMNs build the first
barrier of the host against microorganisms. The phagocytosis of microorganisms by PMNs is dramatically enhanced in the presence of LPS and a good
example of nonspecific immunostimulatory capacity of endotoxin.

PMNs are able to neutralize LPS because they possess enzymes for the degradation (deacylation and dephosphorylation) of LPS and lipid A to nontoxic partial structures *(46,47)*. Furthermore, they contain and release antimicrobial proteins with strong affinity for LPS and thereby neutralize LPS bioactivity. One of these proteins is the cationic 55-kDa bactericidal/permeability-increasing protein (BPI) that is able to kill bacteria by binding to the bacterial surface. Other examples are the PMN proteins CAP 18 and CAP 37, which also have LPS-inactivating potential *(48,49)*.

PMNs contribute to inflammatory reactions after their activation by the release of inflammatory mediators. Furthermore, damage of endothelial cells and thereby destruction of the lining of blood vessels was observed. PMNs also contribute to local inflammatory reactions after penetration of the vessel wall into the tissue.

3.3. B- and T-Lymphocytes

The polyclonal activation of murine B lymphocytes in response to LPS, resulting in proliferation, differentiation, and the secretion of immunoglobulins, seems to be an early defense mechanism, as it leads to the enhanced release of antibodies with various antimicrobial specificities *(50)*.

Human T-lymphocytes (CD4⁻ as well as CD8⁺) are able to proliferate and secrete Th1-type lymphokines upon LPS stimulation *(12)*. Activation of T lymphocytes by LPS depends on accessory monocytes providing costimulatory signals *(51)*.

Murine T cells have also been reported to proliferate in vitro in response to LPS *(52,53)* and moreover it has been shown that LPS-stimulated CD8⁺/CD4⁻ murine T lymphocytes are able to suppress the humoral immune response to bacterial polysaccharides such as pneumococcal type III polysaccharides *(54)*. This suppressing activity by CD8⁺/CD4⁻ murine T lymphocytes can be downregulated by lipid A, leading to increased antipolysaccharide antibody production *(54)*. All these findings support the hypothesis that LPS is indeed involved in cellular immunity against microorganisms.

3.4. Vascular Cells, Epithelial Cells

Vascular cells (endothelial cells and smooth muscle cells) as well as epithelial cells can respond to LPS by the release of cytokines, for example, IL-1, IL-6, and/or IL-8 *(55–58)*. In addition they can produce several other mediators such as prostacyclin, nitric oxide, platelet-activating factor, interferons, and colony-stimulating factors *(59,60)*. Another reaction of vascular cells is the expression of adhesion molecules upon LPS or IL-1 stimulation, which contributes strongly to the regulation of the inflammatory response during infection.

3.5. Hematopoietic Stem Cells

Hematopoietic stem cells are precursor cells, which provide the hematopoietic system with mature cells after multiplication and differentiation. This reactivity of stem cells is under the control of different stem cell growth factors. It has now become evident that stem cells play an active role in innate immunoreactions. The stimulation of CD80-expression on human monocytes as well as the induction of cytokine production and proliferation of T lymphocyte by LPS requires a low number of CD34$^+$ stem cells *(125)*. The mechanism, however, by which these CD34$^+$ stem cells exert their accessory cell function remains to be clarified.

4. Interaction of LPS with Soluble or Membrane Proteins

A prerequisite for the activation of cells by LPS is the interaction of LPS with specific LPS-binding molecules on the surface of LPS responsive target cells. To date, various LPS binding structures have been described in the literature, but physiological relevance has been demonstrated for only a few of them.

The most prominent cell surface protein, which has been shown to be involved in the activation of cells by LPS, is the 55-kDa glycoprotein CD14. CD14 exists as a glycosyl-phosphatidylinositol(GPI)-anchored membrane protein (mCD14) on monocytic cells, polymorphonuclear leukocytes, and on some B lymphocytes.

Binding of LPS to mCD14 on monocytes is necessary for the stimulation of these cells, leading to the production and release of immune mediators. Anti-CD14 antibodies are able to abolish the cytokine production of myeloid cells in response to LPS, which clearly demonstrates the importance of the CD14 molecule. Recently, these findings were supported by a CD14-knockout (CD14$^{-/-}$ko) model, as CD14$^{-/-}$ mice are more resistant to LPS-induced lethal shock in vivo. Furthermore peripheral blood mononuclear cells (PBMC) from CD14$^{-/-}$ mice did not respond to LPS concentrations up to 100 ng/mL *(61)*.

The binding of LPS to CD14 is markedly enhanced by a serum component, termed LPS-binding protein (LBP) *(62,63)*. LBP is a 60-kDa glycoprotein that concentration of which increases during the acute-phase response from 5–10 µg/mL up to 200 µg/mL *(62)*. LBP reduces the concentration of LPS necessary for the activation of monocytes by forming LBP–LPS complexes, which then are recognized by CD14. The function of LBP in inflammatory reactions has disadvantageous as well as beneficial effects: In D-galactosamine-sensitized mice, blocking of LBP by anti-LBP monoclonal antibodies (MAbs) leads to the protection from lethal endotoxic shock induced by a low challenge of LPS *(64)*. On the other hand it was shown in an LBP knockout model that LBP-deficient mice were not able to combat a Gram-negative infection because they are incapable of inducing an inflammatory response to LPS *(65)*.

Although mCD14 is accepted as a key molecule for LPS binding in monocytic cells, the mechanism of cell activation after LPS binding to mCD14 is still unclear. It is obvious that mCD14, lacking a transmembrane domain, is incapable of transmitting a signal through the membrane like a classical hormone receptor. Furthermore, the GPI anchor has been shown to be unnecessary for the activation of cells via mCD14 *(66)*. Therefore, at least one accessory molecule performing the signal transduction is postulated to build a functional LPS receptor together with mCD14. When high doses of LPS are present, monocyte stimulation occurs in an mCD14 independent way and cannot be blocked by anti-CD14 MAbs.

Beside mCD14 several types of soluble CD14 (48, 53, 55 kDa) are present in concentrations of approx 2–6 µg/mL *(67)* in serum, either released by monocytes or secreted as GPI-free forms *(68–70)*. It has been shown that LPS binds directly to sCD14, a process that is markedly facilitated by LBP, although LBP is not present in the generated LPS–sCD14 complexes *(71)*. LPS–sCD14 complexes enable the activation of some mCD14⁻ cells, for example, endothelial cells, fibroblasts, and smooth muscle cells to produce cytokines *(67,72,73)*. Therefore it has been postulated that a specific receptor for the LPS–sCD14 complexes is expressed on these cells which mediates the activation of CD14-negative cells by LPS. A possible candidate for the sCD14-LPS receptor is a 216-kDa protein, which was recently developed by crosslinking experiments *(74)* on the astrocytoma cell line U373. However, the functional involvement of this protein in LPS signaling remains to be elucidated. In conclusion, CD14 as a membrane or soluble receptor plays an important but not essential role as a primary LPS recognition molecule in various cells. Additional membrane molecules, however, seem to be necessary for cell activation.

One of these molecules, recently discovered to be involved in LPS recognition, possibly is the GPI-linked membrane molecule decay accelerating factor (DAF, CD55). CD55 was identified to be the 80-kDa LPS binding membrane protein (LMP80) which was described as a candidate for an LPS-responsive element *(75)*. When monocyte membranes are incubated with LPS, CD55 coprecipitates together with LPS by the use of anti-LPS MAbs, and furthermore CD55-transfected Chinese hamster ovary (CHO) cells are capable of responding to LPS (El-Samalouti et al., *unpublished observations*).

Furthermore, the adhesion molecules of the β-integrin family, which are phagocytic receptors of leukocytes, are able to recognize Gram-negative bacteria by their LPS moiety. β-Integrins are expressed as transmembrane heterodimeric receptors containing the common unit CD18 paired with one of the CD11 subunits CD11a (LFA-1), CD11b (CR3), and CD11c (CR4). A specific LPS-binding region was shown only for CD11b/CD18 *(76)*, but all three receptors were found to be capable of mediating the activation of NFκB in response to LPS in transfected CHO cells when CD14 is absent *(77–79)*. However, the cytosolic domain of CR3 was shown not to be involved in the signaling process *(80)*. Another LPS binding structure on macrophages is the scavenger receptor *(81,82)*, which undisputedly lacks signal-transduction capacity, but is functionally involved in the clearance of LPS from the host.

Most recently, the Toll-like receptor 2 (TLR2) has been reported to function as a signaling receptor in response to LPS. Toll receptors were initially identified in *Drosophila,* where they are involved in embryonic development and in the antifungal response of the adult fly. Five human homologs of Toll (TLR1-5) have been identified to date, which are suggested to participate in the immune response. It was shown that TLR2-transfected embryonic kidney cells are able to respond to LPS by activation of NFκB and that this response is dependent on the presence of LBP and sCD14 and markedly enhanced when the cells are cotransfected with mCD14 *(83,84)*.

We like to suppose the assembly of multiple membrane molecules that may form a LPS receptor domain (LRD). It is well known that GPI-linked molecules accumulate together with cholesterol and glycosphingolipids to form special membrane microdomains called *rafts* or "detergent insoluble glycolipid-enriched domains" (DIGs) *(85–87)*. These domains are the loci of numerous cell functions, from membrane traffic and cell morphogenesis to cell signaling, and a large number of signaling molecules are concentrated here. Indeed, mCD14 was shown to be present in DIGs, associated mainly with the protein tyrosine kinase p53/56^{lyn} *(88)*, but also with GTP-binding proteins and ouabain-inhibitable Na$^+$/K$^+$ATPase in low-density domains of the monocyte membrane *(89)*. The GPI anchor of mCD14 was recently demonstrated to be responsible for the presence of the molecule in DIGs, as a transmembrane form of CD14 was not detected in the Triton X-100 insoluble fraction of cells *(90)*. However, the cellular response mediated by transmembrane CD14 or GPI-linked CD14 was different only partially. Like mCD14, CD55 is involved in transmembrane signaling *(91–93)* and is also present in DIGs, combined mainly with p56^{lck} and p59^{fyn} *(91,94)*.

In conclusion, LPS recognition and signaling involves various known and unknown membrane molecules. We therefore hypothesize that these various membrane molecules form a functional LPS receptor domain composed of different subunits. These may either be located in DIGs or assemble in DIGs after binding of LPS to mCD14 or to other molecules. Our present view of the recognition of LPS by inflammatory cells (e.g., monocytes/macrophages) is demonstrated in **Fig. 3**.

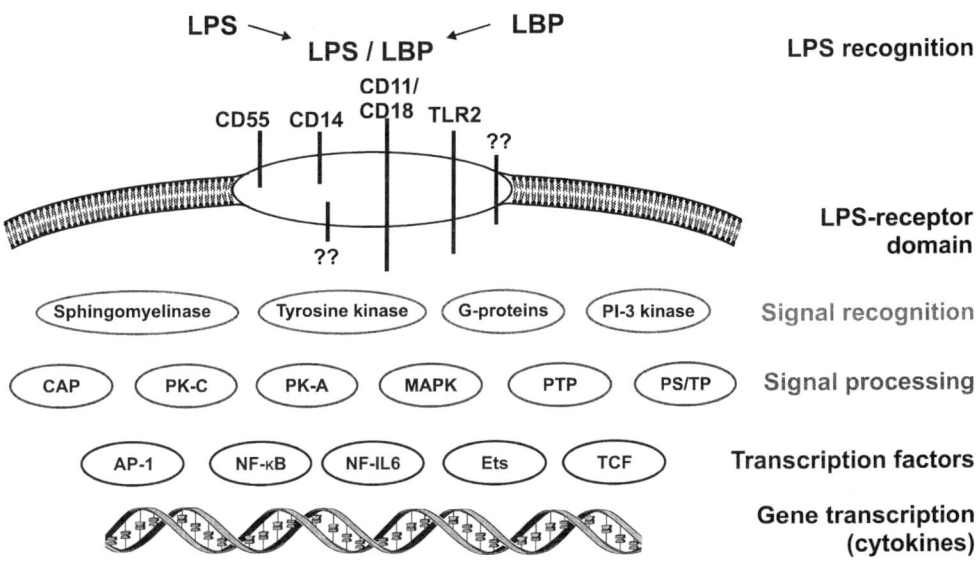

Fig. 3. Endotoxin recognition and signaling in monocytes/macrophages.

5. LPS-Induced Signal Transduction and Signal Processing

The transduction of an extracellular signal, which ultimately results in the transcription of distinct genes, is a complex mechanism, including extracellular recognition of a ligand, transmission of the signal through the membrane, intracellular recognition, further signal processing, activation and translocation of transcription factors, and finally transcription and expression of genes (**Fig. 3**).

A large body of data has been published in the past about signaling cascades activated after LPS binding to CD14, but a continuous order of events from binding of LPS to cytokine release has not been established. The major unresolved question was discussed previously: How does the LPS signal pass through the cell membrane after binding to a GPI-linked receptor? Membrane CD14 has been shown by immunocoprecipitation to be associated with the protein tyrosine kinases p53/56[lyn] *(88)*. Binding of LPS leads to the phosphorylation of p53/56[lyn] and of other src family kinases such as p58/64 [hck] and p59[c-fgr] *(88,95)*. Nevertheless the src kinases seem to play a minor role in LPS-induced cell activation, as knockout mice, deficient in expression of lyn, hck, and fgr respond normally to LPS with regard to TNF-α, IL-2 and IL-6 *(96)*. On the other hand, it was shown in the human system that the addition of the tyrosine kinase inhibitors herbimycin A and genistein prior to LPS stimulation abolished the cytokine release of mononuclear phagocytes *(97)*. Other early signal transduction elements that become activated by LPS are heterotrimeric G-proteins *(98,99)* and the sphingomyelinase *(100–102)*.

Possible downstream targets of the tyrosine kinases are mitogen activated protein (MAP) kinases. Three groups of MAP kinases were described to be rapidly activated by phosphorylation in response to LPS: the extracellular signal regulated kinase (ERK) *(103,104)*, the c-jun N-terminal kinase (JNK) *(105,106)*, and the p38 kinase *(107,108)*. The ERK-1/2 kinases are activated by the MKK-1 (MEK) via the upstream effectors Ras and Raf-1 and the substrates of ERK-1/2 are the transcription factors ELK-1 and c-Myc, phospholipase A_2, and other protein kinases. Although ERK-1/2 are involved in the regulation of TNF-α transcription *(109,110)*, this mechanism only partially mimics the LPS effects.

The other two MAP kinases have a greater impact on the transcription of cytokine genes. The MAP-kinases JNK1 and JNK2 are activated by MKK4 (or SEK-1, JNKK1) which itself is a target of the MEKK1 kinase. Upstream activators of MEKK1 are strongly stimuli and cell type dependent; in some cases the small G-proteins Ras, Rac, or Cdc42 are involved. The substrates of JNK1 and JNK2 are the transcription factors c-jun, activation transcription factor 2 (ATF-2), and the ternary complex factor (TCF). After dimerization these proteins bind to the AP-1 or CRE binding site present in many cytokine promotors.

The p38 kinase plays the most crucial role in the regulation of the cytokine response upon LPS stimulation, as a specific inhibitor of this kinase, SB203580, completely prevents the release of cytokines after LPS stimulation of monocytes *(111)*. P38 is activated by MKK3 or MKK6 after their phosphorylation by MEKK5. Upstream activators may include small G-proteins Rho, Rac, or Cdc42. Several downstream targets of p38 are identified, such as the protein kinases MAPKAPK2/3, MNK1/2, and PRAK and the transcription factors CHOP10 and, MEF2C, the TCF member Sap1, and ATF2, but little is known about the mechanism of transcription regulation by p38. An influence on the AP-1 and CRE activity is conceivable because the activation of MEF2C by p38 should lead to the up-regulation of c-jun transcription *(112)*. More likely, the function of p38 lies in translational regulation rather than the activation of transcription factors, as it was shown that the inhibition of p38 does not affect the mRNA level of cytokines *(113)*.

The inhibition of protein kinase C (PKC) during LPS treatment resulted in a decreased IL-1β and TNF-α secretion *(114,115)*, whereas protein kinase A (PKA) inhibition affects mainly IL-6 release *(97)*. Because inhibitory anti-CD14 MAb did not affect PKC activity it is assumed that PKC is a pathway used by CD14-independent mechanisms *(115)*. Interestingly, LPS-induced activation of PKC is independent of the mobilization of Ca^{2+} and diacylglycerol and involves mainly the atypical PKC isoforms PKC-ξ and PKC-ϵ *(116,117)*. Those atypical PKC isoforms are the downstream targets of the phosphatidylinositol 3-kinase (PI-3 kinase), which was also shown to be activated on LPS stimulation *(118,119)*.

An important transcription factor, which is involved in the regulation of nearly all genes expressed in response to inflammation, is NFκB. NFκB belongs to the NFκB/REL family of dimeric transcription factors, which are sequestered in the cytosol by the inhibitory protein I-κB. Phosphorylation of I-κB leads to its ubiquitinylation and subsequent degradation which results in an unmasked nuclear localization signal of the NFκB dimer and its translocation into the nucleus *(120)*. The upstream activators of NFκB were recently identified for the TNF-α pathway *(121)*. Phosphorylation of I-κB occurs by a high molecular mass protein complex called I-κB kinase (IKK) *(122,123)*. IKK may be activated by NIK, another kinase, in cooperation with Traf2, which is associated with the TNF receptor *(124)*. Interestingly, NFκB can also be activated by MEKK1, which is the JNK activating kinase, through the phosphorylation of the I-κB kinase complex *(122)*.

In summary, we can conclude that a whole panel of signal-transduction pathways are activated by LPS. The picture becomes more complicated by the fact that there is a significant amount of crosstalk between the different pathways, and therefore the elucidation of the initial activating step is a difficult process.

6. Conclusion

Endotoxin is the major component of the outer leaflet of Gram-negative bacteria and has profound immunstimulatory and inflammatory capacity. During severe Gram-negative infections, endotoxin leads to a variety of pathophysiological reactions, possibly resulting in the clinical picture of sepsis. Septic shock syndrome, induced by endotoxin, is associated with a mortality rate of 20–50% and causes approx 100,000 death annually in the United States.

The past years have provided new insights into the biology and chemistry of the endotoxin molecule. Knowledge about the molecular structure of endotoxin, the synthesis of lipid A analogs, and the development of endotoxin antagonists have provided much information about the process of cellular activation by endotoxin. There was also a great deal of progress in knowledge concerning the interplay between endotoxin and soluble or cellular receptors, the signal transduction pathways that are activated by endotoxin, and the cellular response to endotoxin. However, the precise mechanisms of endotoxin activity in the host were proven to be rather complex and remain unclear to date. Nevertheless, these advances provide the basis for the development of better strategies to combat Gram-negative infections in the future.

Acknowledgments

This work was supported by a grant from the Deutsche Forschungs-gemeinschaft (SFB 367, projects C5 to A. J. U.), by the Bundesministerium für Bildung, Wissenschaft, Forschung und Technologie (BMBF Grant 01KI9471 to A. J. U.), and the Fonds der Chemischen Industrie (H.-D. F.)

References

1. Pfeiffer, R. (1892) Untersuchungen über das Choleragift. *Z. Hyg.* **11,** 393–412.
2. Rietschel, E. T. and Brade, H. (1992) Bacterial endotoxins. *Sci. Am.* **267,** 54–61.
3. Rietschel, E. T., Kirikae, T., Schade, F. U., Mamat, U., Schmidt, G., Loppnow, H., Ulmer, A. J., Zähringer, U., Seydel, U., and Di Padova, F. E. (1994) Bacterial endotoxin: molecular relationships of structure to activity and function. *FASEB J.* **8,** 217–225.
4. Schletter, J., Heine, H., Ulmer, A. J., and Rietschel, E. T. (1995) Molecular mechanisms of endotoxin activity. *Arch. Microbiol.* **164,** 383–389.
5. Vogel, S. N. and Hogan, M. M. (1990) Role of cytokines in endotoxin-mediated host responses, in *Immunophysiology: The Role of Cells and Cytokines in Immunity and Inflammation* (Oppenheim, J. J. and Shevach, E. M., eds.), Oxford University Press, New York, pp. 238–258.
6. Nogare, D. (1991) Southwestern Internal Medicine Conference: septic shock. *Am. J. Med. Sci.* **302,** 50–65.
7. Galanos, C., Freudenberg, M., Katschinski, T., Salomao, R., Mossmann, H., and Kumazawa, Y. (1992) Tumor necrosis factor and host response to endotoxin, in *Bacterial Endotoxic Lipopolysaccharides,* Vol. II, *Immunpharmacology and Pathophysiology* (Morrison, D. C. and Ryan, J. L., eds.), CRC, Boca Raton, FL, pp. 75–102.
8. Dinarello, C. A. (1984) Interleukin-1. *Rev. Infect. Dis.* **6,** 51–95.
9. Beutler, B. and Cerami, A. (1998) The biology of catechin/TNF. A primary mediator of the host response. *Annu. Rev. Immunol.* **7,** 625–655.
10. Loppnow, H. (1994) LPS, recIL1 and smooth muscle cell-Il1 activate vascular cells by specific mechanisms, in *Bacterial Endotoxins: Basic Science to Anti-Sepsis Strategies* (Levin, J., Van Deventer, S. J. H., Van der Poll, T., and Sturk, A., eds.), Wiley-Liss, New York, pp. 309–321.
11. Haziot, A., Tsuberi, B. Z., and Goyert, S. M. (1993) Neutrophil CD14: biochemical properties and role in the secretion of tumor necrosis factor-alpha in response to lipopolysaccharide. *J. Immunol.* **150,** 5556–5565.
12. Mattern, T., Thanhäuser, A., Reiling, N., Toellner, K. M., Duchrow, M., Kusumoto, S., Rietschel, E. T., Ernst, M., Brade, H., and Flad, H. D. (1994) Endotoxin and lipid A stimulate proliferation of human T cells in the presence of autologous monocytes. *J. Immunol.* **153,** 2996–3004.
13. Lüderitz, O., Freudenberg, M., Galanos, C., Lehmann, V., Rietschel, E. T., and Shaw, D. (1982) Lipopolysaccharides of gram-negative bacteria, in *Current Topics in Membrane and Transport* (Razin, S. and Rottem, S., eds.), Academic Press, New York, pp. 79–151.

14. Jann, K. and Jann, B. (1994) Structure and biosynthesis of O-antigens, in *Chemistry of Endotoxin* (Rietschel, E. T., ed.), Elsevier Science Publishers, Amsterdam, pp. 138–186.

15. Knirel, Y. A. (1990) Polysaccharide antigens of *Pseudomonas aeruginosa. Crit. Rev. Microbiol.* **17,** 273–304.

16. Rietschel, E. T., Brade, H., Holst, O., Brade, L., Müller-Loennies, S., Mamat, U., Zähringer, U., Beckmann, F., Seydel, U., Brandenburg, K., Ulmer, A. J., Mattern, T., Heine, H., Schletter, J., Loppnow, H., Schonbeck, U., Flad, H.-D., Hauschildt, S., Schade, U. F., Di Padova, F. E., Kusumoto, S., and Schumann, R. R. (1996) Bacterial endotoxin: Chemical constitution, biological recognition, host response, and immunological detoxification. *Curr. Top. Microbiol. Immunol.* **216,** 39–81.

17. Rietschel, E. T., Brade, H., Brade, L., Kaca, W., Kawahara, K., Lindner, B., Lüderitz, T., Tomita, T., Schade, U., and Seydel, U. (1985) Newer aspects of the chemical structure and biological activity of bacterial endotoxins. *Prog. Clin. Biol. Res.* **189,** 31–51.

18. Holst, O. and Brade, H. (1992) Chemical structure of the core region of bacterial lipopolysaccharides, in *Bacterial Endotoxic Lipopolysaccharides* (Morrison, D. C. and Ryan, J. L., eds.), CRC, Boca Raton, FL, pp. 135–170.

19. Unger, F. M. (1981) The chemistry and biological significance of 3-deoxy-2-D-*mann*o-2-octulosonic acid (Kdo). *Adv. Carbohydr. Chem. Biochem.* **38,** 323–387.

20. Di Padova, F. E., Brade, H., Barclay, G. R., Poxton ,I. R., Liehl, E., Schuetze, E., Kocher, H. P., Ramsay, G., Schreier, M. H., and McClelland, D. B. (1993) A broadly cross-protective monoclonal antibody binding to *Escherichia coli* and *Salmonella* lipopolysaccharides. *Infect. Immun.* **61,** 3863–3872.

21. Di Padova, F. E., Gram, H., Barclay, R., Kleuser, B., Liehl, E., and Rietschel, E. T. (1993) New anticore LPS monoclonal antibodies with clinical potential, in *Bacterial Endotoxin: Recognition and Effector Mechanisms* (Levin, J., Alving, C. R., Munford, R. S., and Stuetz, P. L., eds.), Elsevier, Amsterdam, pp. 325–335.

22. Zähringer, U., Lindner, B., and Rietschel, E. T. (1994) Molecular structure of lipid A, the endotoxic center of bacterial lipopolysaccharides. *Adv. Carbohydr. Chem. Biochem.* **50,** 211–276.

23. Galanos, C., Lüderitz, O., Rietschel, E. T., Westphal, O., Brade, H., Brade, L., Freudenberg, M., Schade, U., Imoto, M., and Yoshimura, H. (1985) Synthetic and natural *Escherichia coli* free lipid A express identical endotoxic activities. *Eur. J. Biochem.* **148,** 1–5.

24. Imoto, M., Yoshimura, H., Shimamoto, T., Sakaguchi, N., Kusumoto, S., and Shiba,T. (1987) Total synthesis of *Escherichia coli* lipid A, the endotoxically active principle of cell-surface lipopolysaccharide. *Bull. Chem. Soc. Jpn.* **60,** 2205–2214.

25. Kusumoto, S., Yamamoto, H., and Shiba, T. (1984) Chemical synthesis of lipid X and lipid Y, acylglucosamine-1-phosphates isolated from *Escherichia coli* mutants. *Tetrahedr. Lett.* **25,** 3727–3730.

26. Rietschel, E. T., Brade, L., Brandenburg, K., Flad, H.-D., de Jong-Leuveninck, J., Kawahara, K., Lindner, B., Loppnow, H., Lüderitz, T., and Schade, U. (1987) Chemical structure and biologic activity of bacterial and synthetic lipid A. *Rev. Infect. Dis.* **9(Suppl 5),** S527–S536

27. Rietschel, E. T., Brade, L., Schade, F. U., Seydel, U., Zähringer, U., Kusumoto, S., and Brade, H. (1988) Bacterial endotoxins: properties and structure of biologically active domains, in *Surface of Microorganisms and Their Interactions with the Mammalian Host* (Schrinner, E., Richmont, M. H., Seibert, G., and Schwarz, U., eds.), Verlag Chemie, Weinheim, pp. 1–41.

28. Feist, W., Ulmer, A. J., Musehold, J., Brade, H., Kusumoto, S., and Flad, H.-D. (1989) Induction of tumor necrosis factor-alpha release by lipopolysaccharide and defined lipopolysaccharide partial structures. *Immunobiology* **179,** 293–307.

29. Wang, M. H., Chen, Y. Q., Flad, H.-D., Baer, H. H., Feist, W., and Ulmer, A. J. (1993) Inhibition of interleukin-6 release and T-cell proliferation by synthetic mirror pseudo cord factor analogues in human peripheral blood mononuclear cells. *FEMS Immunol. Med. Microbiol.* **6,** 53–61.

30. Rietschel, E. T., Kirikae, T., Feist, W., Loppnow, H., Zabel, P., Brade, L., Ulmer, A. J., Brade, H., Seydel, U., Zähringer, U., Schlaak, M., Flad, H.-D., and Schade, F. U. (1991) Molecular aspects of the chemistry and biology of endotoxin, in *Molecular Aspects of Inflammation (42. Colloquium Mosbach, 1991)* (Sies, H., Flohe, L., and Zimmer, G., eds.), Springer Verlag, Berlin, pp. 207–231.

31. Loppnow, H., Brade, L., Brade, H., Rietschel, E. T., Kusumoto, S., Shiba, T., and Flad, H.-D. (1986) Induction of human interleukin 1 by bacterial and synthetic lipid A. *Eur. J. Immunol.* **16,** 1263–1267.

32. Wang, M. H., Flad, H.-D., Feist, W., Brade, H., Kusumoto, S., Rietschel, E. T., and Ulmer, A. J. (1991) Inhibition of endotoxin-induced interleukin-6 production by synthetic lipid A partial structures in human peripheral blood mononuclear cells. *Infect. Immun.* **59,** 4655–4664.

33. Kovach, N. L., Yee, E., Munford, R. S., Raetz, C. R., and Harlan, J. M. (1990) Lipid IVA inhibits synthesis and release of tumor necrosis factor induced by lipopolysaccharide in human whole blood ex vivo. *J. Exp. Med.* **172,** 77–84.

34. Loppnow, H., Libby, P., Freudenberg, M., Krauss, J. H., Weckesser, J., and Mayer, H. (1990) Cytokine induction by lipopolysaccharide (LPS) corresponds to lethal toxicity and is inhibited by nontoxic *Rhodobacter capsulatus* LPS. *Infect. Immun.* **58,** 3743–3750.

35. Qureshi, N., Takayama, K., and Kurtz, R. (1991) Diphosphoryl lipid A obtained from the nontoxic lipopolysaccharide of *Rhodopseudomonas sphaeroides* is an endotoxin antagonist in mice. *Infect. Immun.* **59,** 441–444.

36. Christ, W. J., Asano, O., Robidoux, A. L., Perez, M., Wang, Y., Dubuc, G. R., Gavin, W. E., Hawkins, L. D., McGuinness, P. D., and Mullarkey, M. A. (1995) E5531, a pure endotoxin antagonist of high potency. *Science* **268,** 80–83.

37. Homma, J. Y., Matsuura, M., and Kumazawa, Y. (1990) Structure–activity relationship of chemically synthesized nonreducing parts of lipid A analogs. *Adv. Exp. Med. Biol.* **256,** 101–119.

38. Saiki, I., Maeda, H., Murata, J., Takahashi, T., Sekiguchi ,S., Kiso, M., Hasegawa, A., and Azuma, I. (1990) Production of interleukin 1 from human monocytes stimulated by synthetic lipid A subunit analogues. *Int. J. Immunopharmacol.* **12,** 297–305.

39. Kirikae, T., Schade, F. U., Zähringer, U., Kirikae, F., Brade, H., Kusumoto, S., Kusama, T., and Rietschel, E. T. (1994) The significance of the hydrophilic backbone and the hydrophobic fatty acid regions of lipid A for macrophage binding and cytokine induction. *FEMS Immunol. Med. Microbiol.* **8,** 13–26.

40. Galanos, C., Lehmann, V., Lüderitz, O., Rietschel, E. T., Westphal, O., Brade, H., Brade, L., Freudenberg, M. A., Hansen-Hagge, T., and Lüderitz, T. (1984) Endotoxic properties of chemically synthesized lipid A part structures. Comparison of synthetic lipid A precursor and synthetic analogues with biosynthetic lipid A precursor and free lipid A. *Eur. J. Biochem.* **140,** 221–227.

41. Delude, R. L., Savedra, R. J., Zhao, H., Thieringer, R., Yamamoto, S., Fenton, M. J., and Golenbock D. T. (1995) CD14 enhances cellular responses to endotoxin without imparting ligand-specific recognition. *Proc. Natl. Acad. Sci. USA* **92,** 9288–9292.

42. Seydel, U., Labischinski, H., Kastowsky, M., and Brandenburg, K. (1993) Phase behavior, supramolecular structure, and molecular conformation of lipopolysaccharide. *Immunobiology* **187,** 191–211.

43. Brandenburg, K., Mayer, H., Koch, M. H., Weckesser, J., Rietschel, E. T., and Seydel, U. (1993) Influence of the supramolecular structure of free lipid A on its biological activity. *Eur. J. Biochem.* **218,** 555–563.

44. Takayama, K., Mitchell, D. H., Din, Z. Z., Mukerjee, P., Li, C., and Coleman, D. L. (1994) Monomeric Re lipopolysaccharide from *Escherichia coli* is more active than the aggregated form in the Limulus amebocyte lysate assay and in inducing Egr-1 mRNA in murine peritoneal macrophages. *J. Biol. Chem.* **269,** 2241–2244.

45. Freudenberg, M. A., Keppler, D., and Galanos, C. (1986) Requirement for lipopolysaccharide-responsive macrophages in galactosamine-induced sensitization to endotoxin. *Infect. Immun.* **51,** 891–895.

46. Munford, R. S. and Hall, C. L. (1989) Purification of acyloxyacyl hydrolase, a leukocyte enzyme that removes secondary acyl chains from bacterial lipopolysaccharides. *J. Biol. Chem.* **264,** 15,613–15,619.

47. Luchi, M. and Munford R. S. (1993) Binding, internalization, and deacylation of bacterial lipopolysaccharide by human neutrophils. *J. Immunol.* **151,** 959–969.

48. Larrick, J. W., Morgan, J. G., Palings, I., Hirata, M., and Yen, M. H. (1991) Complementary DNA sequence of rabbit CAP18—a unique lipopolysaccharide binding protein. *Biochem. Biophys. Res. Commun.* **179,** 170–175.

49. Pereira, H. A., Erdem, I., Pohl, J., and Spitznagel, J. K. (1993) Synthetic bactericidal peptide based on CAP37: a 37-kDa human neutrophil granule-associated cationic antimicrobial protein chemotactic for monocytes. *Proc. Natl. Acad. Sci. USA* **90,** 4733–4737.

50. Andersson, J., Melchers, F., Galanos, C., and Lüderitz, O. (1973) The mitogenic effect of lipopolysaccharide on bone marrow-derived mouse lymphocytes. Lipid A as the mitogenic part of the molecule. *J. Exp. Med.* **137,** 943–953.

51. Mattern, T., Flad, H.-D., Brade, L., Rietschel, E. T., and Ulmer, A. J. (1998) Stimulation of human T lymphocytes by LPS is MHC unrestricted, but strongly dependent on B7 interactions. *J. Immunol.* **160,** 3412–3418.

52. Milner, E. C., Rudbach, J. A., and Voneschen, K. B. (1983) Cellular responses to bacterial lipopolysaccharide: T cells recognize LPS determinants. *Scand. J. Immunol.* **18,** 21–28.

53. Vogel, S. N., Hilfiker, M. L., and Caulfield, M. J. (1983) Endotoxin-induced T lymphocyte proliferation. *J. Immunol.* **130,** 1774–1779.

54. Baker, P. J. (1993) Effect of endotoxin on suppressor T cell function. *Immunobiology* **187,** 372–381.

55. Pober, J. S. and Cotran, R. S. (1990) The role of endothelial cells in inflammation. *Transplantation* **50,** 537–544.

56. Loppnow, H. and Libby, P. (1989) Adult human vascular endothelial cells express the IL6 gene differentially in response to LPS or IL1. *Cell Immunol.* **122,** 493–503.

57. Loppnow, H. and Libby, P. (1990) Proliferating or interleukin 1-activated human vascular smooth muscle cells secrete copious interleukin 6. *J. Clin. Invest.* **85,** 731–738.

58. Loppnow, H. and Libby, P. (1992) Functional significance of human vascular smooth muscle cell-derived interleukin 1 in paracrine and autocrine regulation pathways. *Exp. Cell Res.* **198,** 283–290.

59. Libby, P., Loppnow, H., Flee, J. C., Palmer, H., Li, H. M., Warner, S. J. C., Salomon, R. N., and Clinton, S. K. (1991) Production of cytokines by vascular cells—an update and implications for artherogenesis, in *Arteriosclerosis—Cellular and Molecular Interactions in the Artery Wall* (Gottlieb, A. L., Langille, B. L., and Federoff, S., eds.), Plenum, New York, pp. 161–169.

60. Mantoviani, A. and Bussolino, F. (1991) Endothelium-derived modulators of leukocyte function., in *VascularEndothelium: Interactions with Circulating Cells.* (Gordon, J. L., ed.), Elsevier, New York, pp. 129–140.

61. Haziot, A., Ferrero, E., Kontgen, F., Hijiya, N., Yamamoto, S., Silver, J., Stewart, C. L., and Goyert, S. M. (1996) Resistance to endotoxin shock and reduced dissemination of gram-negative bacteria in CD14-deficient mice. *Immunity* **4,** 407–414.

62. Schumann, R. R., Leong, S. R., Flaggs, G. W., Gray, P. W., Wright, S. D., Mathison, J. C., Tobias, P. S., and Ulevitch, R. J. (1990) Structure and function of lipopolysaccharide binding protein. *Science* **249,** 1429–1431.

63. Wright, S. D., Ramos, R. A., Tobias, P. S., Ulevitch, R. J., and Mathison, J. C. (1990) CD14, a receptor for complexes of lipopolysaccharide (LPS) and LPS binding protein [see comments]. *Science* **249,** 1431–1433.

64. Gallay, P., Heumann, D., Le, R. D., Barras, C., and Glauser, M. P. (1994) Mode of action of anti-lipopolysaccharide-binding protein antibodies for prevention of endotoxemic shock in mice. *Proc. Natl. Acad. Sci. USA* **91,** 7922–7926.

65. Jack, R. S., Fan, X., Bernheiden, M., Rune, G., Ehlers, M., Weber, A., Kirsch, G., Mentel, R., Furll, B., Freudenberg, M., Schmitz, G., Stelter, F., and Schütt, C. (1997) Lipopolysaccharide-binding protein is required to combat a murine gram-negative bacterial infection. *Nature* **389,** 742–745.

66. Lee, J. D., Kravchenko, V., Kirkland, T. N., Han, J., Mackman, N., Moriarty, A., Leturcq, D., Tobias, P. S., and Ulevitch, R. J. (1993) Glycosyl-phosphatidylinositol-anchored or integral membrane forms of CD14 mediate identical cellular responses to endotoxin. *Proc. Natl. Acad. Sci. USA* **90,** 9930–9934.

67. Frey, E. A., Miller, D. S., Jahr, T. G., Sundan, A., Bazil, V., Espevik, T., Finlay, B. B., and Wright, S. D. (1992) Soluble CD14 participates in the response of cells to lipopolysaccharide. *J. Exp. Med.* **176,** 1665–1671.

68. Bazil, V. and Strominger, J. L. (1991) Shedding as a mechanism of down-modulation of CD14 on stimulated human monocytes. *J. Immunol.* **147,** 1567–1574.

69. Bufler, P., Stiegler, G., Schuchmann, M., Hess, S., Krüger, C., Stelter, F., Eckerskorn, C., Schütt C., and Engelmann H. (1995) Soluble lipopolysaccharide receptor (CD14) is released via two different mechanisms from human monocytes and CD14 transfectants. *Eur. J. Immunol.* **25,** 604–610.

70. Durieux, J. J., Vita, N., Popescu, O., Guette, F., Calzada-Wack, J., Munker, R., Schmidt, R. E., Lupker, J., Ferrara, P., and Ziegler-Heitbrock, H. W. (1994) The two soluble forms of the lipopolysaccharide receptor, CD14: characterization and release by normal human monocytes. *Eur. J. Immunol.* **24,** 2006–2012.

71. Hailman, E., Lichenstein, H. S., Wurfel, M. M., Miller, D. S., Johnson, D. A., Kelley, M., Busse, L. A., Zukowski M. M., and Wright S. D. (1994) Lipopolysaccharide (LPS)-binding protein accelerates the binding of LPS to CD14. *J. Exp. Med.* **179,** 269–277.

72. Pugin, J., Schurer-Maly, C. C., Leturcq, D., Moriarty, A., Ulevitch, R. J., and Tobias, P. S. (1993) Lipopolysaccharide activation of human endothelial and epithelial cells is mediated by lipopolysaccharide-binding protein and soluble CD14. *Proc. Natl. Acad. Sci. USA* **90,** 2744–2748.

73. Loppnow, H. (1994) LPS, recIL1 and smooth muscle cell-IL1 activate vascular cells by specific mechanisms. *Prog. Clin. Biol. Res.* **388,** 309–321.

74. Vita, N., Lefort, S., Sozzani, P., Reeb, R., Richards, S., Borysiewicz, L. K., Ferrara, P., and Labeta M. O. (1997) Detection and biochemical characteristics of the receptor for complexes of soluble CD14 and bacterial lipopolysaccharide. *J. Immunol.* **158,** 3457–3462.

75. Schletter, J., Brade, H., Brade, L., Krüger, C., Loppnow, H., Kusumoto, S., Rietschel, E. T., Flad, H.-D., and Ulmer, A. J. (1995) Binding of lipopolysaccharide (LPS) to an 80-kilodalton membrane protein of human cells is mediated by soluble CD14 and LPS-binding protein. *Infect. Immun.* **63,** 2576–2580.

76. Wright, S. D., Levin, S. M., Jong, M. T., Chad, Z., and Kabbash, L. G. (1989) CR3 (CD11b/CD18) expresses one binding site for Arg-Gly-Asp-containing peptides a second site for bacterial lipopolysaccharide. *J. Exp. Med.* **169,** 175–183.

77. Ingalls, R. R. and Golenbock, D. T. (1995) CD11c/CD18, a transmembrane signaling receptor for lipopolysaccharide. *J. Exp. Med.* **181,** 1473–1479.

78. Ingalls, R. R., Arnaout, M. A., Delude, R. L., Flaherty, S., Savedra, R. J., and Golenbock, D. T. (1998) The CD11/CD18 integrins: characterization of three novel LPS signaling receptors. *Prog. Clin. Biol. Res.* **397,** 107–117.

79. Flaherty, S. F., Golenbock, D. T., Milham, F. H., and Ingalls, R. R. (1997) CD11/CD18 leukocyte integrins: new signaling receptors for bacterial endotoxin. *J. Surg. Res.* **73,** 85–89.

80. Ingalls, R. R., Arnaout, M. A., and Golenbock, D. T. (1997) Outside-in signaling by lipopolysaccharide through a tailless integrin. *J. Immunol.* **159,** 433–438.

81. Wright, S. D. (1991) Multiple receptors for endotoxin. *Curr. Opin. Immunol.* **3,** 83–90.

82. Hampton, R. Y., Golenbock, D. T., Penman, M., Krieger, M., and Raetz, C. R. (1991) Recognition and plasma clearance of endotoxin by scavenger receptors. *Nature* **352,** 342–344.

83. Yang, R. B., Mark, M. R., Gray, A., Huang, A., Xie, M. H., Zhang, M., Goddard, A., Wood, W. I., Gurney, A. L., and Godowski, P. J. (1998) Toll-like receptor-2 mediates lipopolysaccharide-induced cellular signalling [In Process Citation]. *Nature* **395,** 284–288.

84. Kirschning, C. J., Wesche, H., Ayers, M., and Rothe, M. (1998) Human Toll-like receptor 2 confers responsiveness to bacterial LPS. *J. Exp. Med.* **188,** 2091–2097.

85. Harder, T. and Simons, K. (1997) Caveolae, DIGs, and the dynamics of sphingolipid-cholesterol microdomains. *Curr. Opin. Cell Biol.* **9,** 534–542.

86. Brown, D. A. and Rose, J. K. (1992) Sorting of GPI-anchored proteins to glycolipid-enriched membrane subdomains during transport to the apical cell surface. *Cell* **68,** 533–544.

87. Cinek, T. and Horejsi, V. (1992) The nature of large noncovalent complexes containing glycosyl-phosphatidylinositol-anchored membrane glycoproteins and protein tyrosine kinases. *J. Immunol.* **149,** 2262–2270.

88. Stefanova, I., Corcoran, M. L., Horak, E. M., Wahl, L. M., Bolen, J. B., and Horak, I. D. (1993) Lipopolysaccharide induces activation of CD14-associated protein tyrosine kinase p53/56lyn. *J. Biol. Chem.* **268,** 20,725–20,728.

89. Wang, P. Y., Kitchens, R. L., and Munford, R. S. (1995) Bacterial lipopolysaccharide binds to CD14 in low-density domains of the monocyte-macrophage plasma membrane. *J. Inflamm.* **47,** 126–137.

90. Pugin, J., Kravchenko, V. V., Lee, J. D., Kline, L., Ulevitch, R. J., and Tobias, P. S. (1998) Cell activation mediated by glycosylphosphatidylinositol-anchored or transmembrane forms of CD14. *Infect. Immun.* **66,** 1174–1180.

91. Shenoy-Scaria, A. M., Kwong, J., Fujita, T., Olszowy, M. W., Shaw, A. S., and Lublin, D. M. (1992) Signal transduction through decay-accelerating factor. Interaction of glycosyl-phosphatidylinositol anchor and protein tyrosine kinases p56lck and p59fyn 1. *J. Immunol.* **149,** 3535–3541.

92. Davis, L. S., Patel, S. S., Atkinson, J. P., and Lipsky, P. E. (1988) Decay-accelerating factor functions as a signal transducing molecule for human T cells. *J. Immunol.* **141,** 2246–2252.

93. Shibuya, K., Abe, T., and Fujita, T. (1992) Decay-accelerating factor functions as a signal transducing molecule for human monocytes. *J. Immunol.* **149,** 1758–1762.

94. Stefanova, I., Horejsi, V., Ansotegui, I. J., Knapp, W., and Stockinger, H. (1991) GPI-anchored cell-surface molecules complexed to protein tyrosine kinases. *Science* **254,** 1016–1019.

95. Beaty, C. D., Franklin, T. L., Uehara, Y., and Wilson, C. B. (1994) Lipopolysaccharide-induced cytokine production in human monocytes: role of tyrosine phosphorylation in transmembrane signal transduction. *Eur. J. Immunol.* **24,** 1278–1284.

96. Meng, F. and Lowell, C. A. (1997) Lipopolysaccharide (LPS)-induced macrophage activation and signal transduction in the absence of Src-family kinases Hck, Fgr, and Lyn. *J. Exp. Med.* **185,** 1661–1670.

97. Geng, Y., Zhang B., and Lotz M. (1993) Protein tyrosine kinase activation is required for lipopolysaccharide induction of cytokines in human blood monocytes. *J. Immunol.* **151,** 6692–6700.

98. Jakway, J. P. and DeFranco, A. L. (1986) Pertussis toxin inhibition of B cell and macrophage responses to bacterial lipopolysaccharide. *Science* **234,** 743–746.

99. Daniel-Issakani, S., Spiegel, A. M., and Strulovici, B. (1989) Lipopolysaccharide response is linked to the GTP binding protein, Gi2, in the promonocytic cell line U937. *J. Biol. Chem.* **264,** 20,240–20,247.

100. Barber, S. A., Detore, G., McNally, R., and Vogel, S. N. (1996) Stimulation of the ceramide pathway partially mimics lipopolysaccharide-induced responses in murine peritoneal macrophages. *Infect. Immun.* **64,** 3397–3400.

101. Barber, S. A., Perera, P. Y., and Vogel, S. N. (1995) Defective ceramide response in C3H/HeJ (Lpsd) macrophages. *J. Immunol.* **155,** 2303–2305.

102. Joseph, C. K., Wright, S. D., Bornmann, W. G., Randolph, J. T., Kumar, E. R., Bittman, R., Liu, J., and Kolesnick, R. N. (1994) Bacterial lipopolysaccharide has structural similarity to ceramide and stimulates ceramide-activated protein kinase in myeloid cells. *J. Biol. Chem.* **269,** 17,606–17,610.

103. Weinstein, S. L., Gold, M. R., and DeFranco, A. L. (1991) Bacterial lipopolysaccharide stimulates protein tyrosine phosphorylation in macrophages. *Proc. Natl. Acad. Sci. USA* **88,** 4148–4152.

104. Weinstein, S. L., Sanghera, J. S., Lemke, K., DeFranco, A. L., and Pelech, S. L. (1992) Bacterial lipopolysaccharide induces tyrosine phosphorylation and activation of mitogen-activated protein kinases in macrophages. *J. Biol. Chem.* **267,** 14,955–14,962.

105. Hambleton, J., Weinstein, S. L., Lem, L., and DeFranco, A. L. (1996) Activation of c-Jun N-terminal kinase in bacterial lipopolysaccharide-stimulated macrophages. *Proc. Natl. Acad. Sci. USA* **93,** 2774–2778.

106. Sanghera, J. S., Weinstein, S. L., Aluwalia, M., Girn, J., and Pelech, S. L. (1996) Activation of multiple proline-directed kinases by bacterial lipopolysaccharide in murine macrophages. *J. Immunol.* **156,** 4457–4465.

107. Han, J., Lee, J. D., Tobias, P. S., and Ulevitch, R. J. (1993) Endotoxin induces rapid protein tyrosine phosphorylation in 70Z/3 cells expressing CD14. *J. Biol. Chem.* **268,** 25,009–25,014.

108. Han, J., Lee, J. D., Bibbs, L., and Ulevitch, R. J. (1994) A MAP kinase targeted by endotoxin and hyperosmolarity in mammalian cells. *Science* **265,** 808–811.

109. Geppert, T. D., Whitehurst, C. E., Thompson, P., and Beutler, B. (1994) Lipopolysaccharide signals activation of tumor necrosis factor biosynthesis through the ras/raf-1/MEK/MAPK pathway. *Mol. Med.* **1,** 93–103.

110. Reimann, T., Buscher, D., Hipskind, R. A., Krautwald, S., Lohmann-Matthes, M. L., and Baccarini, M. (1994) Lipopolysaccharide induces activation of the Raf-1/MAP kinase pathway. A putative role for Raf-1 in the induction of the IL-1 beta and the TNF-alpha genes. *J. Immunol.* **153,** 5740–5749.

111. Lee, J. C., Laydon, J. T., McDonnell, P. C., Gallagher, T. F., Kumar, S., Green, D., McNulty, D., Blumenthal, M. J., Heys, J. R., and Landvatter, S. W. (1994) A protein kinase involved in the regulation of inflammatory cytokine biosynthesis. *Nature* **372,** 739–746.

112. Han, J., Jiang, Y., Li, Z., Kravchenko, V. V., and Ulevitch, R. J. (1997) Activation of the transcription factor MEF2C by the MAP kinase p38 in inflammation. *Nature* **386,** 296–299.

113. Perregaux, D. G., Dean, D., Cronan, M., Connelly, P., and Gabel, C. A. (1995) Inhibition of interleukin–1 beta production by SKF86002: evidence of two sites of in vitro activity and of a time and system dependence. *Mol. Pharmacol.* **48,** 433–442.

114. Coffey, R. G., Weakland, L. L., and Alberts, V. A. (1992) Paradoxical stimulation and inhibition by protein kinase C modulating agents of lipopolysaccharide evoked production of tumour necrosis factor in human monocytes. *Immunology* **76,** 48–54.

115. Shapira, L., Takashiba, S., Champagne, C., Amar, S., and Van, D. T. (1994) Involvement of protein kinase C and protein tyrosine kinase in lipopolysaccharide-induced TNF-alpha and IL-1 beta production by human monocytes. *J. Immunol.* **153,** 1818–1824.

116. Herrera-Velit, P., Knutson ,K. L., and Reiner, N. E. (1997) Phosphatidylinositol 3-kinase-dependent activation of protein kinase C-zeta in bacterial lipopolysaccharide-treated human monocytes. *J. Biol. Chem.* **272,** 16,445–16,452.

117. Shapira, L., Sylvia, V. L., Halabi, A., Soskolne, W. A., Van, D. T., Dean, D. D., Boyan, B. D., and Schwartz, Z. (1997) Bacterial lipopolysaccharide induces early and late activation of protein kinase C in inflammatory macrophages by selective activation of PKC-epsilon. *Biochem. Biophys. Res. Commun.* **240,** 629–634.

118. Herrera-Velit, P. and Reiner N. E. (1996) Bacterial lipopolysaccharide induces the association and coordinate activation of p53/56lyn and phosphatidylinositol 3-kinase in human monocytes. *J. Immunol.* **156,** 1157–1165.

119. Park, Y. C., Lee, C. H., Kang, H. S., Chung, H. T., and Kim, H. D. (1997) Wortmannin, a specific inhibitor of phosphatidylinositol–3-kinase, enhances LPS-induced NO production from murine peritoneal macrophages. *Biochem. Biophys. Res. Commun.* **240,** 692–696.

120. Baeuerle, P. A. and Baltimore, D. (1996) NF-kappa B: ten years after. *Cell* **87,** 13–20.
121. Maniatis, T. (1997) Catalysis by a multiprotein IkappaB kinase complex. *Science* **278,** 818–819.
122. Lee, F. S., Hagler, J., Chen, Z. J., and Maniatis, T. (1997) Activation of the IkappaB alpha kinase complex by MEKK1, a kinase of the JNK pathway. *Cell* **88,** 213–222.
123. Regnier, C. H., Song, H. Y., Gao ,X., Goeddel, D. V., Cao, Z., and Rothe, M. (1997) Identification and characterization of an IkappaB kinase. *Cell* **90,** 373–383.
124. Malinin, N. L., Boldin ,M. P., Kovalenko, A. V., and Wallach, D. (1997) MAP3K-related kinase involved in NF-kappaB induction by TNF, CD95 and IL-1. *Nature* **385,** 540–544.
125. Mattern, T., Girroleit, G., Flad, H.-D., Rietschel, E. M., and Ulmer, A. J. (1999) CD34-positive haematopoietic stem cells exert accessory function in LPS-induced T-cell proliferation and CD80 expression on monocytes. *J. Exp. Med.* **189,** 693–700.

17

Matrix-Assisted Laser Desorption/Ionization Time-of-Flight Mass Spectrometry of Lipopolysaccharides

Buko Lindner

1. Introduction

The outer membrane of Gram-negative bacteria, which include many human pathogens, contains various proteins, polysaccharides, and glycolipids. Of these, lipopolysaccharides (LPS) are of particular microbiological, immunological, and medical importance. As the major amphiphilic components of the outer leaflet of the outer membrane, LPS fulfill a vital role for the organization and function of the outer membrane (e.g., effective permeation barrier to harmful substances). Furthermore, LPS represent the main surface antigen (O-antigen) harboring binding sites for antibodies and are thus involved in the specific recognition by the host organism's defense system. When released from bacteria, for example, during multiplication, death, or lysis, LPS induce in mammalis a broad spectrum of physiological and pathological activities such as stimulation of cytokine production and act as potent bacterial toxins responsible for the toxic manifestation of Gram-negative infections (e.g., septic shock). To emphasize these activities LPS have also been termed *endotoxins*.

Chemically, endotoxins consist of a hydrophilic polysaccharide and a covalently bound hydrophobic lipid component, called lipid A, anchoring LPS in the outer membrane. In *Enterobacteriaceae* the polysaccharide can be subdivided into two structurally distinct regions, a core oligosaccharide and a long heteropolysaccharide chain, the O-specific chain, which is generally composed of a sequence of varying numbers of identical oligosaccharides, the repeating units. The nature and type of linkage or substitution as well as the number and sequence of the monosaccharides within the repeating units are characteristic

From: *Methods in Molecular Biology, vol. 145: Bacterial Toxins: Methods and Protocols*
Edited by: O. Holst © Humana Press Inc., Totowa, NJ

for a given LPS. The O-chains comprise an enormous compositional and structural variability among different serotypes and species. In rough mutant strains and in some nonenterobacterial species the O-specific chain is absent. The core region of LPS which often consists of a branched heterooligosaccharide can formally be subdivided into an outer core consisting of neutral hexoses and an inner core mainly composed of L-*glycero*-D-*manno* heptose (L,D-Hep) and 3-deoxy-D-*manno*-oct-2-ulopyranosonic acid (Kdo), to which phosphate or 2-aminoethyl-(pyro)phosphate groups may be linked. Lipid A structures of different bacterial origin are built up according to a conserved and common architecture consisting of a β-1,6-linked disaccharide backbone containing amino sugars that carries up to seven fatty acids in ester and amid linkages (mainly 3-hydroxy and acyloxyacyl fatty acids) and one or two partially substituted phosphate groups at positions 1 and 4' of the disaccharide backbone. Lipid A has been shown to constitute the "endotoxic principle" of LPS although many of its biological activities are enhanced by the Kdo-containing inner core. For details on biological activity and chemical structure the reader is referred to Chapter 16 of this volume and to comprehensive reviews *(1–3)*.

It is well known that small alterations of the chemical structure of endotoxin may lead not only to dramatic changes of their biological activity but even to antagonistic effects *(4)*. Therefore, a detailed knowledge of the chemical composition of the most often heterogeneous natural or derivatized LPS samples is important. To characterize the complex chemical structures various diagnostic tools such as chemical compositional analysis, nuclear magnetic resonance (NMR) spectroscopy, and modern soft desorption/ionization mass spectrometric techniques need to be combined (for review *see [5]*).

This chapter concentrates on the application of matrix-assisted laser desorption time-of-flight mass spectrometry (MALDI-TOF MS *[6]*) for the structural elucidation of endotoxins. In contrast to peptides and proteins, the mass spectrometric analysis of LPS is hampered by their thermolability, heterogeneity, and their intrinsic amphiphilic character, that is, the presence of hydrophobic and hydrophilic regions. Particularly the last property, which leads to spontaneous aggregation of the glycolipid in either organic nonpolar or aqueous solution, requires special efforts to obtain homogeneous microcrystalline sample matrix layers, a prerequisite for reproducibility and sensitivity of MALDI-MS. Therefore, special attention is given to the description of routinely used sample preparation methods and modes of mass spectrometric analysis to provide detailed information on (1) exact molecular masses to confirm structural proposals; (2) type, number, and distribution of fatty acid residues on the lipid A backbone; and (3) intrinsic biological heterogeneity within the lipid A, core, and O-chain polysaccharides of native, unmodified LPS.

2. Materials

2.1. Endotoxins

1. Smooth and rough forms of LPS can be isolated from bacteria by the phenol–water method *(7)* and the phenol–chloroform–light petroleum ether (PCP) method *(8)*, respectively, and subsequently dialyzed and lyophilized.
2. De-*O*-acylated LPS can be prepared from the crude LPS by treatment with anhydrous hydrazine (37°C, 30 min) *(9)*; *see* Chapter 19.
3. Free lipid A was obtained by acetate buffer treatment (1 *M*, 100°C, 1.5 h) of LPS and converted to the triethylamine (TEN) salt form *(10)*. Dephosphorylated lipid A was obtained by treatment of LPS with 48% hydrofluoric acid (48 h, 4°C) followed by acetate buffer treatment *(11)*.

2.2. Chemicals and Separation Media

Matrix compounds for MALDI-MS:

1. 2,4,6-Trihydroxyacetophenone (THAP, Aldrich, Steinheim, Germany).
2. 2,5-Dihydroxybenzoic acid (gentisic acid, DHB, Aldrich, Steinheim, Germany).
3. 3-Hydroxypicolinic acid (PA Aldrich, Steinheim, Germany).
4. 3,5-Dimethoxy-4-hydroxycinnamic acid (sinapic acid, SA, Fluka, Deisenhofen, Germany).

Other:

5. TEN.
6. High-performance liquid chromatography (HPLC)-grade acetonitrile (Sigma, Deisenhofen, Germany).
7. Cation-exchanger Amberlite-120 (H$^+$-form)
8. Methanol.
9. Chloroform.
10. Silica gel 60 thin-layer chromatography (TLC) aluminium sheets (Merck, Darmstadt, Germany).

3. Methods

3.1. Preparation of Isolated Sample Material

In this **Subheading** a numerically ordered protocol for the different steps of sample preparation is given. It should be mentioned that only those methods are described that seem in our hands be best suited to routinely detect ions from intact molecules with high sensitivity.

1. Solubilization of LPS and lipid A samples: About 1 nmol of crystalline sample material is dispersed in 7.5 μL of deionized water in an 0.5-mL Eppendorf tube. For the disaggregation of the amphiphilic glycolipids *(10)*, 5 μL of 0.36 M TEN are added and the dispersion is vortex-mixed and sonicated for 2 min at 40°C in

an ultrasonic bath (Sonorex RK100, Bandelin Electronics, Berlin, Germany). For possible alterations and critical points *see* **Notes 1–3**.

2. Removal of disturbing cations: Small droplets ($\approx$100 µL) of cation exchanger (e.g., Amberlite-120) are deposited onto a piece of Parafilm™ (American National Can, Greenwich, CT) and excess fluid is absorbed by tissue paper (Kimwipes™, Kimberly-Clark). The sample solution is pipetted onto the remaining cation exchange resin beads and after some seconds of exposure transferred to a new vial for mixing with the matrix solution (*see* **Note 4**).

3. Matrix solution: 75 mg of THAP are dissolved in 1 mL of methanol containing 200 µL of 0.1% trifluoroacetic acid and 100 µL of acetonitrile (for further explanation *see Notes 5* and *6*). Matrix solution has to be prepared freshly for daily use.

4. Mixing of sample and matrix solution and deposition on sample target: Approximately 2 µL of matrix solution and the same volume of sample solution are placed into a 0.5-mL Eppendorf tube and mixed carefully for a few seconds with a vortex mixer (*see* **Note 7**). Then, small droplets (aliquotes of 0.5–1 µL) of the resulting mixture are deposited on the metallic surface of the spectrometer's sample stage (*see* **Note 8**). The droplets are dried at 20–25°C in a stream of air. When the liquid has completely evaporated, the sample can be loaded into the mass spectrometer.

3.2. Preparation from TLC Spots

TLC is a frequently used sensitive analytical method to prove the purity or heterogeneity of isolated lipid A. To reliably identify the chemical structure of TLC-separated spots and for detailed fragmentation analysis (*see* **Subheading 4.2.2.**), we have developed a method to obtain MALDI-MS spectra from such spots.

1. TLC-separation: Approximately 15 µg of crude lipid A is deposited in two bands of 1 cm length onto the ready-to-use silica gel 60 aluminum plate (*see* **Note 9**). The TLC plate is developed in a solvent system consisting of chloroform–methanol–water (100:75:15, by vol). The spots of one lane are visualized by charring after dipping in sulfuric acid reagent or by exposure to iodine vapors. The unstained spots of the other lane are scraped off and transferred to an Eppendorf tube.

2. Extraction, purification, and sample preparation: The sample is extracted from the silica gel and desalted by the addition of approx 30 µL of the TLC solvent mixture, vortex-mixed vigorously for about 1 min, and then transferred to a Micropure™ 0.45 µm separator (Millipore, Eschborn, Germany) filled with a thin layer of ion-exchange resin beads. The separator is placed into a filtrate vial (1.5-mL Eppendorf tube) and centrifuged for 30 s, followed by a second centrifugation after addition of 30 µL of solvent. The chloroform of the TLC solvent in the filtrate vial is evaporated in a stream of nitrogen (*see* **Note 10**) and is solubilized as described under **Subheading 3.2.**, **step 1** and can then be directly mixed with matrix solution and deposited on the probe tip (*see* **Subheading 3.1.**, **step 4**).

3.3. MALDI-TOF MS

3.3.1. Instrumentation

Mass spectrometric analyses shown in this report were performed with a Bruker-Reflex III (Bruker-Franzen Analytik, Bremen, Germany) that can be operated both with a linear (LIN-) and a two-stage reflectron (RE-)TOF mass analyzer with either continuous or delayed ion extraction for increased mass resolution and accuracy. A nitrogen laser (wavelength $\lambda = 337.1$ nm, pulse duration $d = 3$ ns) is applied for laser desorption. In RE-TOF configuration fragment ion spectra can be acquired originating from metastable decomposition of a preselected parent ion within the time-of-flight tube (post-source decay, PSD).

Unless otherwise stated, the ions were accelerated to 20 keV and 28 keV in the negative and positive ion mode, respectively. Mass spectra shown are the sum of at least 50 laser shots. Mass scale calibration was performed externally and in some cases internally with glycolipids and lipid A derivatives of known chemical structure.

3.3.2. Mass Spectrometric Analysis

The characteristics of the mass spectra obtained under different modes of instrumental operation are outlined in this **Subheading** and the structural information that can be extracted is exemplarily demonstrated for a few mass spectra obtained from native smooth and rough form LPS and lipid A samples, respectively. Additional examples of MALDI-MS for glycolipids are given in the literature *(12–20)*.

3.3.2.1. CHARACTERISTICS OF MASS SPECTRA OF NATIVE LPS

Owing to their negatively charged phosphate and carboxyl groups, underivatized LPS and lipid A are more sensitively detected in the negative ion mode (as singly charged deprotonated molecular ions, $[M-H]^-$) than in the positive ion mode. In this mode multiple cationization ($[M+H +n(X–H)]^+$, $X = Na, K, n = 0,1,2,3$) and loss of phosphate groups add an undesirable level of complexity, decrease sensitivity, and reduce mass resolution, thus, sometimes, preventing a straight forward unequivocal interpretation of the mass spectra.

In contrast to peptides and proteins which normally do not give rise to prompt fragmentation, the mass spectra of LPS samples exhibit in the LIN-TOF configuration under continuous or delayed ion extraction, significant fragment ion peaks generated either during the laser-induced desorption process or by decomposition within the ion source. The fact that the relative abundances of these ion peaks are dependent on the laser irradiance, with lowest intensities near the threshold for ion formation, proves that they are not generated from

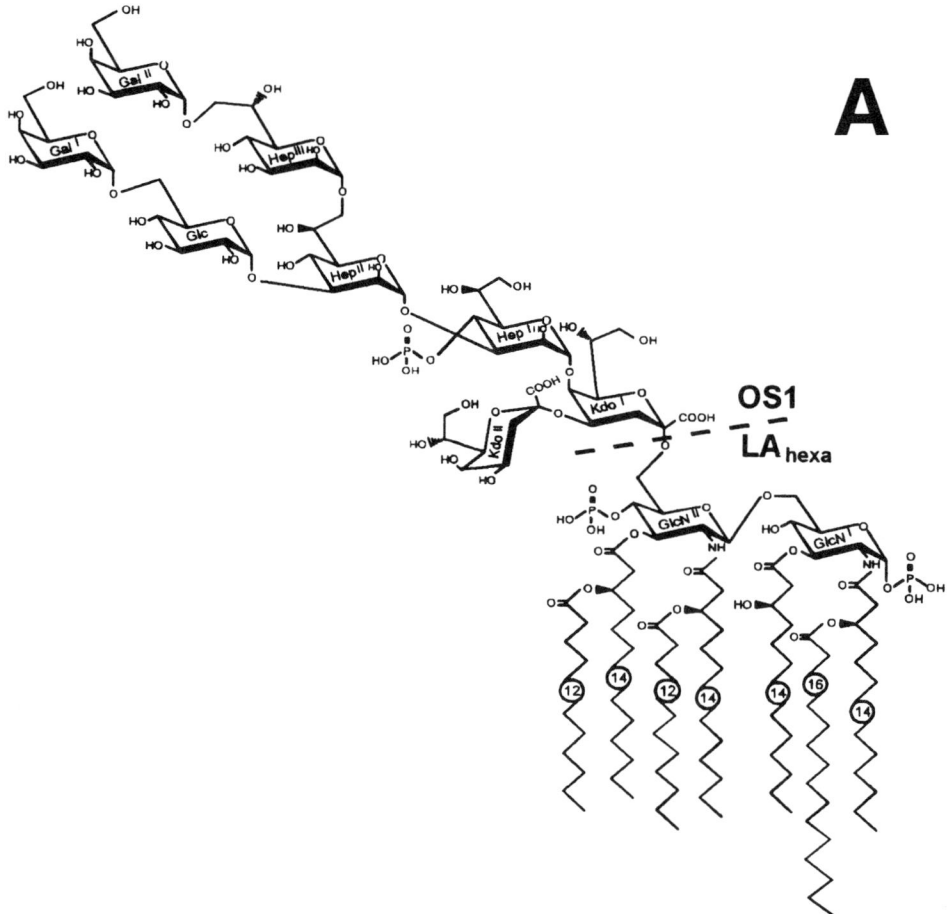

Fig. 1. Chemical structure **(A)** of the most abundant LPS (OS1$_{hexa}$) **(14)** and negative ion LIN-TOF MALDI mass spectrum of native LPS isolated from *Erwinia
carotovora FERM* P-7576 acquired under continuous ion extraction **(B)**.

molecular species present in a heterogeneous LPS sample but are LPS fragments
originating from the rupture between Kdo and the lipid A. The cleavage of this
very acid labile glycosidic bond yield peaks related to either the polysaccharide
portion (which is—according to the nomenclature of Domon and Costello
[21]—characterized as a B-type oligosaccharide fragment), or to the lipid A component where the glycosidic oxygen remains at the lipid A moiety (*see* **Fig. 1**).
As an example we show the negative ion MALDI mass spectrum of the complete
native LPS isolated from *Erwinia carotovora* generated in LIN-TOF configuration under continuous ion extraction (*see* **Note 11**). The spectrum reflects in its

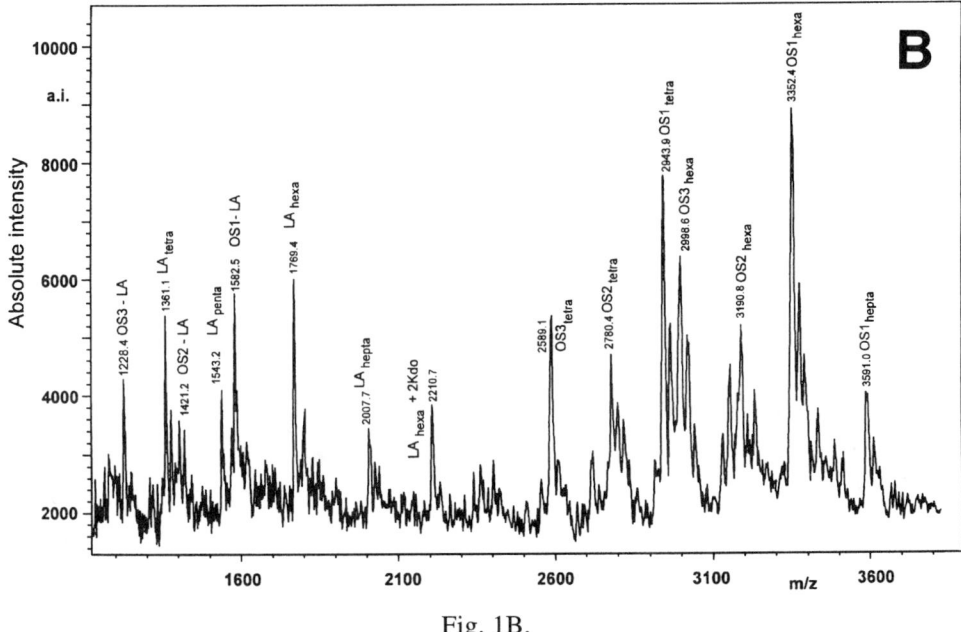

Fig. 1B.

complexity the heterogeneity of the complete LPS with respect to the core oli-gosaccharide (OS1–OS3) and with respect to the number of primary and sec-ondary *O*-linked fatty acid residues within lipid A *(22)*. For instance, $OS1_{hepta}$ stands for the quasimolecular ion $[M-H]^-$ of a hepta-*N,O*-acyl derivative of OS1. Fragment ions (*m/z* 1100–2300) originating from the cleavage of lipid A show on the one hand the heterogeneity in the degree of acylation of lipid A (LA_{treta}–LA_{hepta}) and on the other hand the heterogeneity within the core oli-gosaccharide (OS1-LA–OS3-LA) (*see* **Note 12**). These data fit well with the results of various chemical and NMR analyses on the chemical structure of the core region and free lipid A elucidated by Fukuoka et al. *(14,22)*.

Mass spectra of the native or de-*O*-acylated S-form LPS with a long polysac-charide O-chain exhibit clusters of ions up to *m/z* >10,000, indicating the pres-ence of a large number of components with different molecular masses. Although these clusters of ions are not well resolved, the mass differences between the main components indicate the size of the repeating unit *(12)*. In **Fig. 2** part of the negative ion mass spectrum of the S-form of LPS from *Sal-monella enterica* sv. Minnesota is given yielding a mass difference of *m/z* 648 corresponding to a repeating unit consisting of four hexoses (*see* **Note 13**).

3.3.2.2. Characteristic of Mass Spectra of Free Lipid A

Negative ion mass spectra of free lipid A samples in LIN-TOF and in RE-TOF configuration exhibit nearly the same pattern of abundant molecular ions

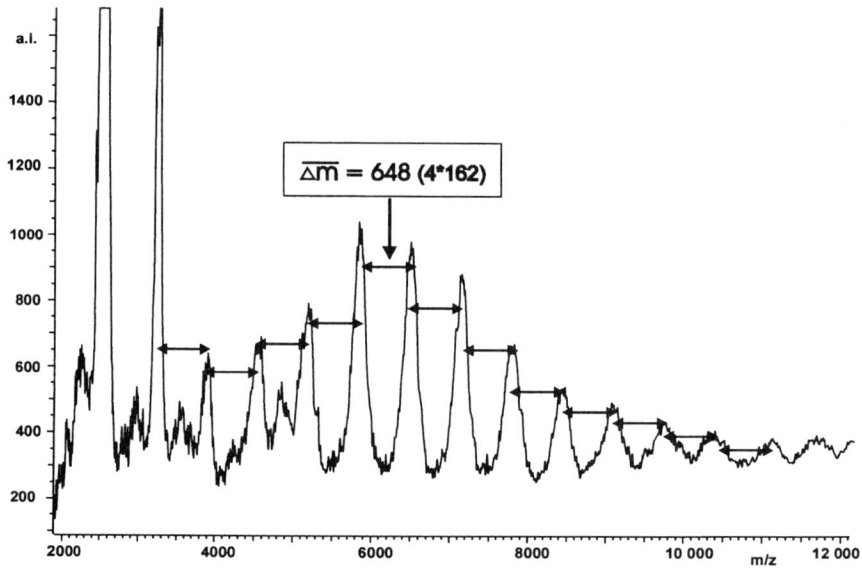

Fig. 2. Part of the negative ion LIN-TOF MALDI mass spectrum of the S form of LPS from a *Salmonella enterica* svar. Minnesota strain acquired under continuous ion extraction.

showing no or only minor fragmentation as demonstrated in **Fig. 3** for a free lipid A preparation isolated from *S. enterica* sv. Minnesota strain R595. The isolated lipid A mainly consists of four molecular species, a hepta- and a hexa-acylated diphosphoryl lipid A both carrying a 4-amino-4-deoxy-L-arabinopyranose (L-Ara*p*4N) in nonstoichiometric amounts. The inset shows an enlargement of the molecular ion peak of the diphosphoralyted hexaacyl component acquired in the RE-TOF mode under delayed ion extraction (*see* **Note 14**).

In the positive ion mode, however, abundant prompt laser-induced fragment ions are induced originating from the cleavage of phosphate substituents such as L-Ara*p*4N and phosphorethanolamine, the phosphate group at position C1, and the rupture of the glycosidic bond leading to the formation of an oxonium ion of the distal glucosaminyl residue (GlcN II) (*see* **Fig. 4**, scheme of the chemical structure). The latter fragment ion is of special diagnostic importance, since it allows the determination of the fatty acid distribution on the reducing and nonreducing glucosamine directly from underivatized diphosphoryl lipid A. If the oxonium ions cannot be assigned unequivocally because of the heterogeneity of a lipid A sample with respect to the number and type of fatty acids, separation of the different components is necessary. As an example, the negative and positive ion LIN-TOF mass spectrum of the TLC-separated

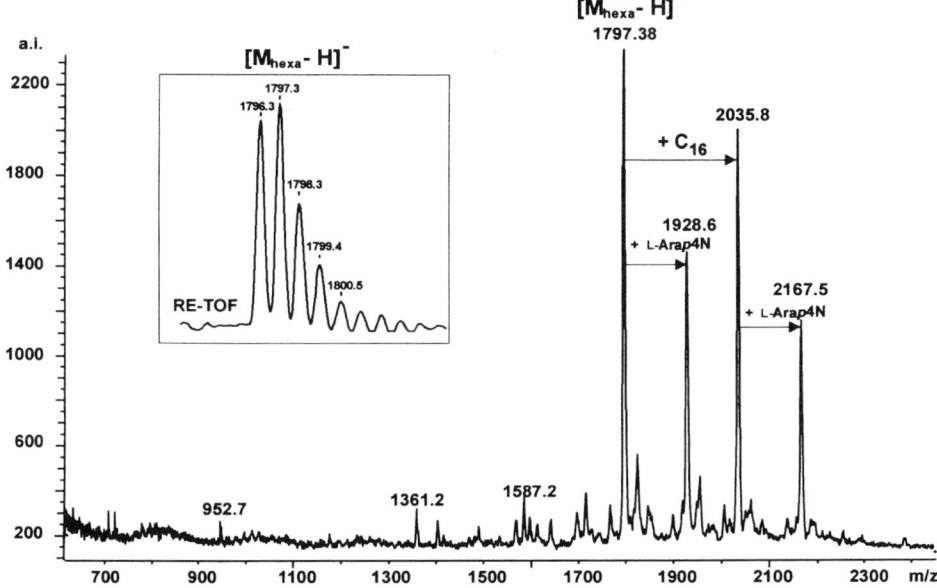

Fig. 3. Negative ion LIN-TOF MALDI mass spectrum of free lipid A isolated from *S. enterica* sv. Minnesota strain R595. Inset shows enlargement of the molecular ion peak at *m/z* 1797 acquired in RE-TOF configuration under delayed ion extraction.

hexaacyl diphosphoryl lipid A component of *S. enterica* sv. Minnesota (as described in **Subheading 3.2.**) is shown. The oxonium ion at *m/z* 1087 provides clear evidence that this hexaacyl lipid A has an asymmetric fatty acid distribution (four fatty acid residues are linked to the nonreducing GlcN II and two fatty acids to the reducing GlcN I).

By stepwise reduction of the reflector voltage, fragment ion spectra can be acquired originating from the metastable decomposition of a preselected parent molecular ion in the first field-free drift region of the analyzer (PSD analysis [23]). The capability of PSD-MS to provide detailed information on the type, number, and linkage of the fatty acids in heterogeneous lipid A samples is demonstrated in **Fig. 5**, showing the positive ion RE-TOF mass spectrum of a crude, heterogeneous lipid A-HF isolated from *E. coli* strain F515 (**Fig. 5A**), the spectrum obtained with gating switched on to select the pentaacyl component at *m/z* 1451 as parent ion (**Fig. 5B**), and the respective PSD daughter ion spectrum (**Fig. 5C**). All abundant fragment ion peaks originated exclusively from cleavage of ester-linked fatty acid residues as indicated in the fragmentation scheme. Furthermore, fragment ions are identified as originating from amide-linked 3-hydroxy fatty acids. These ions are generated by the rupture of

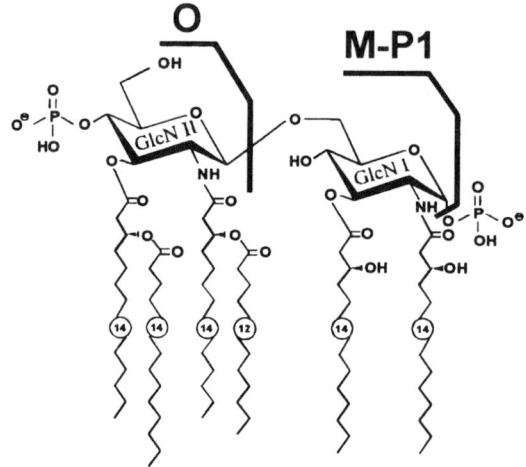

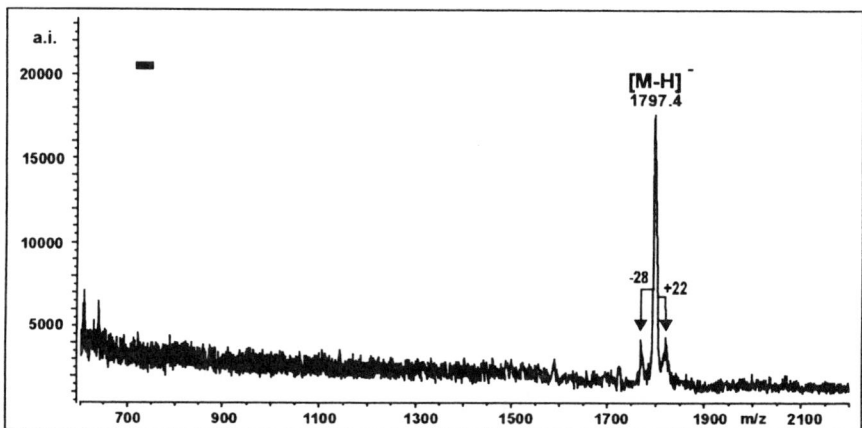

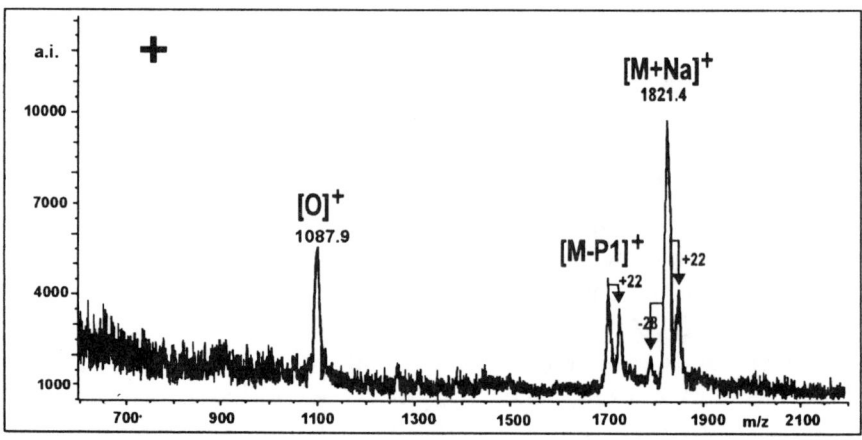

the bond between C_2 and C_3 within the fatty acid chain (c) and are detectable only when the ester-linked fatty acid in a neighboring position of the glucosamine (C3) is already cleaved (*see* **Note 15**). However, PSD mass spectra of lipid A components do not provide any fragment ion of intensity significant enough to allow the determination of the number of fatty acids linked to the reducing and nonreducing glucosamine (e.g., oxonium, or other fragment ion as observed in MS–MS applying high-energy collision-induced decomposition [CID] *[16]* or laser desorption *[24,25]*).

4. Notes

1. For purified lipid A and Re-type LPS the amount of material needed for sample preparation can be reduced to approx 50 pmol (corresponding to approx 100 ng) to still obtain mass spectra with satisfying signal-to-noise ratio.
2. For the S-form LPS with long polysaccharide O-chain the amount of TEN might be reduced.
3. Use this solution immediately: Storage for longer than 1 d might cause degradation of LPS and lipid A by cleavage of ester-bound fatty acid residues.
4. If only very small amounts of material are available, desalting could also be performed after the addition of matrix solution *(19)* or directly on the sample stage *(26)*.
5. Experiments with other matrices showed that as in the positive as well as in the negative ion mode THAP matrix gives best results with respect to sensitivity, resolution, and low level of prompt fragmentation within the ion source, followed by super DHB, a 9:1 (v/v) mixture of 2,5-dihydroxybenzoic acid and 2-hydroxy-5-methoxybenzoic acid.
6. The properties of a particular matrix compound for MALDI-MS may vary from batch to batch and from company to company. Therefore, it might be necessary to recrystallize the matrix compound from deionized water to improve the matrix-assisted desorption process.
7. During mixing the solution should be carefully observed for changes in the solvent composition indicative of precipitation of matrix or sample. In this case, either some microliters of solvent should be added, or mixing should be repeated by using only half of the sample solution.
8. It is good practice to prepare more then one dried droplet of a particular compound on the probe tip, as crystallization may not always be optimal. Furthermore, samples of appropriate standard alone and—if necessary—in a mixture with the compound under investigation should be prepared for external and internal calibration, respectively.

Fig. 4. *(previous page)* Negative and positive ion LIN-TOF MALDI mass spectrum of a hexaacyl lipid A prepared from a TLC-separated spot of crude lipid A isolated from LPS of *S. enterica* svar. Minnesota strain R595 acquired under delayed ion extraction.

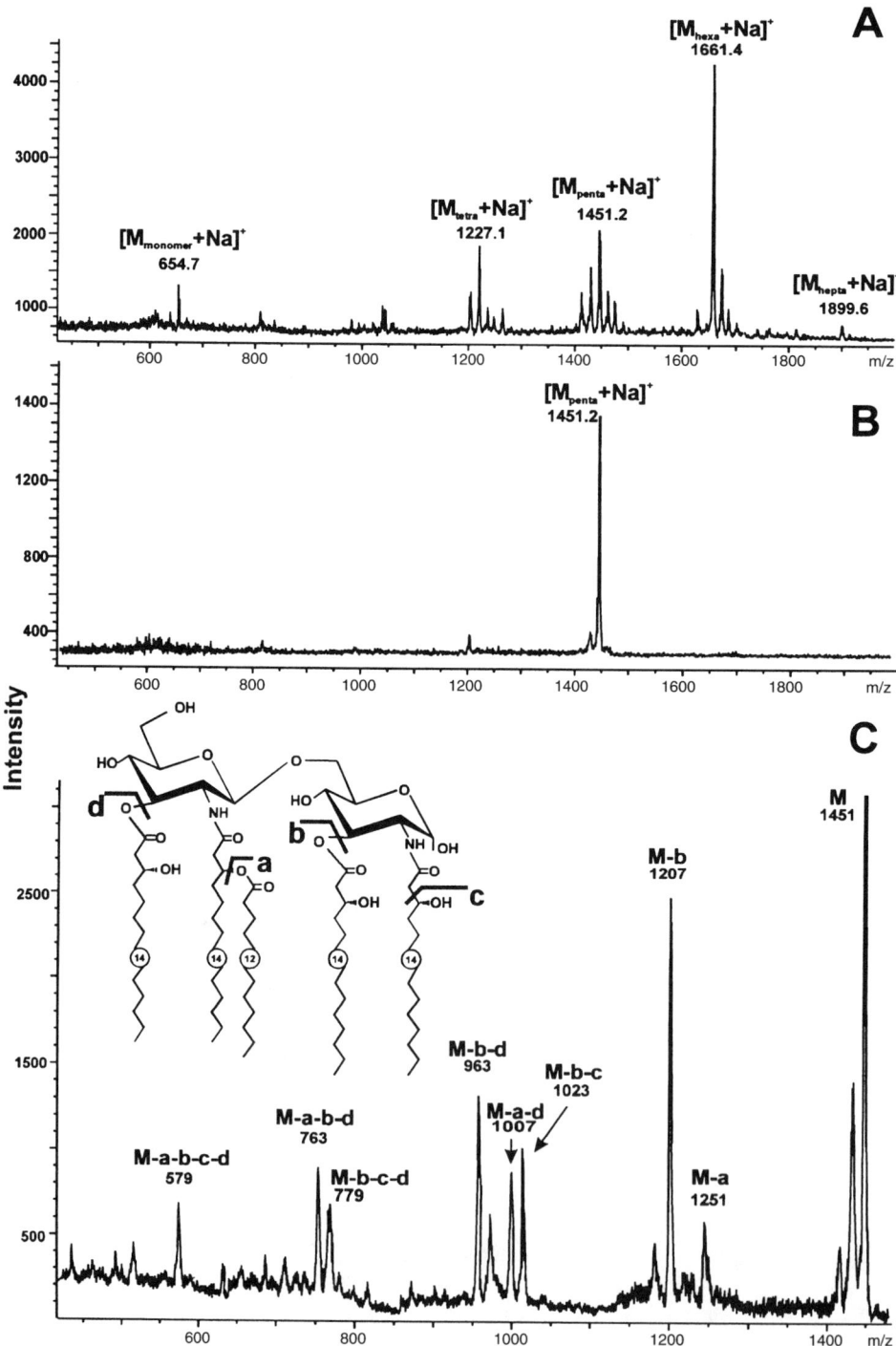

9. Because sensitivity depends on the degree of heterogeneity of the crude lipid A preparation the minimal amount needed might be decreased. For lipid A isolated from *E. coli* strain F515 the method was successful even when only 5 µg/band were deposited on the TLC plate.

10. It is worth mentioning that extraction from silica gel and desalting of the sample could also be performed without the use of the Micropure™ separator in two separate steps, although recovery is decreased.

11. Under delayed ion extraction identical mass peaks are registered at clearly improved mass resolution and accuracy but with decreased relative intensities of the molecular ions. Spectra of LPS analyzed in the RE-TOF configuration comprise mainly fragment ions and molecular ions of very low intensity indicating that the molecules rapidly decay within the time-of-flight tube.

12. To unequivocally identify the origin of different peaks in heterogeneous LPS samples, it could be advantageous to analyze also the de-*O*-acylated LPS. The removal of the *O*-linked fatty acids reduces in most cases the heterogeneity within the lipid A moiety *(14)*. Furthermore, as mentioned by Gibson et al. *(19)*, de-*O*-acylated LPS are more soluble and therefore more likely to cocrystallize with the matrix compound, resulting in better sensitivity and mass resolution.

13. It should be mentioned that for some S-form LPS we measured very poor signal-to-noise ratios in the high mass region, making signal smoothing and accumulation of a greater number of single shot spectra necessary. De-*O*-acylation might help to decrease signal-to-noise ratios (*see* **Note 12**).

14. It should be mentioned that in LIN-TOF configuration the molecular ions of the L-Ara*p*4N-containing compounds are detected with slightly higher relative abundances.

15. Similar information on fatty acids can be obtained also from negative ion mode PSD mass spectra of diphosphoryl lipid A components. However, the additional loss of one or both phosphoryl groups adds an undesirable complexity.

Acknowledgments

I would like to thank H. Lüthje for excellent technical assistance; O. Holst, U. Zähringer, and S. Fukuoka for their collegial collaboration; and K. Brandenburg for proofreading the manuscript.

References

1. Zähringer, U., Lindner, B., and Rietschel, E. T. (1994) Molecular structure of lipid A, the endotoxic center of bacterial lipopolysaccharides. *Adv. Carbohydr. Chem. Biochem.* **50,** 211–276.

Fig. 5. (*previous page*) Metastable decay analysis of lipid A-HF isolated from LPS from *E. coli* strain F515. (**A**) Positive ion RE-TOF MALDI mass spectrum of the heterogeneous lipid A sample; (**B**) spectrum of a preselected pentaacyl parent ion; (**C**) composite PSD daughter ion spectrum.

2. Holst, O. (1999) Chemical structure of the core region of lipopolysaccharides, in *Endotoxins in Health and Disease* (Brade, H., Morrison, D. C., Opal, S., and Vogel, S., eds.), Marcel Dekker, New York, pp. 115–154.

3. Rietschel, E. T., Brade, H., Holst, O., Brade, L., Müller-Loennies, S., Mamat, U., Zähringer, U., Beckmann, F., Seydel, U., Brandenburg, K., Ulmer, A. J., Mattern, T., Heine, H., Schletter, J., Loppnow, H., Schönbeck, U., Flad, H.-D., Hausschildt, S., Schade, U. F., Di Padova, F., Kusumoto, S., and Schumann, R. R. (1996) Bacterial endotoxin: chemical constitution, biological recognition, host response, and immunological detoxification, in *Pathology of Septic Shock* (Rietschel, E. T. and Wagner, H., eds.), Springer-Verlag, Berlin, pp. 39–81.

4. Rietschel, E. T., Kirikae, T., Schade, F. U., Mamat, U., Schmidt, G., Loppnow, H., Ulmer, A., Seydel, U., Di Padova, F. E., Schreier, M., and Brade, H. (1994) Bacterial endotoxin: molecular relationships of structure to activity and function. *FASEB J.* **8**, 217–225.

5. Zähringer, U., Lindner, B., and Rietschel, E. T. (1999) Chemical structure of lipid A. Recent methodical advances towards the complete structural analysis of a biologically active molecule, in *Endotoxin in Health and Disease* (Brade, H., Morrison, D. C., Opal, S., and Vogel, S., eds.), Marcel Dekker, New York, 93–114.

6. Karas, M. and Hillenkamp, F. (1988) Laser desorption ionization of proteins with molecular masses exceeding 10 000 daltons. *Anal. Chem.* **60**, 2299–2301.

7. Westphal, O., Lüderitz, O., and Bister, F. (1952) Über die Extraktion von Bakterien mit Phenol/Wasser. *Z. Naturforsch.* **7**, 148–155.

8. Galanos, C., Lüderitz, O., and Westphal, O. (1969) A new method for the extraction of R-lipopolysaccharides. *Eur. J. Biochem.* **9**, 245–249.

9. Haishima, Y., Holst, O., and Brade, H. (1992) Structural investigation on the lipopolysaccharide of *Escherichia coli* rough mutant F653 representing the R3 core type. *Eur. J. Biochem.* **203**, 127–134.

10. Galanos, C. and Lüderitz, O. (1975) Electrodialysis of lipopolysaccharides and their conversion to uniform salt forms. *Eur. J. Biochem.* **54**, 603–610.

11. Helander, I. M., Lindner, B., Brade, H., Altman, K., Lindberg, A. A., and Rietschel, E. T. (1988) Chemical structure of the lipopolysaccharide of *Haemophilus influenzae* strain I69 Rd⁻/b⁺: description of a novel deep-rough chemotype. *Eur. J. Biochem.* **177**, 483–492.

12. Jachymek, W., Petersson, C., Helander, A., Kenne, L., Lugowski, C., and Niedziela, T. (1995) Structural studies of the O-specific chain and a core hexasaccharide of *Hafnia alvei* strain 1192 lipopolysaccharide. *Carbohydr. Res.* **269**, 125–138.

13. Kaltashov, I. A., Doroshenko, V., Cotter, R. J., Takayama, K., and Qureshi, N. (1997) Confirmation of the structure of lipid A derived from the lipopolysaccharide of *Rhodobacter sphaeroides* by a combination of MALDI, LSIMS, and tandem mass spectrometry. *Anal. Chem.* **69**, 2317–2322.

14. Fukuoka, S., Knirel, Y. A., Lindner, B., Moll, H., Seydel, U., and Zähringer, U. (1997) Elucidation of the structure of the core region and the complete structure of the R-type lipopolysaccharide of *Erwinia carotovora* FERM P- 7576. *Eur. J. Biochem.* **250**, 55–62.

15. White, K. A., Kaltashov, I. A., Cotter, R. J., and Raetz, C. R. H. (1997) A mono-functional 3-deoxy-D-*manno*-octulosonic acid (Kdo) transferase and a Kdo kinase in extracts of *Haemophilus influenzae*. *J. Biol. Chem.* **272,** 16555–16563.

16. Qureshi, N., Kaltashov, I., Walker, K., Doroshenko, V., Cotter, R. J., Takayama, K., Sievert, T. R., Rice, P. A., Lin, J. S., and Golenbock, D. T. (1997) Structure of the monophosphoryl lipid A moiety obtained from the lipopolysaccharide of *Chlamydia trachomatis*. *J. Biol. Chem.* **272,** 10,594–10,600.

17. Guo, L., Lim, K. B., Gunn, J. S., Bainbridge, B., Darveau, R. P., Hackett, M., and Miller, S. I. (1997) Regulation of lipid A modifications by *Salmonella typhimurium* virulence genes phoP-phoQ. *Science* **276,** 250–253.

18. Rahman, M. M., Guard-Petter, J., and Carlson, R. W. (1997) A virulent isolate of *Salmonella enteritidis* produces a *Salmonella typhi*-like lipopolysaccharide. *J. Bacteriol.* **179,** 2126–2131.

19. Gibson, B. W., Engstrom, J. J., John, C. M., Hines, W., and Falick, A. M. (1997) Characterization of bacterial lipooligosaccharides by delayed extraction matrix-assisted laser desorption ionization time-of-flight mass spectrometry. *J. Am. Soc. Mass Spectrom.* **8,** 645–658.

20. Juhasz, P. and Costello, C. E. (1992) Matrix-assisted laser desorption ionization time-of-flight mass spectrometry of underivatized and permethylated gangliosides. *J. Am. Soc. Mass Spectrom.* **3,** 785–796.

21. Domon, B. and Costello, C. E. (1988) Oligosaccharide fragmentation nomenclature. *Glycoconjugate J.* **5,** 397–409.

22. Fukuoka, S., Kanishima, H., Nagawa, Y., Nahanishi, H., Ishihawa, K., Niwa, Y., Tamiya, E., and Karube, I. (1992) Structural characterization of the lipid A component of *Erwinia carotovora* lipopolysaccharide. *Arch. Microbiol.* **157,** 311–318.

23. Kaufmann, R. (1995) Matrix-assisted laser desorption ionization (MALDI) mass spectrometry: a novel analytical tool in molecular biology and biotechnology. *J. Biotechnol.* **41,** 155–175.

24. Cotter, R. J., Honovich, J., Qureshi, N., and Takayama, K. (1987) Structural determination of lipid A from Gram-negative bacteria using laser desorption mass spectrometry. *Biomed. Environm. Mass Spectrom.* **14,** 591–598.

25. Lindner, B., Zähringer, U., Rietschel, E. T., and Seydel, U. (1990) Structural elucidation of lipopolysaccharides and their lipid A component: application of soft ionization mass spectrometry, in *Analytical Microbiology Methods: Chromatography and Mass Spectrometry* (Fox, A., Morgan, S. L., Larsson, L., and Odham, G., eds.), Plenum, New York, pp. 149–161.

26. Suzuki, H., Müller, O., Guttman, A., and Karger, B. (1997) Analysis of 1-aminopyrene-3,6,8-trisulfonate-derivatized oligosaccharides by capillary electrophoresis with MALDI time-of-flight mass spectrometry. *Anal. Chem.* **69,** 4554–4559.

18

Applications of Combined Capillary Electrophoresis–Electrospray Mass Spectrometry in the Characterization of Short-Chain Lipopolysaccharides

Haemophilus influenzae

Pierre Thibault and James C. Richards

1. Introduction

Lipopolysaccharide (LPS) is an essential component of the outer membrane of all Gram-negative bacteria (*1*). This complex class of lipoglycans can trigger a cascade of immunological responses in mammals including endotoxic effects and serum antibody production. LPSs have been found to exhibit a common molecular architecture consisting of at least two distinct regions: a carbohydrate-containing region and a lipid moiety referred to as lipid A (*2*). In enteric bacteria (e.g., *Escherichia coli*, *Salmonella* strains), the carbohydrate-containing region consists of a high molecular mass O-specific polysaccharide that is covalently linked to a low molecular mass core oligosaccharide (*3*). Other bacteria, including *Haemophilus* and *Neisseria* spp., produce only short-chain LPS in which the carbohydrate region typically contains mixtures of low molecular mass but structurally diverse oligosaccharide components (*4*). Short-chain LPS is often referred to as lipooligosaccharide (LOS). *Haemophilus influenzae* expresses heterogeneous populations of these low molecular mass LPSs which exhibit extensive antigenic diversity among multiple oligosaccharide epitopes. This pathogen remains a major cause of disease worldwide. Six capsular serotypes and an indeterminate number of nontypable (i.e., acapsular) strains of *H. influenzae* are recognized. In the developed world, nontypable strains are the second major cause of otitis media infections in children, while serotype b

From: *Methods in Molecular Biology, vol. 145: Bacterial Toxins: Methods and Protocols*
Edited by: O. Holst © Humana Press Inc., Totowa, NJ

capsular strains are associated with invasive diseases, including meningitis and pneumonia *(5)*. The carbohydrate regions of *H. influenzae* LPS molecules provide targets for recognition by host immune responses, and expression of certain oligosaccharide epitopes is known to contribute to disease pathogenesis. Molecular structural studies of LPS from a number of different strains has resulted in a structural model in which a conserved L-*glycero*-D-*manno*-heptose (Hep)-containing inner-core trisaccharide moiety is attached via a phosphorylated 3-deoxy-D-*manno*-oct-2-ulosonic acid (Kdo) residue to the lipid A component *(6–12)*. In this structural model, each of the Hep residues within the triad can provide a point for further oligosaccharide chain elongation. The addition of phosphate-containing substituents, which include free phosphate (P), phosphoethanolamine (PEtn), pyrophosphoethanolamine (PPEtn), and phosphocholine (PCho), also contributes to the structural variability of these molecules. Moreover, *H. influenzae* LPS can undergo phase variation between defined oligosaccharide structures, which creates the possibility of an extensive repertoire of oligosaccharide epitopes in a single strain *(13,14)*. The structural diversity arising from phase variation has complicated the study of the molecular features of these molecules and their role in commensal and pathogenic behavior in the host. The availability of the complete sequence of the *H. influenzae* strain Rd genome *(15)* has led to significant progress in identifying the genes that are responsible for LPS expression in this pathogen *(16)*. The heterogeneity and structural complexity of short-chain LPS within and between *H. influenzae* strains pose significant analytical challenges.

Over the last decade, mass spectrometry (MS) has played an increasingly important role in the characterization of carbohydrate-containing regions of short-chain LPS. For example, liquid secondary ion mass spectrometry (LSIMS) and tandem mass spectrometry were used to characterize oligosaccharide samples obtained following mild acid hydrolysis of LPS from *H. influenzae* and *Neisseria gonorrheae* *(6,17)* and *H. ducreyi* *(18)*. More recently, electrospray mass spectrometry (ESMS) has been shown to be a sensitive analytical technique for profiling the structural heterogeneity of de-*O*-acylated LPS samples from *Haemophilus* and *Neisseria* strains *(9–12,19–23)*. These investigations have generally relied on direct infusion of sample into the mass spectrometer, following prior purification by either gel-filtration or ion-exchange chromatography. While providing valuable analytical data, information on the nature and distribution of LPS glycoform (i.e., molecular species differing in the number of sugar resides) and isoform (i.e., molecular species having the same composition but differing in the distribution of sugar residues) populations is often lacking. During the last few years, we have devoted considerable effort to the development of on-line electrophoretic separation techniques that are compatible with ESMS for the analysis of isoform and glycoform distributions of short-chain LPSs *(24–26)*. We have found that coupling capillary electrophoresis to electrospray mass spectrometry (CE–ESMS) provides unparalleled resolution for the identification of glycoform populations owing to differing oligosaccharide compositions and phosphate substitution patterns *(26)*. We have recently developed an adsorption preconcentration approach for on-line enrichment of de-*O*-acylated LPS samples for subsequent CE separation and ESMS identification of glycoform patterns from as few as five single bacterial colonies *(27)*. Applications of CE–ESMS involving mixed scan functions and selective ion monitoring provide a specific and sensitive method for characterizing LPS glycoforms that display substantial phosphorylated groups. In conjunction with tandem mass spectrometry (MS-MS), these applications provide a powerful approach for obtaining structural information on short-chain LPS.

2. Materials
2.1. Growth of Bacterial Strains

1. Bacterial strains: *Haemophilus influenzae* strains RM118, RM153, RM7004, and 319 (*see* **Note 1**).
2. Chocolate agar plates from Quelab of Montreal, Canada.
3. Liquid growth media: 3.7% brain heart infusion (BHI, w/v), 10 mg/L of hemin, 2 mg/L of nicotinamide adenine dinucleotide (NAD).
4. Phosphate–buffered saline (PBS): 6.7 mM Potassium phosphate, pH 7.4, containing 150 mM NaCl and 0.02% NaN$_3$ (w/v).
5. Bacterial killing solution: 0.5% Phenol (w/v) in PBS.

2.2. Extraction of Cell-Wall LPS

1. Phenol.
2. Water bath.
3. Omni mixer with stainless steel container.
4. Refrigerated centrifuge.
5. Ultracentrifuge.

2.3. Preparation of De-O-Acylated LPS

1. LPS preparation: *H. influenzae* purified LPS (*see* **Subheading 3.2.**); lyophilized powder of LPS-containing digested cells (*see* **Subheading 3.5.**).
2. De-*O*-acylation reagent: Anhydrous hydrazine.
3. Acetone.
4. Eppendorf centrifuge.

2.4. Profiling the Distribution of LPS Glycoforms by On-Line CE–ESMS

1. Elga water filtration system was used to obtain deionized water: all buffer solutions were filtered through a Millipore 0.45-μm filter.

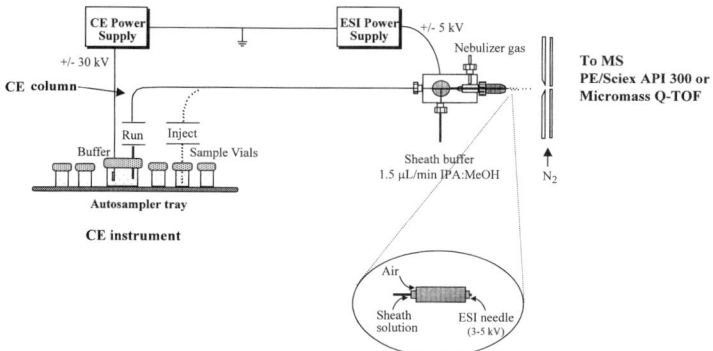

Fig. 1. Schematic representation of the coaxial CE–ESMS interface. The electro-spray needle (27-gauge) is butted against a low dead volume tee that enables the delivery of the sheath solutions to the end of the capillary column. Separations are typically obtained on 90 cm lengths of bare fused silica with the appropriate CE separation buffer for anionic or cationic separation. A voltage of 30 kV is applied at the injection end of the capillary. The outlet of the fused-silica capillary (185 μm o.d.) is tapered to 75 μm o.d. The effective voltage across the capillary is typically 25 kV.

2. CE instrument coupled to tandem mass spectrometer using a coaxial sheath flow interface (*see* **Fig. 1**).
3. Sheath buffer solution: Isopropanol–methanol (7:3, v/v).
4. Capillary columns: 90 cm length ×50 μm i.d. of bare fused silica.
5. CE separation buffers: 30 mM morpholine–acetate, pH 9.0 containing 5% methanol (v/v, for negative ion detection); 30 mM aqueous ammonium acetate, pH 8.5, containing 5% methanol (for positive ion detection).

2.5. Preparation of De-O-acylated LPS from Single Colonies of Bacteria

1. 1.5-mL Polypropylene tubes.
2. Proteinase K solution: 25 μg/mL in water.
3. Ammonium acetate buffer: 20 mM, pH 7.5.
4. Deoxyribonuclease I (DNase)–ribonuclease (RNase) solution: Ammonium acetate buffer containing 10 μg/mL of DNase and 5 μg/mL of RNase.
5. Eppendorf centrifuge.

2.6. CE–ESMS Analysis of De-O-Acylated LPS from Single Colonies of Bacteria by On-Line Preconcentration

1. Fabrication of preconcentrator: 180 mm i.d. polytetrafluorethylene (PTFE) tubing; poly (styrene-divinylbenzene) membrane (SDB-XC); 7 and 80 cm lengths of fused silica.
2. Elution buffer: 10% (v/v) 0.1 M Formic acid in acetonitrile.

3. Methods

3.1. Growth of Bacterial Strains (See Note 2)

1. Resuscitate bacterial strains from frozen stocks on chocolate agar plates and incubate at 37°C for 16 h.
2. Select colonies from plates and cultivate in 10 L-batches of BHI broth supplemented with hemin and NAD at 37°C for 20 h *(28)*.
3. Harvest cells by low-speed centrifugation (5000*g*).
4. Decant centrifugate.
5. Resuspend cell pellet in bacterial killing solution and stir for 16 h.
6. Collect bacterial cell mass by centrifugation (5000*g*).

3.2. Extraction of Cell-Wall LPS (See Note 3)

1. Prepare an approx 90% phenol solution by adding 50-mL of water to 500 g of commercial grade phenol. Heat on high for 3–4 min in a microwave oven to liquefy.
2. Preheat deionized water to 65–70°C in a water bath.
3. Add bacterial cell mass (approx 20 g wet mass) to stainless steel container. Add 50 mL each of preheated water and phenol solution.
4. Stir vigorously by mechanical mixing for 20–30 min.
5. Place stainless steel container in ice bucket to cool mixture to below 10°C.
6. Separate aqueous and phenol phases by low-speed centrifugation (5000*g*) at 5°C for 30 min. Collect water phase.
7. Add 50 mL of warm water (65–70°C) to phenol phase and repeat **steps 4–6**.
8. Combine water phase extracts (approx. 100 mL) and collect LPS by following **steps 9–10** or **steps 11–12**.
9. Dialyze against running tap water for 2–3 d.
10. Collect LPS by lyophilization.
11. Precipitate LPS from water phase (**step 8**) by addition of four volumes of ethanol.
12. Collect LPS by low-speed centrifugation and lyophilization.
13. Dissolve LPS in water to a final concentration of 1–2%, and ultracentrifuge (105,000*g*) at 4°C for 5 h.
14. Decant water and repeat step 13 (2×).
15. Collect purified LPS by suspending in water and lyophilization.

3.3. Preparation of De-O-Acylated LPS (See Note 4)

1. Suspend 0.5–1 mg of purified LPS preparation in 200 µL of anhydrous hydrazine and incubate at 37°C for 1 h with constant stirring.
2. Cool reaction mixture in ice (0°C) and slowly add 600 µL of cold acetone to destroy excess hydrazine.
3. Obtain precipitated product by centrifugation.
4. Wash pellet with 600 µL of acetone (2×); and, then with 500 µL of a mixture of acetone–water (4:1, v/v).
5. Dissolve in water and lyophilize to obtain de-O-acylated LPS as a powder.

3.4. Profiling the Distribution of LPS Glycoforms by On-Line CE–ESMS (See Note 5)

1. Condition CE column by rinsing sequentially with 1 M NaOH (15 min), deionized water (20 min), and the CE separation buffer (15 min). The sheath buffer solution is delivered at 1.5 µL/min to the back tee of the CE–ESMS interface.
2. Redissolve de-O-acylated LPS in CE separation buffer to give a concentration of approx 100 µL/mL (calculated using purified LPS).
3. Using an autosampler, inject 40 nL of de-O-acylated LPS solution onto a CE column.
4. Collect negative ion spectra at an orifice potential of 50 V by scanning the first quadrupole (Q1).

3.5. Preparation of De-O-acylated LPS from Single Colonies of Bacteria (See Note 6)

1. Scrape bacterial colonies from agar plate into a polypropylene tube and suspend in 100 µL of water.
2. Lyophilize cell suspension and resuspend in 90 µL of water.
3. Add 10 µL of proteinase K solution and incubate at 37°C for 90 min.
4. Stop the reaction by raising the temperature to 65°C for 10 min, then cool to 20–22°C and lyophilize.
5. Add 200 µL of DNase/RNase solution and incubate at 37°C for 4 h.
6. Lyophilize to obtain powder of digested cells containing free LPS.
7. Convert to de-O-acylated LPS as described in **Subheading 3.3.** Dissolve product in 40 µL of water for CE–ESMS analysis.

3.6. CE–ESMS Analysis of De-O-Acylated LPS from Single Colonies of Bacteria by On-Line Preconcentration (See Note 7)

1. Prepare preconcentrator (cPC) by inserting a piece of SDB-XC membrane midpoint in a 2 cm length of PTFE tubing and activate by rinsing with methanol for 30 min.
2. Install a cPC device between a 7 cm fused silica capillary transfer line and a 80 cm CE column.
3. Install a cPC-CE column using the same experimental setup shown in **Fig. 1.**
4. Condition a cPC-CE column sequentially with five column volumes each of elution buffer and CE separation buffer (*see* **Subheading 2.4., step 5**).
5. Inject 4.5 µL of de-O-acylated LPS solution from colony extracts (*see* **Subheading 3.4.**) at a constant pressure of 2000 mbar for 3 min. Rinse column with four column volumes of CE separation buffer and elute sample from the stationary phase with small plug (approx 25 nL) of elution buffer at 300 mbar for 0.1 min. A small positive pressure is then applied to prevent readsorption of the analyte.
6. Collect spectra as described in **Subheading 3.4.**

3.7. Identification of Phosphate-Containing Substituents by High/Low Orifice Voltage Stepping (See Note 8)

1. Equilibrate the CE–ESMS system, prepare the solution of de-*O*-acylated LPS, and inject sample exactly as described in **Subheading 3.4.**
2. Collect negative or positive ion spectra using high (120 V)/low (30 V) orifice voltage stepping in conjunction with alternating selected ion monitoring and full mass scan acquisition modes.

3.8. Characterization of De-O-acylated LPS by CE–ESMS-MS (See Note 9)

1. Inject 4.5 μL of de-*O*-acylated LPS onto CE column using conditions identical to those described in **Subheading 3.4.**
2. Select ions for collisional activation and collect mass spectra in positive ion mode.

4. Notes

1. *H. influenzae* strains RM118, RM153, and RM7004 are from the culture collection of Professor E. R. Moxon (Oxford University, U.K.). *H. influenzae* RM118 is a capsular deficient serotype d strain (referred to as strain Rd⁻), obtained from the same source as the strain used in the *Haemophilus* genome sequencing project *(15)*. Strains RM153 (Eagan) and RM 7004 are encapsulated serotype b disease isolates from the United States *(29)*. Strain 319 is a Hex₄ phase variant of strain Eagan, obtained from Dr. J. Weiser (University of Pennsylvania).

2. It is important to carry out growth and manipulation of *H. influenzae* bacteria under level II containment to ensure proper biosafety. Once bacteria are killed by stirring cells with a phenol-containing solution (bacterial killing solution), LPS can be extracted using the precautions normally followed in the analytical chemistry laboratory.

3. For large-scale LPS preparation (approx 20 g wet mass of bacterial cells), material is obtained by the hot aqueous phenol extraction procedure *(30)*. We have found that high yields of LPS can be achieved when the bacterial cell mass is dried before extraction by phenol–water, a procedure that is described elsewhere *(31)*. LPS is obtained from the aqueous phase following extensive dialysis against running tap water or by precipitation with ethanol, and then purified by ultracentrifugation *(10,16)*. Although, lower yields are obtained by the ethanol precipitation procedure, these LPS preparations have been found to contain less RNA as determined by sugar analysis. Also, the latter preparations contain higher populations of LPS in which the Kdo residues are substituted by pyrophosphoethanolamine groups instead of phosphate at the O–4 position *(10)*.

4. Using the mild de-*O*-acylation conditions described, anhydrous hydrazine provides an excellent method for solubilizing short-chain LPS samples though the removal of ester-linked fatty acids from the lipid A region of the molecule. It is important to recognize, however, that the procedure also effects release of O-acetyl substituents, which may be present in the core region of the mol-

ecule (cf *32*). This procedure is now well established for probing the heterogeneity of LPS from *H. influenzae (21)*. Samples of de-*O*-acylated LPS are
readily separated and amenable to analysis by CE–ESMS (*see* **Note 5**) as they
typically carry a net negative charge due to the acidic Kdo residue and phosphate-containing substituents.

5. The method provided was developed using a Crystal Model 310 CE instrument
interfaced to a API 300 triple quadrupole mass spectrometer (Perkin-Elmer/
Sciex) via a coaxial interface configuration *(27,33)* (*see* **Fig. 1**). For high-resolution experiments, a Micromass hybrid quadrupole-time-of-flight (Q-TOF) instrument was employed for mass spectral analysis. TOF analysis offers superior
resolution and enables collection of a wide mass range with excellent sensitivity.
For rapid screening of the glycoform and isoform distributions of de-*O*-acylated
LPS samples, spectra are generally obtained in the negative ion mode. As with all
high-resolution separation techniques, it is important to filter all aqueous solution through a 0.45-μm filter before use. We have found that superior resolution
of de-*O*-acylated LPS glycoforms can be achieved when a morpholine buffer
system is used compared to ammonium acetate or ammonium formate separation
buffers *(26,27)*. Methanol at concentrations of 2–5% is added to the separation
buffer to minimize adsorption of de-*O*-acylated LPS on the capillary surface and
thereby enhance sensitivity. Detection limits of the order of 10 μg/mL (30 ng
on-column injection) of de-*O*-acylated LPS can routinely be achieved with an
aqueous solution of 30 *mM* morpholine containing 5% methanol. An example of
the CE separation and sensitivity achievable with a triple quadrupole instrument
is shown in **Fig. 2** for a de-*O*-acylated LPS sample from the *H. influenzae* type b
strain RM7004. It is noteworthy that a 10–20-fold increase in detection limit can
be achieved with a Q-TOF instrument *(26)*. Analysis of the de-*O*-acylated LPS
from RM7004 was carried out with quadrupole detection following injection of
20 ng of sample. This type b *H. influenzae* strain elaborates a complex mixture of
LPS glycoforms in which one to ten hexose residues are attached to the inner-
core triheptose element (*also see 23*). The most abundant of LPS glycoforms
contain four, five, or six hexose residues. The two-dimensional depiction of *m/z*
vs time shown in **Fig. 2B** provides a useful method to identify closely related
families of LPS glycoforms. Horizontal groupings of molecular species (or
isoforms) arise from different distributions of substituents on the inner core element. The diagonal patterns arise from progressive additions of hexose residues
to the inner core element, resulting in increases in molecular mass with concomitant decreases in electrophoretic mobility. Another diagonal grouping of
glycoforms is observed, which has similar glycan distributions, but carries an
additional PEtn group. This additional PEtn group is most likely present as a
PPEtn moiety, attached to the O–4 position of Kdo, a structural feature that has
been firmly established for the type b strain, Eagan *(10)*. The effect that phosphate substituents can have on electrophoretic mobility of de-*O*-acylated LPS is
illustrated for the Eagan strain 319 in **Fig. 3**. This type b strain expresses a limited number of glycoforms in which those containing four hexose residues pre-

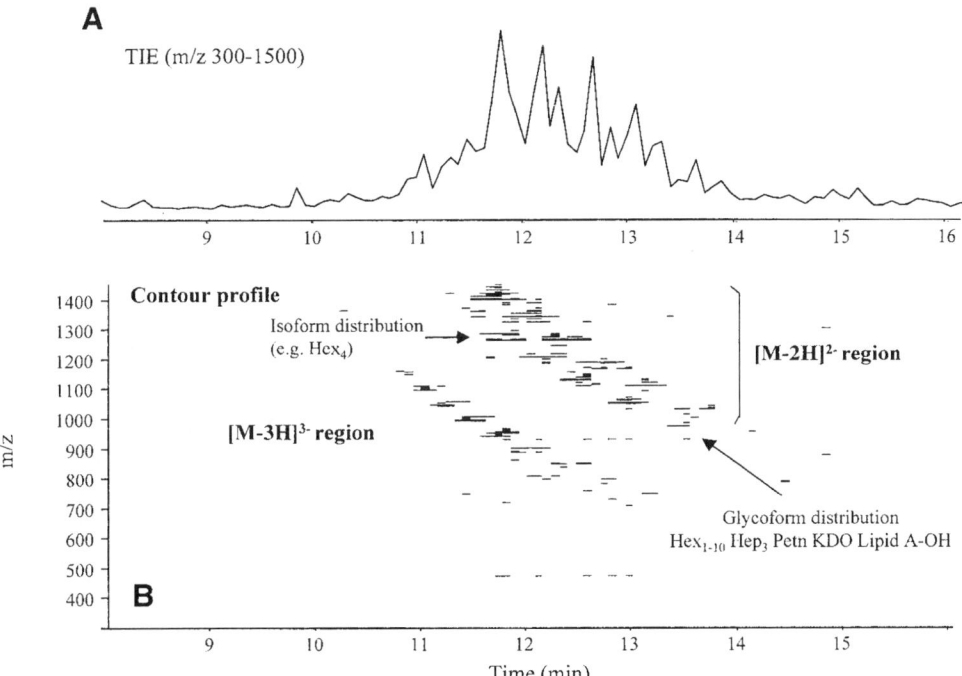

Fig. 2. Negative ion CE–ESMS of de-*O*-acylated LPS from *H. influenzae* RM7004.
(**A**) Total ion electropherogram (*m/z* 300–1500). (**B**) Contour plot of *m/z vs* time showing families of closely related glycoforms as diagonally disposed groupings of molecular species. The doubly and triply deprotonated regions are indicated and the inset
shows the predicted range in composition. Isoforms having identical mass, but different distributions of hexose residues, or phosphate-containing substituents are indicated for the Hex₄ glycoform. The data were acquired using 30 ng of sample with
triple quadrupole analysis.

dominate. Glycoforms which differ as a result of addition of a PEtn group on
Kdo–P (i.e., Kdo–PPEtn) occur at 9.18 min (M_r: 2600.0) and 8.86 min (M_r:
2723.0), respectively. The lower mobility of the glycoforms containing Kdo–
PPEtn is consistent with the increase in molecular mass without significant
change in net change at the pH (9.0) of the CE separation buffer. A matching pair
of Hex₄ glycoforms that carry an additional phosphate group are observed at
10.20 and 10.05 min, respectively, owing to their corresponding increases in net
negative charge. Minor glycoforms, substituted by PCho, are also observed
(**Fig. 3**). Expression of PCho on *H. influenzae* LPS has been found to be phase
variable, the extent of which can vary among different strains *(11,12,14,23)*. PCho
groups can be readily identified by tandem mass spectral analysis (precursor ion
monitoring) *(34)* or by high/low orifice voltage stepping (*see* **Note 8**).

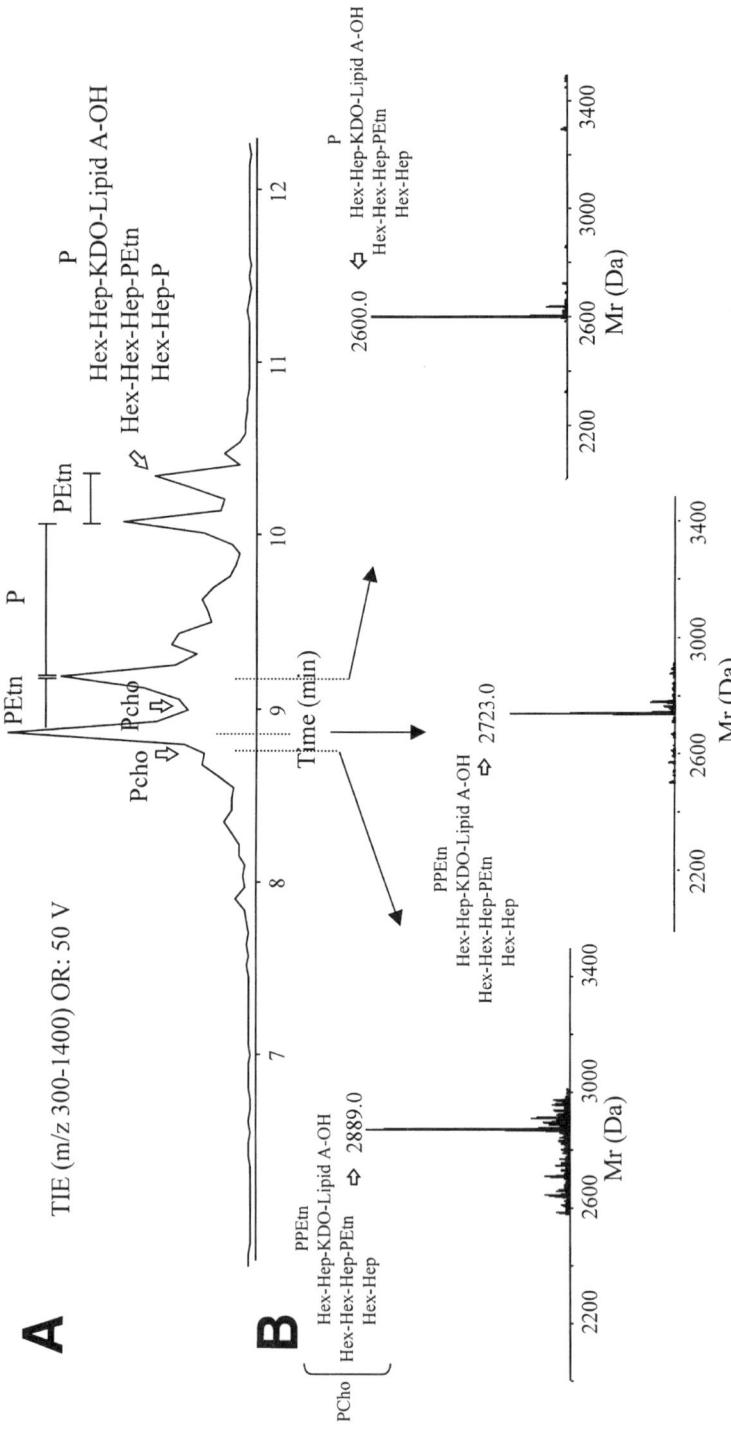

Fig. 3. Negative ion CE–ESMS of de-O-acylated LPS from H. influenzae strain 319. (A) Total ion electropherogram (m/z 300–1400). (B) Reconstructed mass profiles for major Hex$_4$ glycoforms are indicated for peaks at 8.76, 8.86, and 9.18 min. The data were obtained by injection of 20 ng of total de-O-acylated LPS on a triple quadrupole instrument.

6. A micro-LPS extraction technique can be used for analysis of single colonies from plate-grown cells *(27)*. For example, the release of LPS can be achieved by successive treatment of lyophilized cells (approx 200 μg) obtained by excising approx 5 colonies from chocolate agar plates (*see* **Subheading 3.1.**) followed by proteinase K, DNase, and RNase digestion. Nucleic acid depolymerization is required to reduce the viscosity of solutions. Care should be taken when excising bacterial colonies to avoid scraping agar into the vials. Anhydrous hydrazine is added (*see* **Subheading 3.3.**) to the lyophilized cells containing free LPS to generate de-*O*-acylated material (*see* **Note 4**) for CE–ESMS analysis (*see* **Note 7**). An advantage of this extraction technique is that all steps can be carried out in a single vial, minimizing losses due to sample transfer.

7. We have found that on-line preconcentration of de-*O*-acylated LPS samples by adsorption onto C_{18} chromatography particles or SDB-XC membranes can lead to a 40-fold improvement in concentration detection limits of total de-*O*-acylated LPS *(27)*. Adsorption onto SDB-XC membranes generally gives better reproducibility and linear responses for these glycolipid samples compared to C_{18} absorbant. We have found this technique to be particularly suitable for the analysis of trace amounts of de-*O*-acylated LPS mixtures from bacterial colonies excised from agar plates. As an example, **Fig. 4** shows cPC–CE–ESMS analysis of de-*O*-acylated LPS from five colonies of *H. influenzae* strain RM153. The mass spectra arising from the most prominent glycoforms were detected at 8.5 and 9.1 min. These correspond to the Hex_4 LPS glycoform (*m/z* 1299; $[M-2H]^{2-}$; 866, $[M-3H]^{3-}$) shown in the insert and its phosphorylated analogue (*m/z* 1340, $[M-2H]^{2-}$; 893, $[M-3H]^{3-}$) (*see* **Note 6**). The lack of resolution in CE separation using the preconcentration technique (**Fig. 4**) is due to the concurrent application of a small inlet pressure during the separation. This technique can be used in conjunction with mixed scan functions (*see* **Note 8**) and selective ion monitoring to detect specific molecular species present in complex mixtures of glycoforms. In addition, this technique is applicable for probing the glycoform distribution of de-*O*-acylated LPS from bacteria (approx 10^8 bacteria) obtained from in vitro (e.g., tissue culture) or in vivo sources.

8. A mixed scan function in which a high/low orifice voltage stepping function is incorporated into the mass spectral acquisition parameters has proved effective to probe for phosphate-containing residues, such as PPEtn and PCho *(26)*. The example illustrated in **Fig. 5** is for positive ion detection of PCho residues appended to a hexose residue in the outer core region of de-*O*-acylated LPS from *H. influenzae* RM118 *(12)*. The experiment is designed to acquire fragment ions at *m/z* 328 (Hex–PCho) under selected ion monitoring at an orifice voltage of 120 V (150 ms duration) together with the full mass scan (*m/z* 300–1400) at a lower orifice voltage of 50 V (3.5 s duration). This method can also be employed in the negative ion mode for the detection of PPEtn groups (which produce a fragment ion of *m/z* 220) *(26)*, as well as terminal Neu5Ac groups (*m/z* 290) *(35)*. Although good resolution of de-*O*-acylated LPS glycoforms are obtained using the morpholine separation buffer (*see* **Fig. 5**), this buffer system is not completely

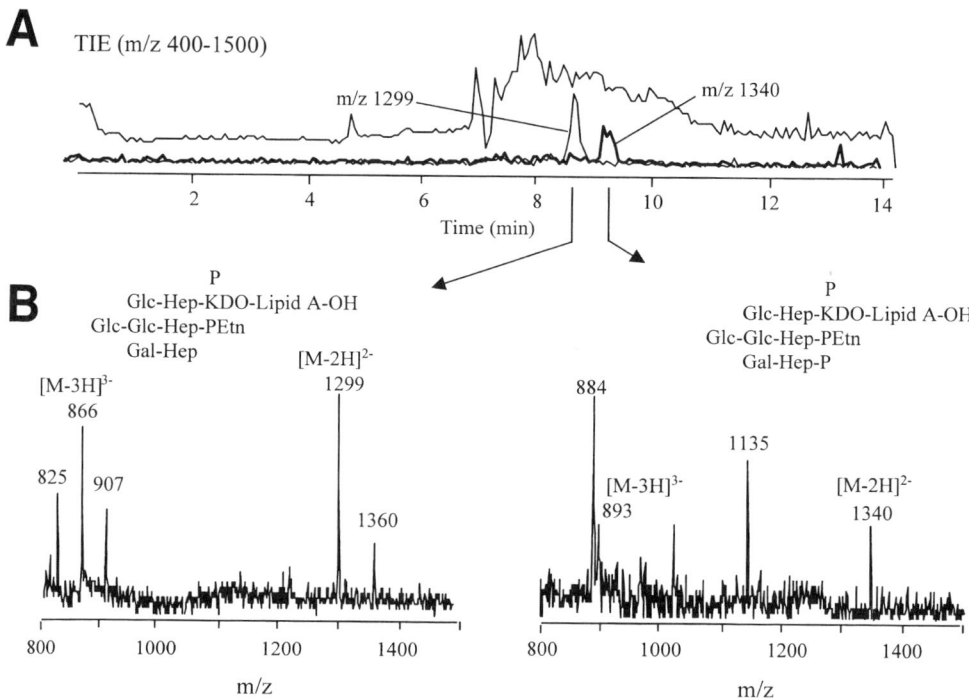

Fig. 4. Negative ion CE–ESMS of de-*O*-acylated LPS extracted from five colonies of *H. influenzae* strain RM135 adopted from *(27)* using the proconcentration device. **(A)** Total ion electropherogram (*m/z* 400–1500). **(B)** Extracted mass spectra for the major Hex$_4$ glycoforms eluting at 8.5 and 9.3 min are indicated together with the proposed structures *(10)*.

compatible with positive ion detection. This is due to the propensity of morpholine to form multimolecular adducts with multiply protonated species *(26)*. This problem can be overcome by using ammonium acetate buffers (*see* **Subheading 2.4.**), although this reduces electrophoretic resolution (**Fig. 5B**).

9. Useful structural information can be obtained for de-*O*-acylated LPS by tandem mass spectrometry of the doubly or triply charged molecular ions. Generally ions selected in Q1 are focussed on the RF-only quadrupole (Q2) where collisional activation is induced with argon or nitrogen gas (typically at a collision energy of 75 eV in the laboratory frame of reference). Fragment ions, so obtained, are separated and recorded by scanning the third quadrupole (Q3) of a triple quadrupole instrument or the TOF analyzer of a Q-TOF instrument. Fragmentation in the negative ion mode generally gives only limited structural information because cleavage occurs primarily between the Kdo-α-D-GlcN bond to afford a de-*O*-acylated lipid A fragment at *m/z* 951 *(24–26)* (**Fig. 6A**). Information concerning the nature of the phosphate substituent on the Kdo moiety (P or PPEtn) can also

A

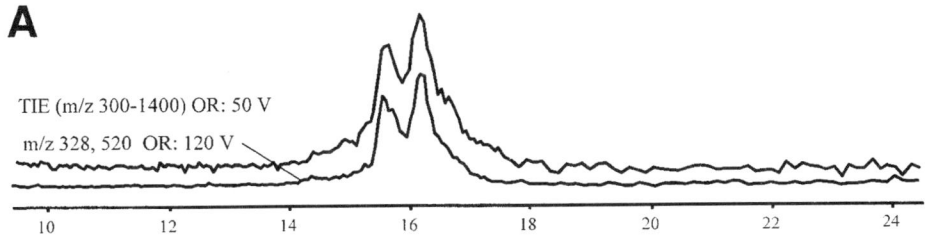

B

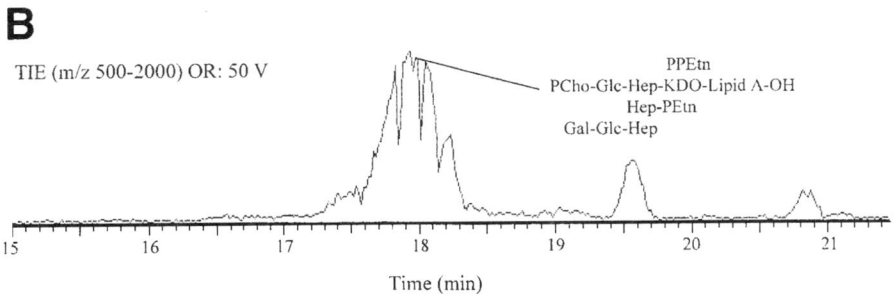

Fig. 5. Positive ion CE–ESMS of de-*O*-acylated LPS from *H. influenzae* strain RM118. **(A)** Mixed scan acquisition in the morpholine buffer system showing total ion electropherogram (300–1400) and single-ion monitoring at *m/z* 328 and 520. **(B)** Total ion electropherogram (500–2000) in ammonium acetate buffer system. The de-*O*-acylated LPS sample was obtained from a Hex$_3$ phase variant of this *H. influenzae* strain in which the major glycoforms carry PCho groups as revealed by SIM for fragment ion at *m/z* 328. The presence of two peaks resolved at 15.5 and 17.0 min for the Hex$_3$ glycoforms using the morpholine buffer system and due to an additional PEtn moiety in the former.

be obtained from negative ion MS–MS experiments due to diagnostic fragment ions at *m/z* 97 or 220 *(10)*, and this provides the basis for a selective ion monitoring detection technique (*see* **Note 8**). Tandem mass spectrometry of oligosaccharides in the positive ion mode generally produce abundant fragment ions arising from cleavage at the glycosidic bonds. The formation of these B and Y type ions can provide valuable sequencing information. For example, the tandem mass spectrum of the Hex$_3$ glycoform from the PCho-containing de-*O*-acylated LPS of the *H. influenzae* RM118 (**Fig. 6B**) provides considerable structural information. Diagnostic ions at *m/z* 520 and 328 arising from PCho–Hex–Hep and PCho–Hex confirm the location of the PCho moiety on the first branching heptose. Evidence for further substitution originating from the terminal heptose of the inner-core element was obtained when an oligosaccharide fragment ion (*m/z* 1351) produced at high cone voltage (70 V) was selected in Q1 for collisional activation (**Fig. 6B**). The second generation of fragment ions thus obtained produces sequence infor-

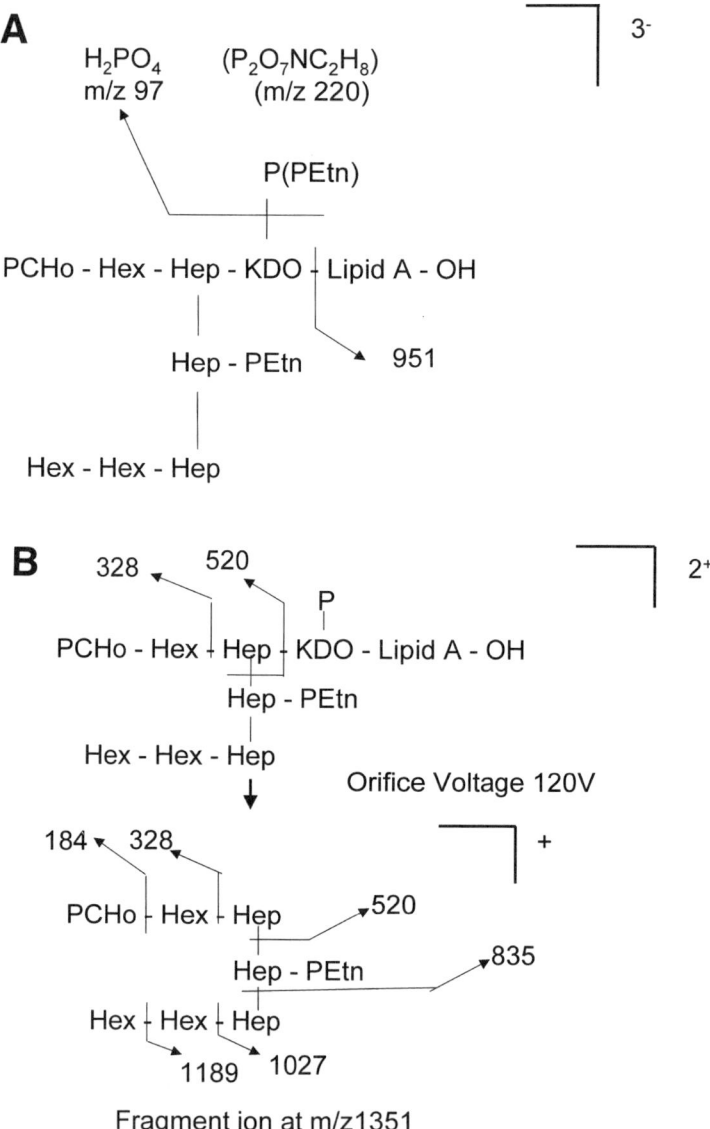

Fig. 6. Fragmentation of de-*O*-acylated LPS from Hex₄ glycoforms of *H. influenzae* RM118 observed in negative and positive ion ES MS-MS. **(A)** Major fragments observed, arising from triply deprotonated ions containing Kdo–P (*m/z* 866.8) or Kdo–PPEtn (*m/z* 907.9) moieties *(12)*. **(B)** Major fragments observed arising from the doubly protonated ion ([M+2H]²⁺, *m/z* 1302) and from the singly charged fragment ion (*m/z* 1351) promoted using an orifice voltage of 120 V following CE separation under the conditions shown in **Fig. 5B**.

mation consistent with that obtained from detailed structural analysis *(12)*. The detection of positive ions from de-*O*-acylated LPS species that are normally anionic in solution opens the door to valuable structural information not normally obtainable from their negative ion counterparts.

Acknowledgments

We thank our colleagues and collaborators without whom this work would not have evolved: Adele Martin, Don Krajcarski, Dr. Jianjun Li, and Dr. Andrew Cox at the National Research Council of Canada; Professor E. Richard Moxon and Dr. Derek Hood from Oxford University; Dr. Elke Schweda from the Karolinska Institute; and Dr. Jeff Weiser from the University of Pennsylvania.

References

1. Rietschel, E. Th., Brade, L., Schade, U. F., Seydel, U., Zähringer, U., Kusumoto, S., and Brade, H. (1988) Bacterial endotoxins: properties and structure of biologically active domains, in *Surface Structures of Microorganisms and Their Interactions with the Mammalian Host* (Schrinner, E., Richmond, M. H., Seibert, G., and Schwarz, U., eds.), Verlag Chemie, Weinheim, Germany, pp. 1–41.
2. Raetz, C. R. H. (1990) Biochemistry of endotoxins. *Annu. Rev. Biochem.* **59,** 129–170.
3. Holst, O. and Brade, H. (1992) Chemical structure of the core region of Lipopolysaccharides, in *Bacterial Endotoxic Lipopolysaccharides* (Morrison, D. C. and Ryan, J. L., eds.), CRC, Boca Raton, Fl, pp. 135–170.
4. Holst, O., Brade, H., and Kosma, P. (1992) GLC-MS of reduced, acetylated, and methylated (2 → 4)- and (2 → 8)-linked disaccharides of 3-deoxy-D-manno-octulopyranosonic acid (Kdo). *Carbohydr. Res.* **231,** 65–71.
5. Turk, D. C. (1981) *Haemophilus influenzae*, epidemiology, immunology and prevention of disease, in *Anonymous* (Sell, S. H. and Wright, P. F., eds.), Elsevier, New York, pp. 3–9.
6. Phillips, N. J., Apicella, M. A., Griffiss, J. M., and Gibson, B. W. (1992) Structural characterization of the cell surface lipooligosaccharides from a nontypable strain of *Haemophilus influenzae. Biochemistry* **31,** 4515–4526.
7. Schweda, E. K. H., Hegedus, O. E., Borrelli, S., Lindberg, A. A., Weiser, J. N., Maskell, D. J., and Moxon, E. R. (1993) Structural studies of the saccharide part of the cell envelope lipopolysaccharide from *Haemophilus influenzae* strain AH1–3 (lic3+). *Carbohydr. Res.* **246,** 319–330.
8. Schweda, E. K. H., Jansson, P.-E., Moxon, E. R., and Lindberg, A. A. (1995) Structural studies of the saccharide part of the cell envelope lipooligosaccharide from *Haemophilus influenzae* strain galEgalK. *Carbohydr Res.* **272,** 213–224.
9. Phillips, N. J., McLaughlin, R., Miller, T. J., Apicella, M. A., and Gibson, B. W. (1996) Characterization of two transposon mutants from *Haemophilus influenzae* Type b with altered lipooligosaccharide biosynthesis. *Biochemistry* **35,** 5937–5947.
10. Masoud, H., Moxon E. R., Martin, A., Krajcarski, D., and Richards, J. C. (1997) Structure of the variable and conserved lipopolysaccharide oligosaccharide

epitopes expressed by *Haemophilus influenzae* serotype b strain Eagan. *Biochemistry* **36**, 2091–2103.

11. Risberg, A., Schweda, E. K. H., and Jansson, P.-E. (1997) Structural studies of the cell-envelope oligosaccharide from the lipopolysaccharide of *Haemophilus influenzae* strain RM.118–28. *Eur. J. Biochem.* **243**, 701–707.

12. Risberg, A., Masoud, H., Martin, A., Richards, J. C., Moxon, E. R., and Schweda, E. K. H. (1999) Structural analysis of the lipopolysaccharide oligosaccharide epitopes expressed by a capsule-deficient strain of *Haemophilus influenzae* Rd. *Eur. J. Biochem.* **261**, 171–180.

13. Kimura, A. and Hansen, E. (1986) Antigenic and phenotypic variations of *Haemophilus influenzae* type b lipopolysaccharide and their relationship to virulence. *Infect. Immun.* **51**, 69–79.

14. Weiser, J. N., Shchepetov, M., and Chong, S. T. H. (1996) Decoration of lipopolysaccharide with phosphorylcholine: a phase-variable characteristic of *Haemophilus influenzae*. *Infect Immun.* **65**, 943–950.

15. Fleischmann, R. D., Adams, M. D., White, O., Clayton, R. A., Kirkness, E. F., Kerlavage, A. R., Bult, C. J., Tomb, J.-F., Dougherty, B. A., Merrick, J. M., McKenney, K., Sutton, G., FitzHugh, W., Fields, C., Gocayne, J. D., Scott, J., Shirley, R., Liu, L.-I., Glodek, A., Kelley, J. M., Weidman, J. F., Phillips, C. A., Spriggs, T., Hedblom, E., Cotton, M. D., Utterback, T. R., Hanna, M. C., Nguyen, D. T., Saudek, D. M., Brandon, R. C., Fine, L. D., Fritchman, J. L., Fuhrmann, J. C., Geoghagen, N. S. M., Gnehm, C. L., McDonald, L. A., Small, K. V., Fraser, C. M., Smith, H. O., and Venter, J. C. (1995) Whole-genome random sequencing and assembly of *Haemophilus influenzae* Rd. *Science* **269**, 496–498.

16. Hood, D. W., Deadman, M. E., Allen, T., Masoud, H., Martin, A., Brisson, J.-R., Fleischmann, R., Venter, J. C., Richards, J. C., and Moxon, E. R. (1996) Use of the complete genome sequence information of *Haemophilus influenzae* strain Rd to investigate lipopolysaccharide biosynthesis. *Mol. Microbiol.* **22**, 951–965.

17. Phillips, N. J., John, C. M., Reinders, L. G., Griffiss, J. M., Apicella, M. A., and Gibson, B. W. (1990) Structural models for the cell surface lipooligosaccharide (LOS) of *Neisseria gonorrhoeae* and *Haemophilus influenzae*. *Biomed. Environ. Mass Spectrom.* **19**, 731–745.

18. Melaugh, W., Phillips, N. J., Campagnari, A. A., Karalus, R., and Gibson, B. W. (1992) Partial characterization of the major lipooligosaccharide from a strain of *Haemophilus ducreyi*, the causative agent of chancroid, a genital ulcer disease. *Am. Soc. Biochem. Mol. Biol.* **267**, 13,434–13,439.

19. Gibson, B. W., Melaugh, W., Phillips, N. J., Apicella, M. A., Campagnari, A. A., and Griffiss, J. M. (1993) Investigation of the structural heterogeneity of lipooligosaccharides from pathogenic *Haemophilus* and *Neisseria* species and R-type lipopolysaccharides from *Salmonella typhimurium* by electrospray mass spectrometry. *J. Bacteriol.* **175**, 2702–2712.

20. Phillips, N. J., Apicella, M. A., Griffiss, J. M., and Gibson, B. W. (1993) Structural studies of the lipooligosaccharides from *Haemophilus influenzae* type b strain A2. *Biochemistry* **32**, 2003–2012.

21. Gibson, B. W., Phillips, N. J., Melaugh, W., and Engstrom, J. J. (1996) Determining structures and functions of surface glycolipids in pathogenic *Haemophilus* bacteria by electrospray ionization mass spectrometry, in *Biochemical and Biotechnological Applications of Electrospray Ionization Mass Spectrometry* (Snyder, A. P., ed.), American Chemical Society, Washington, DC, pp. 166–184.

22. Schweda, E. K. H., Jonasson, J. A., and Jansson, P.-E. (1995) Structural studies of lipooligosaccharides from *Haemophilus ducreyi* ITM 5535, ITM 3147, and a fresh clinical isolate, ACY1: evidence for intrastrain heterogeneity with the production of mutually exclusive sialylated or elongated glycoforms. *J. Bacteriol.* **177,** 5316–5321.

23. Weiser, J. N., Pan, N., McGowan, K. L., Musher, D., Martin, A., and Richards, J. C. (1998) Phosphorylcholine on the lipopolysaccharide of *Haemophilus influenzae* contributes to persistence in the respiratory tract and sensitivity to serum killing mediated by C-reactive protein. *J. Exp. Med.* **187,** 631–640.

24. Kelly, J., Masoud, H., Perry, M. B., Richards, J. C., and Thibault, P. (1996) Separation and characterization of lipooligosaccharides from *Morexella catarrhalis* using capillary electrophoresis-electrospray mass spectrometry and tandem mass spectrometry. *Anal. Biochem.* **233,** 15–30.

25. Auriola, S., Thibault, P., Sadovskaya, I., Altman, E., Masoud, H., and Richards, J. C. (1996) Structural characterization of lipopolysaccharides from *Pseudomonas aeruginosa* using capillary electrophoresis-electrospray ionization mass spectrometry and tandem mass spectrometry, in *Biochemical and Biotechnological Application of Electrospray Ionization Mass Spectrometry* (Snyder, A. P., ed.), American Chemical Society, Washington, D.C., pp. 149–165.

26. Thibault, P., Li, J., Martin, A., Richards, J. C., Hood, D. W., and Moxon, E. R. (1999) Electrophoretic and mass spectrometric strategies for the identification of lipopolysaccharides and immunodeterminants in pathogenic strains of *Haemophilus influenzae*; application to clinical isolates, in *Mass Spectrometry in Medicine and Biology* (Carr, S., Bowers, M. T., and Burlingame, A., eds.), Humana Press, Totowa, NJ, pp. 439–462.

27. Li, J., Thibault, P., Martin, A., Richards, J. C., Wakarchuk, W. W., and vander Wilp, W. (1998) Development of an on-line preconcentration method for the analysis of pathogenic lipopolysaccharide using capillary electrophoresis-electrospray mass spectrometry: application to small colony isolates. *J. Chromatogr. A* **817,** 325–336.

28. High, N. J., Deadman, M. E., and Moxon, E. R. (1993) The role of the repetitive DNA motif (5' -CAAT–3') in the variable expression of the *Haemophilus influenzae* lipopolysaccharide epitope αGal(1–4)βGal. *Mol. Microbiol.* **9,** 1275–1282.

29. Anderson, P., Peter, G., Johnston, R. B., Wetterlow, H., and Smith, D. H. (1972) Immunization of humans with polyribophosphate, the capsular antigen of *Haemophilus influenzae* type b. *J. Clin. Invest.* **51,** 39–44.

30. Westphal, O., Lüderitz, O., and Bister, F. (1952) Über die Extraktion von Bakterien mit Phenol/Wasser. *Z. Naturforsch.* **7b,** 148–155.

31. Masoud, H., Perry, M. B., and Richards, J. C. (1994) Characterization of the lipopolysaccharide of *Moraxella catarrhalis*. Structural analysis of the lipid A from *M. catarrhalis* serotype A lipopolysaccharide. *Eur. J. Biochem.* **220,** 209–216.

32. Wakarchuk, W. W., Gilbert, M., Martin, A., Wu, Y., Brisson, J.-R., Thibault, P., and Richards, J. C. (1998) Structure of an α-2,6-sialylated lipopolysaccharide from *Neisseria meningitidis* immunotype L1. *Eur. J. Biochem.* **254,** 626–633.

33. Kelly, J., Locke, S. J., Ramaley, L., and Thibault, P. (1996) Development of electrophoretic conditions for the characterization of protein glycoforms by capillary electrophoresis-electrospray mass spectrometry. *J. Chromatogr.* **720,** 409–427.

34. Cox, A. D., Howard, M. D., Brisson, J.-R., van der Zwan, M., Thibault, P., Perry, M. B., and Inzana, T. J. (1998) Structural analysis of the phase variable lipopolysaccharide from *Haemophilus somnus* strain 738. *Eur. J. Biochem.* **253,** 507–516.

35. Hood, D. W., Makepeace, K., Deadman, M. E., Rest, R. F., Thibault, P., Martin, A., Richards, J. C., and Moxon, E. R. (1999) Sialic acid in the lipolysaccharide of *Haemophilus influenzae*: strain distribution, influence on serum resistance and structural characterization. *Mol. Microbiol.* **33,** 679–692.

19

Deacylation of Lipopolysaccharides and Isolation of Oligosaccharide Phosphates

Otto Holst

1. Introduction

Two types of lipopolysaccharides (LPS) exist: smooth (S) and rough (R) forms *(1–3)*. Both LPS forms are found in wild-type bacteria. They consist of a lipid moiety, lipid A, which comprises a (phosphorylated) disaccharide of glucosamine or 2,3-diamino-2,3-dideoxy-D-glucose that is acylated by ester- and amide-bound fatty acids, and of the core region *(4)* which is covalently linked to lipid A. Only in S-form LPS is this core region substituted further, that is, by the O-specific polysaccharide (O-antigen). Because mutants that are not able to synthesize a minimal core structure are not viable, the core region and lipid A represent the common structural principle of all LPS.

There is one structural element that is present in all core regions, namely 3-deoxy-D-*manno*-oct-2-ulopyranosonic acid (Kdo). This sugar links the core region to lipid A. A majority of core regions possess in addition L-*glycero*-D-*manno*-heptopyranose and are substituted by (di)phosphate residues and/or (di)phosphate esters. The structural analysis of the core region was hampered for a long time by the difficult chemistry of Kdo and, also, by the phosphate substitution of sugars. An appropriate methodology for the analysis of the Kdo-containing inner core region was reported in the mid-1980s *(5,6)*; however, a methodology for the complete analysis of the phosphate substitution of the core region has been developed only recently. We have contributed to the last by developing a method that allows the complete deacylation of LPS *(7,8)*. Its application on R-form LPS results in the isolation of oligosaccharide phosphates, the structural analysis of which identifies in most cases the complete phosphorylated LPS carbohydrate backbone.

From: *Methods in Molecular Biology, vol. 145: Bacterial Toxins: Methods and Protocols*
Edited by: O. Holst © Humana Press Inc., Totowa, NJ

Herein, the deacylation of R-form LPS and the isolation of pure oligosaccharide phosphates that can readily be analyzed by nuclear magnetic resonance (NMR) spectroscopy and mass spectrometry are described. It should be noted that phosphodiester, diphosphate, diphosphodiester (all resulting in a phosphate residue that substitutes the sugar), acetyl, and carbamoyl groups are cleaved under the conditions used and their positions thus cannot be identified. Methods to identify the location of these substituents are outside the scope of this chapter.

2. Materials

The water used in the preparation of solvents is purified by passage through a Milli-Q Water System (Millipore, Bedford, MA).

2.1. Chemicals and Separation Media

All chemicals should possess p.a. quality.

1. Absolute hydrazine (Kodak, Rochester, NY).
2. Acetone.
3. Ammonium hydrogencarbonate.
4. Bio-Gel P2 (Bio-Rad, Munich, Germany).
5. CarboPac PA100 analytical column (4 mm × 250 mm, Dionex, Idstein, Germany).
6. CarboPac PA1 semipreparative column (9 mm × 250 mm, Dionex, Idstein, Germany).
7. Dichloromethane.
8. Helium.
9. Hydrochloric acid.
10. KOH pellets.
11. LPS of *Escherichia coli* strain F515–140 *(5)* or of any *E. coli* Re-chemotype.
12. Nitrogen.
13. Phosphorpentoxide (P_2O_5).
14. Sephadex G-10 (Pharmacia, Freiburg i. Br., Germany).
15. Silica gel 60 thin-layer chromatography (TLC) plates (on aluminium, 20 cm × 20 cm, Merck, Darmstadt, Germany).
16. Sodium acetate.
17. TSK HW40 (S) (Merck, Darmstadt, Germany).

2.2. Buffers and Solutions

1. 4 *M* aqueous KOH.
2. 4 *M* aqueous HCl.
3. 50% Aqueous NaOH (Mallinckrodt Baker, Deventer, The Netherlands).
4. 0.1 *M* aqueous NaOH (stored under helium atmosphere).
5. 1 *M* aqueous sodium acetate, pH 6.0 (stored under helium atmosphere).
6. 1 *M* aqueous sodium acetate in 0.1 *M* aqueous NaOH (stored under helium atmosphere).
7. 10 m*M* aqueous ammonium hydrogencarbonate, degassed.
8. 20% Ethanolic sulfuric acid.

3. Methods

3.1. De-O-Acylation of LPS

To de-*O*-acylate LPS (cleavage of the ester-linked fatty acids), it is incubated with absolute hydrazine (*see* **Notes 1** and **2**).

1. Transfer the LPS (0.5–50 mg, for analytical or preparative purpose, respectively) and a magnetic bar to a vial (10–50 mL, Macherey & Nagel, Düren, Germany), and dry it and an aluminum-Teflon cap in a desiccator over P_2O_5 for 18 h.
2. Add absolute hydrazine (20 mg of LPS/mL) to the sample and seal the vial carefully with the cap.
3. Incubate the sample under stirring in a water bath at 37°C for 30 min.
4. Pour acetone (15 volumes of the hydrazine in **step 2**) into a beaker and cool it in ice. In another beaker, keep approx 10 mL of acetone in ice.
5. Transfer the vial to a fume hood and let the sample cool down to 20–22°C.
6. Open the vial, remove the solution using a (Pasteur) pipet, and drop it carefully into the larger volume of cooled acetone. The hydrazine is destroyed during this process and the de-*O*-acylated LPS precipitates.
7. Wash the vial using the smaller volume of cooled acetone and transfer the washing to the larger volume. Leave the combined acetone phases in ice for 30–60 min.
8. Transfer the acetone with the de-*O*-acylated LPS to a glass centrifuge tube and centrifuge at 2500*g* and 4°C for 10–15 min.
9. After centrifugation, the de-*O*-acylated LPS forms a pellet. Transfer the supernatant carefully with a pipet to a separate beaker and store it until the yield of de-*O*-acylated LPS is known.
10. Wash the precipitate 3–4× with cold acetone (2500*g*, 4°C, 10–15 min). Add the washings to the stored acetone of **step 9**.
11. After the last washing, dry the precipitate in the centrifuge tube in a water bath at 37°C and determine the yield of de-*O*-acylated LPS (usually between 50% and 70% of the LPS).

3.2. De-N-Acylation of De-O-Acylated LPS

For de-*N*-acylation of the de-*O*-acylated LPS sample, a strong alkaline hydrolysis is performed (*see* **Note 3**).

1. Transfer the de-*O*-acylated LPS to a vial (10 mL, as described in **Subheading 3.1.**) and dissolve it in 4 *M* aqueous KOH (20–30 mg/mL).
2. Flush the sample gently with a stream of nitrogen (15 min) and then seal the vial (*see* **Note 4**).
3. Incubate the sample in a heating block at 120°C for 16 h.
4. Take the sample out of the heating block and let it cool to 20–22°C, then to 4°C (ice). A precipitate is formed that contains part of the free fatty acids.
5. Centrifuge the sample (2500*g*, 4°C, 15 min).
6. Transfer the supernatant to a fresh vial and cool it in ice/acetone for 10 min.

7. Adjust the pH carefully to 6 with 4 M aqueous HCl.
8. The sample is now ready for the extraction of free fatty acids. Add a similar volume of dichloromethane and vortex-mix the sample vigorously. After phase separation, the dichloromethane (bottom) phase (containing the fatty acids) is removed and discarded.
9. Repeat **step 8** twice.
10. Evaporate the aqueous phase with a rotary evaporator to remove traces of dichloromethane.
11. The sample is now ready for desalting using gel-permeation chromatography. Prepare a column (approx 80 cm × 1 cm) of Sephadex G-10 in 10 mM aqueous ammonium hydrogencarbonate and allow it to equilibrate (*see* **Note 5**).
12. Dissolve the sample in a maximum of 500 µL of 10 mM aqueous ammonium hydrogencarbonate and transfer it to the column. For separation, a pump may be used, and for detection, a refractometer is highly recommended.
13. Check all fractions for their sugar content by spotting aliquots (2 µL) on a silica gel 60 TLC plate and charring with 20% sulfuric acid in ethanol at 150°C.
14. Sugar-containing fractions are combined and lyophilized to determine the yield of oligosaccharide phosphates (usually 50–70% of the de-O-acylated LPS).

3.3. HPAEC

The desalted oligosaccharide phosphates are now ready for separation using HPAEC. For principles of HPAEC, please read reference (*9*). We use a Dionex DX 300 chromatogaphy system equipped with a pulsed amperometric detector (Dionex, Idstein, Germany; E1, +0.05 V; E2, +0.65 V; E3, –0.65 V) and a Spectra-SYSTEM AS3500 autosampler (Thermo Separation Products, Fremont, CA) (*see* **Note 6**). It is very important to keep the chromatography system and all needed solutions under a helium atmosphere in order to prevent entry of CO_2 from air.

In analytical HPAEC, we use both a CarboPac PA 100 column (4 mm × 250 mm, Dionex, Idstein, Germany) that is eluted at 1 mL/min, using a gradient program of 30–70% 1 M sodium acetate in 0.1 M aqueous NaOH over 20 min, and a CarboPac PA 1 column (4 mm × 250 mm, Dionex, Idstein, Germany) that is eluted at 1 mL/min, using a gradient program of 20 mM–600 mM aqueous sodium acetate, pH 6.0, and post-column addition of 1.5 M NaOH for pulsed amperometric detection over 80 min. For semipreparative HPAEC, a CarboPac PA 1 column (9 mm × 250 mm, Dionex, Idstein, Germany) is eluted at 4 mL/min, using a gradient program of 20 mM–600 mM (or less, according to the results of analytical HPAEC) aqueous sodium acetate, pH 6.0. Here, fractions are detected by spotting 2-µL aliquots on silica gel 60 TLC plates and charring with 20% ethanolic sulfuric acid. The purity of all positive fractions is then investigated using analytical HPAEC and only those fractions that contain pure compounds are combined and then desalted (*see* **Subheading 3.2.**, *see* **Note 7**) and lyophilized.

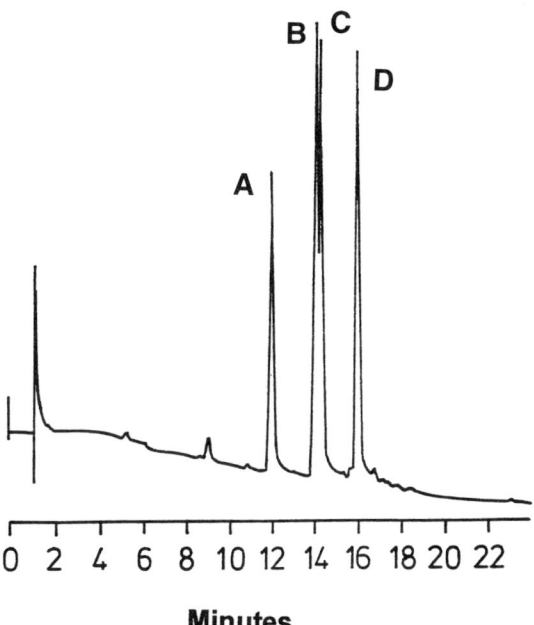

Fig. 1. HPAEC chromatogram of oligosaccharide bisphosphates A-D isolated from LPS of recombinant *E. coli* F515-140 *(9)*. The gradient program was 30–70% 1 *M* sodium acetate in 0.1 *M* aqueous NaOH over 20 min.

3.3.1. Application: The Isolation of Oligosaccharide Bisphosphates Possessing the Core Region of LPS from Chlamydia psittaci (10)

For this purpose, we used the LPS of a recombinant rough-type mutant of *E. coli* (strain F515-140) that expresses the Kdo transferase of *C. psittaci (11)* and thus possesses two types of LPS, that is, the parent Re-type LPS with two Kdo residues linked to lipid A and the chlamydia core containing LPS. After deacylation of the LPS, four oligosaccharide bisphosphates were detected in analytical HPAEC (**Fig. 1**), the structures of which (**Fig. 2**) could be determined by NMR spectroscopy *(10)*. The separation of the oligosaccharide bisphosphates B and C, which differ only in the linkage position of the terminal Kdo (linked to O8 of the second Kdo residue in B and to O4 in C), seemed not to be possible by HPAEC, but could be achieved by affinity chromatography *(10)*. However, we were able to develop an appropriate gradient program later on, by which the separation of B and C is possible (**Fig. 3**):

1. Use a CarboPac PA 1 column (4 mm × 250 mm) at 1 mL/min.
2. For pulsed amperometric detection, use post column addition of 1.5 *M* NaOH at 0.6 mL/min.

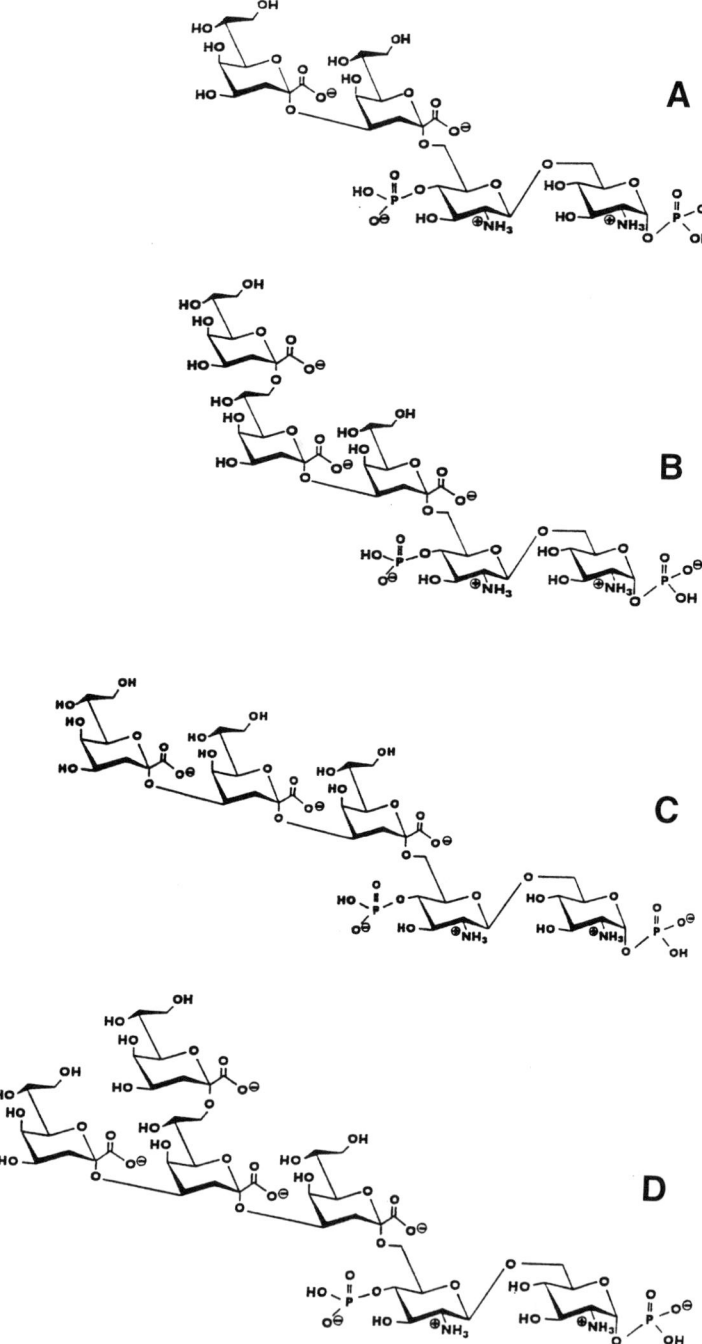

Fig. 2. Structures of oligosaccharide bisphosphates A-D.

3. After injection of the sample, use a gradient program of 30% aqueous sodium acetate, pH 6.0, for 12 min, then 33% for 13 min, and finally again 30% for 5 min.
4. Characterize each oligosaccharide bisphosphate by ¹H- and ¹³C-NMR spectroscopy.

4. Notes

1. Application of the deacylation procedure is useful only if the reducing end of the LPS carbohydrate backbone is substituted, for example, by a phosphate or substituted phosphate residue.
2. Some safety precautions are needed to work with absolute hydrazine, which is highly toxic and an explosive. The bottle with hydrazine is stored in a desiccator over P_2O_5. Any experiment with hydrazine must be performed in a fume hood. After use, the pipets are immediately washed with acetone to destroy the hydrazine. Sample vials that contain hydrazine and are incubated in a water bath must be sealed tightly.
3. Application of the deacylation procedure described herein has limitations. First, if 4-substituted uronic acids are present in the core region, β-elimination processes occur that result in the presence of hex-4-enuronic acids which terminate the isolated oligosaccharide phosphate. The eliminated products cannot be identified, and for the characterization of the complete carbohydrate backbone other degradation procedures must be applied. If D-galacturonic acid is present, a trans-elimination proceeds immediately, leading to a quantitative yield of L-*threo*-hex-4-enuronic acid *(12)*. As shown for the core region of LPS from *Acinetobacter calcoaceticus* strain NCTC 10303 *(13)*, a partial cis-elimination occurs at 4-substituted D-glucuronic acid under the harsh alkaline conditions used. In addition, smaller quantities of Kdo-di- and trisaccharide residues in LPS are cleaved to Kdo-di- and monosaccharide residues, respectively, as shown by treatment of pure oligosaccharide B (**Fig. 2**) with 4 *M* KOH at 120°C *(14)*. Finally, if phosphodiesters are present, phosphate migration may occur *(15)*.
4. This important step reduces significantly the production of artefacts occurring from treatment with hot alkali.
5. Alternative gel-permeation chromatography media are: Bio-Gel P2 (Bio-Rad) in water and, TSK HW 40 (S) (Merck) in water. We usually shift to one of these gels if a separation using Sephadex G-10 is not satisfactory. Further, the use of a microdialysis system (Sialomed, Columbia, MD) with a molecular mass cutoff of either 500 Daltons or even 100 Daltons (dialysis of monosaccharides is possible) is highly recommended for deacylation of small quantities *(14)*.
6. The use of Dionex columns is important. If there is no Dionex chromatography system available, any good HPLC chromatography system will do. Make sure that your pump heads and all tubing are sufficiently rinsed with water (18 h) after chromatography.
7. The use of the Sialomed dialysis system (*see* **Note 5**) is not recommended if sodium acetate is used as the eluting agent. However, if ammonium acetate is used instead, desalting by dialysis works well.

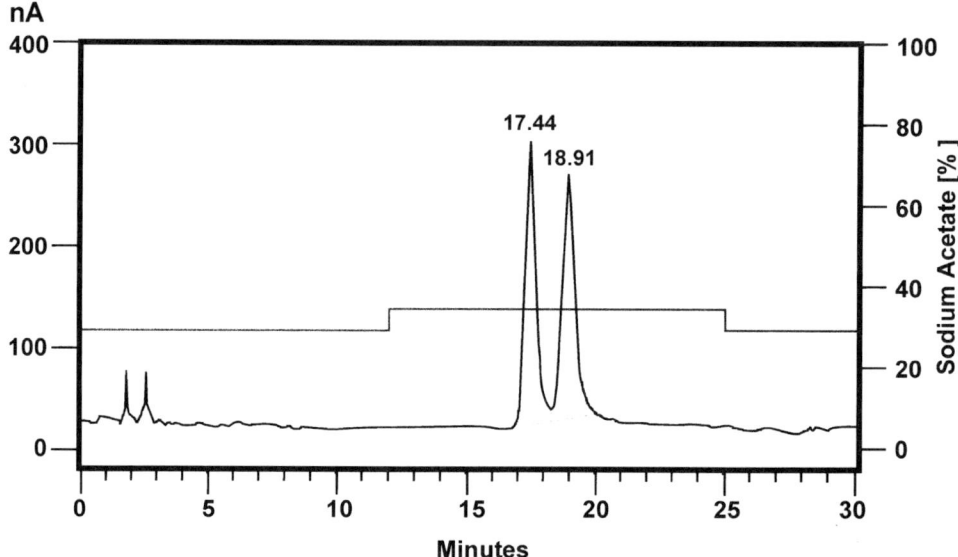

Fig. 3. HPAEC chromatogram and gradient program of the separation of oligosaccharide bisphosphates B and C.

Acknowledgments

I am most grateful to Regina Engel and Veronika Susott for expert technical assistance and help in the development of deacylation protocols and gradient programs in HPAEC. I thank the Deutsche Forschungsgemeinschaft (Grant SFB-470/B1) for financial support.

References

1. Rietschel, E. T., Brade, H., Holst, O., Müller-Loennies, S., Mamat, U., Zähringer, U., Beckmann, F., Seydel, U., Brandenburg, K., Ulmer, A. J., Mattern, T., Heine, H., Schletter, J., Loppnow, H., Schönbeck, U., Flad, H.-D., Hauschildt, S., Schade, U. F., Di Padova, F., Kusumoto, S, and Schumann, R. R. (1996) Bacterial endotoxin: chemical constitution, biological recognition, host response, and immunological detoxification. *Curr. Top. Microbiol. Immunol.* **216,** 39–81.
2. Mamat, U., Seydel, U., Grimmecke, D., Holst, O., and Rietschel, E. T. (1998) Lipopolysaccharides, in *Comprehensive Natural Products Chemistry* (Pinto, M., ed.), Elsevier Science, Amsterdam, Netherlands.
3. Wilkinson, S. G. (1996) Bacterial lipopolysaccharides—themes and variations. *Prog. Lipid Res.* **35,** 283–343.
4. Holst, O. (1999) Chemical structure of the core region of lipopolysaccharides, in Endotoxin in Health and Disease (Brade, H., Morrison, D. C., Opal, S. and Vogel, S., eds.), Marcel Dekker, New York, NY, pp. 115–154.

5. Brade, H. and Rietschel, E. T. (1984) A α-2 → 4-interlinked 3-deoxy-D-*manno*-2-octulosonic acid-disaccharide: a common constituent of enterobacterial lipopolysaccharides. *Eur. J. Biochem.* **145,** 231–236.

6. Tacken, A., Rietschel, E. T., and Brade, H. (1986) Methylation analysis of the heptose/3-deoxy-D-*manno*-2-octulosonic acid-region (inner core) of the lipopolysaccharide from *Salmonella minnesota* rough mutants. *Carbohydr. Res.* **149,** 279–291.

7. Holst, O., Broer, W., Thomas-Oates, J. E., Mamat, U., and Brade, H. (1993) Structural analysis of two oligosaccharide bisphosphates from the lipopolysaccharide of a recombinant strain of *Escherichia coli* F515 (Re chemotype) expressing the genus-specific epitope of *Chlamydia* lipopolysaccharide. *Eur. J. Biochem.* **214,** 703–710.

8. Holst, O., Thomas-Oates, J. E., and Brade, H. (1994) Preparation and structural analysis of oligosaccharide monophosphates from the lipopolysaccharide of recombinant strains of *Salmonella minnesota* and *Escherichia coli* expressing the genus-specific epitope of *Chlamydia* lipopolysaccharide. *Eur. J. Biochem.* **222,** 183–194.

9. DIONEX (1993) Analysis of carbohydrates by high performance anion exchange chromatography with pulsed amperometric detection (HPAE-PAD), Technical Note No. 20.

10. Holst, O., Bock, K., Brade, L., and Brade, H. (1995) The structures of oligosaccharide bisphosphates isolated from the lipopolysaccharide of a recombinant *Escherichia coli* strain expressing the gene gseA [3-deoxy-D-*manno*-2-octulopyranosonic acid (Kdo) transferase] of *Chlamydia psittaci 6BC*. *Eur. J. Biochem.* **229,** 194–200.

11. Mamat, U., Baumann, M., Schmidt, G., and Brade, H. (1993) The genus-specific lipopolysaccharide epitope of *Chlamydia* is assembled in *C. psittaci* and *C. trachomatis* by glycosyltransferases of low homology. *Mol. Microbiol.* **10,** 935–941.

12. Süsskind, M., Müller-Loennies, S., Nimmich, W., Brade, H., and Holst, O. (1995) Structural investigation on the carbohydrate backbone of the lipopolysaccharide from *Klebsiella pneumoniae* R20/O1⁻. *Carbohydr. Res.* **269,** C1–C7.

13. Vinogradov, E. V., Petersen, B. O., Thomas-Oates, J. E., Duus, J. Ø., Brade, H., and Holst, O. (1998) Characterization of a novel branched tetrasaccharide of 3-deoxy-D-*manno*-oct-2-ulosonic acid (Kdo). The structure of the carbohydrate backbone of the lipopolysaccharide from *Acinetobacter baumannii* strain NCTC 10303 (ATCC 17904). *J. Biol. Chem.* **273,** 28,122–28,131.

14. Rund, S., Lindner, B., Brade, H., and Holst, O. (1999) Structural analysis of the lipopolysaccharide from *Chlamydia trachomatis* strain L2. *J. Bio. Chem.* **274,** 16,819–16,824.

15. Brabetz, W., Müller-Loennies, S., Holst, O., and Brade, H. (1997) Deletion of the heptosyltransferase genes *rfaC* and *rfaF* in *Escherichia coli* K-12 results in an Re-type lipopolysaccharide with a high degree of 2-aminoethanol phosphate substitution. *Eur. J. Biochem.* **247,** 716–724.

20

Electrophysiological Measurements on Reconstituted Outer Membranes

Andre Wiese and Ulrich Seydel

1. Introduction

Membranes, in general, constitute the boundary between a cell or a cell compartment and its environment. They are composed of (glyco)lipids and (glyco)proteins, function as permeability barriers, maintain constant ion gradients across the membrane, and guarantee a controlled steady state of fluxes in the cell. Furthermore, the vast majority of cell membranes carry recognition sites for components of the immune system and for interaction/communication with other cells.

For these functions to work properly, a particular lipid composition on each side and distribution between both sides of the lipid bilayer is required. Thus, a membrane is built up from a large variety of lipids, differing in their charge and fatty acid substitution (length and degree of saturation), and these lipids are in a delicate equilibrium providing a suitable environment for protein function and membrane permeability. By a complex interaction of passive and active transport processes—by diffusion through the lipid matrix or protein-aligned transmembrane channels and by energy-dependent ion pumps and transport proteins, respectively—ion gradients are built up, which contribute, together with the charge distribution of the lipids on the two leaflets of the membrane, to a transmembrane potential.

Very particular lipid bilayer matrices with respect to the lipid composition as well as their distribution to the two leaflets are found in bacteria. For example, the outer leaflet of the outer membrane of Gram-negative bacteria is composed solely of a glycolipid, the lipopolysaccharide (LPS), and the inner of a phospholipid mixture (PL) (*1*). We do not describe the chemical nature of

From: *Methods in Molecular Biology, vol. 145: Bacterial Toxins: Methods and Protocols*
Edited by: O. Holst © Humana Press Inc., Totowa, NJ

LPS here and rather refer the reader to a comprehensive review by Rietschel et al. *(2)*.

The inner leaflet of the outer membrane consists of a mixture of phosphatidylethanolamine (PE), phosphatidylglycerol (PG), and diphosphatidylglycerol (DPG)—in the case of *Salmonella enterica* svar. Typhimurium in a molar ratio of PE/PG/DPG = 81:17:2 *(3)*.

This lipid asymmetry provokes a potential difference between the two surfaces of the bilayer membrane, the intrinsic membrane potential. A main component of this potential arises from the difference in the surface charge densities of the two leaflets. The LPS component of the membrane, at its simplest, consists of Re chemotype-like molecules. In this case, each molecule carries four negative charges (referred to LPS Re from *E. coli* F515). For the calculation of the surface charge densities of each leaflet, that is, the number of charges per unit area, the molecular area occupied by the respective lipid molecules has to be considered, which is by a factor of two larger for LPS Re (1.23 nm^2) than for phospholipids (0.55 nm^2). Thus, the surface charge density of the LPS Re leaflet is by a factor of 10 higher than that of the PL leaflet, in which at neutral pH only the PG molecules carry negative charges (one per molecule). From these surface charge densities, the surface potential, the height of which is one determinant for the cell function as well as for the interaction of drugs with the membrane, can be calculated according to the Gouy equation *(4)*. Further determinants are the height and the profile (inner membrane potential difference $\Delta\Phi$) of the potential wall, the latter of which can be determined experimentally *(5)* (*see* **Subheading 3.6.4.**).

From this brief description of the complex architecture of the outer membrane, it becomes obvious that a detailed characterization of its various functions is feasible only with simpler reconstitution systems, in a first step of the unmodified lipid matrix. In such a model system, for example, the influence of the glycolipids on the function of transmembrane proteins may be studied by reconstitution of the proteins into the bilayer and, furthermore, the role of LPS in its potential interaction with membrane active substances such as drugs, detergents, and components of the immune system may be characterized.

The influence of externally applied proteins or drugs on the membrane can be manifold. Their adsorption to the membrane leaflet facing the side of their addition may influence the membrane potential profile owing to changes in the electrostatic environment of the lipid bilayer. Furthermore, the interaction with the lipids may influence the state of order of the acyl chains of membrane lipids, resulting in their fluidization or rigidification depending, among other parameters, on the depth of intercalation of the molecules into the membrane and on the functional groups involved in the interaction. These interactions may, in turn, result in membrane permeabilization either by disturbance of the lamellar structure of the bilayer or by the formation of transmembrane pores. In this short methodological instruction, we will focus only on those events that induce changes in the electrical properties of the membrane.

In 1962, Mueller et al. *(6)* described for the first time a method for forming planar lipid bilayer membranes separating two aqueous phases: A dispersion of phospholipids in a nonpolar solvent such as *n*-decane is spread beneath the surface of an aqueous phase over an aperture of up to several millimeters diameter separating two plastic compartments. The lipid thins out in the center of the aperture until it forms a bilayer that is optically black when viewed in incident light (black lipid membrane, BLM). The membrane is essentially impermeable to ions, and thus application of a voltage across the pure lipid bilayer does not result in any detectable electrical current. Induced membrane disturbances can, therefore, be monitored *via* current measurements (for review *see [7]*). This method of reconstructing biological membranes is obviously not suitable for the reconstitution of asymmetric lipid matrices such as that of the outer membrane of Gram-negative bacteria, because the experimentator has no influence on the arrangement of the different lipids in regard to the specific—asymmetric—composition (for details of this method *see* Chapter 10).

This difficulty could be overcome by techniques introduced by Montal and Mueller *(8)* and Schindler *(9)*. In both methods, bilayers are formed over a small aperture (diameter ≤200 μm) from two lipid monolayers on top of bathing solutions in two compartments separated by the aperture. In the case of the Montal–Mueller technique, lipid solutions in a highly volatile solvent (e.g., chloroform) are spread on the air–water interface, whereas in the case of the Schindler technique the monolayers are built from vesicles added to the bathing solution. When monolayer formation is completed—either after solvent evaporation or after vesicle fusion at the air-water interface—the monolayers are successively raised over the aperture to form the bilayer membrane. Asymmetric membranes can thus be obtained, if different lipids or lipid mixtures are used on the two sides of the aperture (for review *see [10]*). For obvious reasons, the Montal–Mueller and Schindler techniques are most suitable for studying bacterial outer membrane function, the Schindler technique having the advantage of being absolutely solvent free, but the disadvantage of the presence of lipids in the bathing solution. We focus here on the description of the preparation of Montal–Mueller membranes and their electrical characterization as performed in our laboratory to reconstitute the lipid matrix of the outer membrane of Gram-negative bacteria and on the electrical measurements to characterize various interactions and protein functions.

2. Materials
2.1. Lipids

1. Enterobacterial rough mutant LPS (extracted by the phenol–chloroform–petro-leum ether method *[11]*, purified, lyophilized, and transformed into the triethy-lamine salt form *[12]*) dissolved (2.5 mg/mL) in chloroform–methanol (10:1, v/v) at 95°C for 2 min.
2. Phospholipids: PE from bovine brain (type I), PG from egg yolk lecithin (sodium salt), and DPG from bovine heart (sodium salt) from Sigma (Deisenhofen, Ger-many). All phospholipids are used without further purification, dissolved in chlo-roform (2.5 mg/mL), and mixed in a molar ratio of 81:17:2 resembling the PL composition of the inner leaflet of the outer membrane of *Salmonella enterica* sv. Typhimurium *(3)*.

2.2. Buffers and Chemicals

Electrolyte solution (bathing solution): different *pro analysi* grade salts (e.g., NaCl, KCl, LiCl, $MgCl_2$; Merck, Darmstadt, Germany) dissolved in ion-exchanged water (specific conductivity < 0.06 µS/cm). Adjustment of pH (*see* **Note 1**):

pH >8: 5 mM 2-(*N*-cyclohexylamino)-ethanesulfonic acid (CHES)
pH 7:   5 mM *N*-2-hydroxyethylpiperazine-*N'* 2- ethansulfonic acid (HEPES)
pH <6: 5 mM sodium citrate or succinate.

3. Methods
3.1. Mechanical Setup

The mechanical setup of the apparatus for the formation of asymmetric pla-nar bilayer membranes is shown in **Fig. 1**. The apparatus consists of a Faraday cage, a closed cylinder made from aluminum, that contains the measuring chamber in the bottom and the electrical amplifier (headstage) on top. This cage is placed on an electrically controlled heating plate on top of a mag-netic stirrer (**Fig. 1A**). The measuring chamber (**Fig. 1B**) consists of two conical semicircled Teflon compartments carrying various bores. The major bore ($\varnothing = 11$ mm), which forms the preparation chamber with a cylindrical opening ($\varnothing = 5$ mm) facing the planar side of the compartment, is connected via vertical bores ($\varnothing = 1.5$ mm) with the bore ($\varnothing = 5$ mm) for the electrodes. Two smaller bores ($\varnothing = 2$ mm), which are used for the adjustment of the levels of the bathing solution, are oblique with regard to the preparation chamber. The two compartments are pressed tightly together by a conical steel ring (outer diameter $d = 43$ mm) with their planar faces only separated by a Teflon septum containing the aperture for membrane formation.

The lower part of the Faraday cage contains two bores for the hose connec-tion to the syringes to adjust the levels of the bathing solutions. Two further

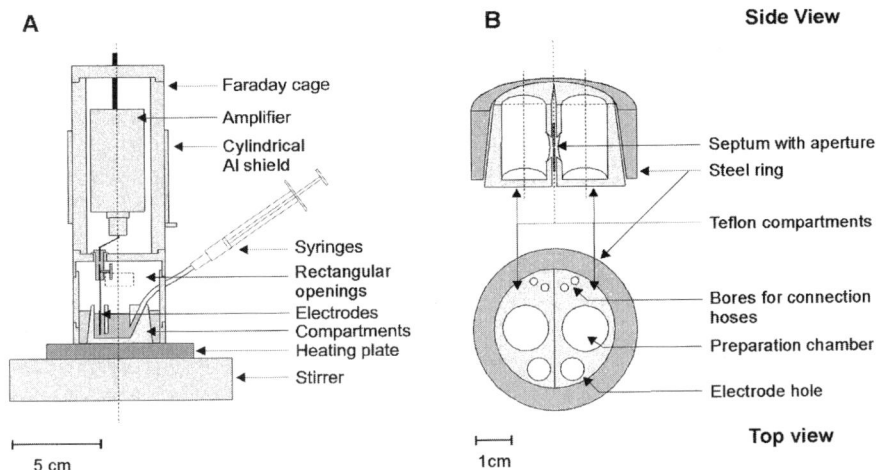

Fig. 1. Mechanical setup of the apparatus for the formation of asymmetric planar lipid bilayer membranes. (**A**) *Side view* on the Faraday cage containing the amplifier and measuring chanber and placed on top of the heating plate and stirrer. (**B**) *Side* and *top views* of the measuring chamber.

rectangular openings at the upper rim on the front and back sides are used for optical control of the membrane (stereo magnifier, Wild 3MZ, Heerbrugg, Switzerland) and its illumination (light source with flexible light guide, KL 1500, Schott, Wiesbaden, Germany). These openings can be closed after membrane formation by a moveable cylindrical shield around the Faraday cage. The headstage contains the amplifier of an L/M-PCA patch clamp amplifier (List-Medical, Darmstadt, Germany). As the amplifier housing is connected to the signal output, it is very important to pay caution to a correct shielding between the outer side of the amplifier and the Faraday cage. For the connection between the electrodes and the output connectors of the amplifier, the bottom of the headstage housing carries polytetrafluorethylene (PTFE)-insolated brass throughputs to which the amplifier outputs on one side, and the Ag/AgCl electrodes (type IVM E255, IN VIVO METRIC, Healdsburg, CA) on the other—easily exchangeable with screws—are connected. To allow a connection between the amplifier and the controller unit of the patch clamp amplifier, the top of the Faraday cage has a small slit (22 mm × 6 mm), which can be closed with an Al slide after insertion of the amplifier into the Al housing. The Faraday cage is connected to the ground connector of the controller unit.

To reduce mechanical vibrations the whole setup for the formation of planar bilayers is placed on a vibration isolation table (T 250, Physik Instrumente GmbH & Co., Waldbronn, Germany)

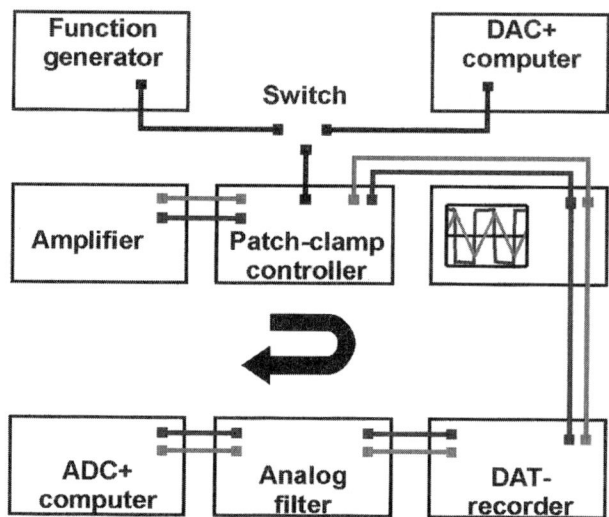

Fig. 2. Diagram of the electrical setup for the measurement of membrane conductivity/current and capacitance.

Of course, you can adjust the geometry and the electrical setup of your apparatus to fit your needs more closely by choosing alternative devices (*see* **Note 2**).

3.2. Electrical Setup

Figure 2 gives an overview of the electrical setup of the apparatus. The amplifier is connected to the controller of the L/M-PCA patch clamp amplifier. The controller is the main unit containing the power supply, signal processing electronics, and all of the controls. The controller is equipped with a capacitance compensation to avoid capacitive current flow as well as with a series resistance compensation to correct errors arising from an access resistance between the electrodes and the bathing solution. It can also be used in current clamp mode. Different amplification ranges (from 0.5 mV/pA to 1000 mV/pA) can be chosen and a built-in six-pole Bessel deep-pass filter can be used for filtering the output signal. The bandwidth of the amplifier is in the range of 50–80 kHz, and the equivalent input noise at 10 kHz is below 0.16 pA (gain 100 mV/pA). The clamp voltage can be changed continuously or stepwise between ±10 mV, ±20 mV, ±50 mV, and ±100 mV from the built-in power supply or by an external voltage source attached to the stimulus input connector. For this purpose, either a function generator (Model 4432:20 MHz, Enertec/Schlumberg, St. Etienne, France) or a digital analog converter (DAC)

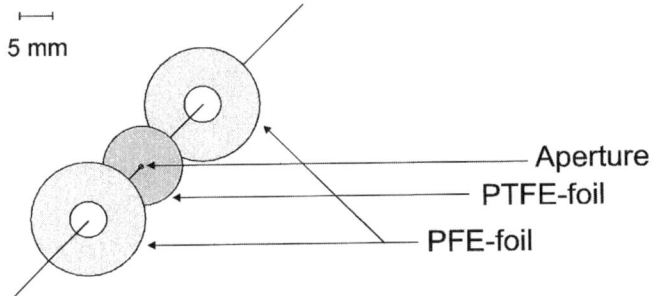

5 mm

Aperture
PTFE-foil
PFE-foil

Fig. 3. Expanded view of a three-layered septum (the aperture is introduced into the centre foil after sintering).

interface (PCI 20428W-1, Intelligent Instrumentation, Leinfelden-Echterdingen, Germany) connected to a PC can be used. Input voltage and output current are displayed on a doubletrace oscilloscope (TDS 220, Tektronix, Beaverton, OR) and are stored with a digital tape recorder (DTR 1202, Biologic, Claix, France) with a sampling frequency of 48 kHz. Before being read into the computer via an analog-digital converter (ADC) interface (PCI 20428W-1, Intelligent Instrumentation, Leinfelden-Echterdingen, Germany), the stored signals can further be deep-pass filtered using a four-pole bessel filter (Ithaco Model 4302, Ithaca, NY). To control the data acquisition and the output of the DAC, the software "Visual Designer 4.0" (Intelligent Instrumentation, Leinfelden-Echterdingen, Germany) is used (*see* **Notes 2** and **3**).

3.3. Preparation of Septa

Membranes are formed over a small aperture (diameter between 100 and 200 µm) in the center of a three-layered foil septum (**Fig. 3**). The inner PTFE foil (Breitenborn GmbH, Wuppertal, Germany) has a thickness of 12 µm, whereas the outer PFE foils (Breitenborn GmbH, Wuppertal, Germany) are 30 µm thick and provide mechanical stability. The outer ring-shaped foils have an outer diameter of 16 mm and an inner diameter of 5 mm; the diameter of the PTFE foil is 11 mm.

Details of septum preparation (*see* **Note 4**) are as follows:

1. Punch the circular foils from the respective sheets.
2. Arrange the foils in the proper way (**Fig. 3**).
3. Place the foils between two aluminum blocks (40 mm × 60 mm × 5 mm).
4. Press the blocks moderately together with two screws.
5. Heat the block in a muffle furnace for about 2.5 min at 600°C.
6. Transfer the block to a water bath (5°C) for approx 1 min.

7. Remove the septum from the block.
8. Place the septum between the electrodes of a spark discharge (the primary coil of a car ignition coil is connected to a 60 V voltage supply, the secondary coil to the electrodes, and the discharge is triggered by a push button connecting the power supply and the primary coil).
9. Control the final size of the aperture with an optical microscope.
10. Clean the septum in chloroform–methanol (2:1, v/v).
11. Dry and store the septa before use in the presence of silica gel to reduce air humidity.

3.4. Calibration for Capacity Measurements

The correct formation of the lipid bilayer is checked by determining the membrane capacitance as a measure for membrane thickness. To this end, a triangular voltage with an amplitude of 15 mV and a frequency of 50 Hz is applied to the membrane. As the ohmic conductivity of the pure bilayer (before the addition of drugs, proteins, etc.) can be neglected, a capacitive output current is generated with an amplitude proportional to the capacitance of the lipid bilayer. To determine the proportionality constant, the setup has to be calibrated (*see* **Note 5**):

1. Connect the function generator to the stimulus input connector of the controller.
2. Apply a triangular signal with an amplitude of 150 mV and a frequency of 50 Hz (the input is scaled down by the "stim.scaling" factor which has to be chosen to 0.1). This setup is always used for the determination of the capacitance of the bilayer.
3. Place different capacitors (capacitance range from 10 to 200 pF) between the electrode junctions of the headstage.
4. Plot the various capacitances vs the amplitudes of the membrane current.

3.5. Formation of Planar Bilayer Membranes and Measurement—General Procedure

In this section, a numerically ordered protocol for the reconstitution of the lipid matrix of the outer membrane is given. Examples are described for the investigation of protein–lipid interactions and for the characterization of the intrinsic membrane potential.

1. Hydrophobization of septa: The success of membrane formation is significantly increased, if the septum is pretreated with a mixture of hexadecane/hexane (1:20, v/v). To this end, the septum is gently moved in the solvent mixture and afterwards thoroughly shaken in air to dry.
2. Assembly of the measuring chamber: The planar sides of the Teflon compartments are first covered with a thin film of medium viscous silicon paste (keep the grease carefully away from the openings). The septum is placed on the planar side of one compartment with its aperture right in the center of the circular opening. The two compartments are then pushed against each other with the septum

between them and firmly pressed together with the conical ring. The measuring chamber is placed on the heating plate and the lower Al tube is placed over it. Now the connection hoses from the syringes are placed into the respective bore holes and the magnetic stirrers are put into the preparation chambers. To ensure an electrolyte connection between the electrodes and the preparation chambers, the bathing solutions (1.8 mL on each side) are filled into the compartments via the electrode holes.

3. Formation of the lipid monolayers: On each side, 2.2 μL of lipid solution are spread on top of the bathing solutions in the preparation chambers, for example, LPS solution on the front side and the PL mixture on the back side. By removal of bathing solution (via syringes), the monolayers are adjusted to a position approx 1 mm below the lower rim of the aperture. Before the preparation of the membranes, the solvent has to be allowed to evaporate for 15 min. During this time, the bathing solution can adjust to the prefixed temperature.

4. Electrical setup: To control membrane formation, the function generator has to be connected to the input connector of the controller for the determination of membrane capacitance (*see* **step 1** and **2** in **Subheading 3.4.**). The headstage is placed on top of the lower aluminum tube in a way that the electrodes fit into the respective bores. The output signal of the amplifier should be a rectangular capacitive current induced by the capacitance of the septum (about 40 pF). An overload signal on the controller would be indicative of a shortcut (leak) in the septum or between the septum and the compartments, *that is*, the experiment failed (proceed with **step 8**).

5. Membrane formation: After the evaporation of the solvent, the two monolayers are successively raised over the aperture by adding bathing solution from the syringes. Best results are obtained by raising the LPS monolayer first. If the formation of the membrane was successful, the capacitive membrane current should have increased to an appropriate value (*see* **Note 5**). A smaller value would be indicative of too thick a membrane, because of, for example, multilayer formation (retry again) or because of silicon grease in the aperture (abort the experiment and proceed with **step 8**). The overload signal indicates an unsuccessful bilayer formation. In this case, lower the monolayer below the aperture and try again to form a stable membrane. Be careful that no lipid remains in the aperture. After several unsuccessful attempts, the experiment should be aborted (proceed with **step 8**, *see* **Note 6**).

6. The measurement: The measurement starts after a further waiting period (5 min) for equilibration of the membrane. The triangular voltage source (for determination of membrane capacitance) is disconnected, and the data acquisition is started (*see* **step 7**). Drugs or proteins are added to the bathing solution and the magnetic stirrer is switched on for 15 s. The Faraday cage is closed by turning the outer aluminum ring over the observation slits. The electrical noise can be reduced further by removing the connection hoses between the syringes and the measuring chamber. A low clamp voltage (10–20 mV) is applied to check whether the incorporation of the external molecules takes place in the absence of any external forces possibly dragging the molecules into the membrane.

7. Data acquisition: Data acquisition starts after membrane equilibration. The digital tape recording is started, and, at the same time, the data are sampled on the hard disk. The "Visual Designer" software package allows the simultaneous data acquisition and controlling of an output voltage via ADC and DAC. To avoid the loss of information, the built-in filter of the controller should be switched to a high cutoff frequency (e.g., 20 kHz). If necessary, a lower cutoff frequency can be chosen for the analog filter between the DAT recorder and ADC. This setup offers the possibility to resample the data from the DAT recorder to the computer at higher sampling rates and cutoff frequencies, if this appears to be necessary at a later time after completion of the measurement.

8. Final procedures: At the end of the experiment, when membrane current exceeds the range of the amplifier—either because of spontaneous membrane rupture or to the action of added membrane-active substances—the clamp voltage is switched to zero, and the apparatus is disassembled. To obtain optimal results, each septum should be used for only one experiment. The silicon paste has to be removed from the Teflon compartments with a tissue paper, and the compartments have to be rinsed according to the cleaning procedure described in **step 9** before reuse. The connection hoses and the syringes can be used several times before replacement. If no further membrane preparation is planned for several hours, the electrodes must be removed and stored in saturated KCl solution.

9. Cleaning procedure: To guarantee a complete removal of all membrane-active substances and of all waste products that might adhere to the compartments, a thorough cleaning in several steps has to be performed. For this, the compartments are first put into ethanol, then into 1 M O_3, and HN finally into 1 M NaOH for at least 30 min in each step. All solutions are heated to 60°C, and the compartments are shaken in the solutions several times. After the complete cleaning procedure, the compartments are thoroughly rinsed in deionized water for 1 h and dried. The syringes and hoses are cleaned with deionized water after each experiment.

3.6. Applications

3.6.1. Determination of Function of Membrane Proteins, Here: Porins

To characterize the channel characteristics of membrane proteins, for example, porins, which are the major membrane proteins in the outer membrane of Gram-negative bacteria and facilitate the diffusion of nutrients and metabolites across this membrane, purified proteins are dissolved in detergents and added to the bathing solution of the PL side *(13,14)*. The incorporation of porins into the bilayer induces a stepwise increase in membrane current (**Fig. 4A**). The channel diameter d can be estimated from the heights of the current steps under the assumption of a circular pore geometry according to the simple equation $\Lambda = (\pi \cdot \sigma \cdot d^2)/4l$ (Λ = single channel conductivity, σ = specific conductance of the bathing solution; l =length of the pore corresponding to membrane thickness). Channel gating can be investigated by connecting the "stimulus input"

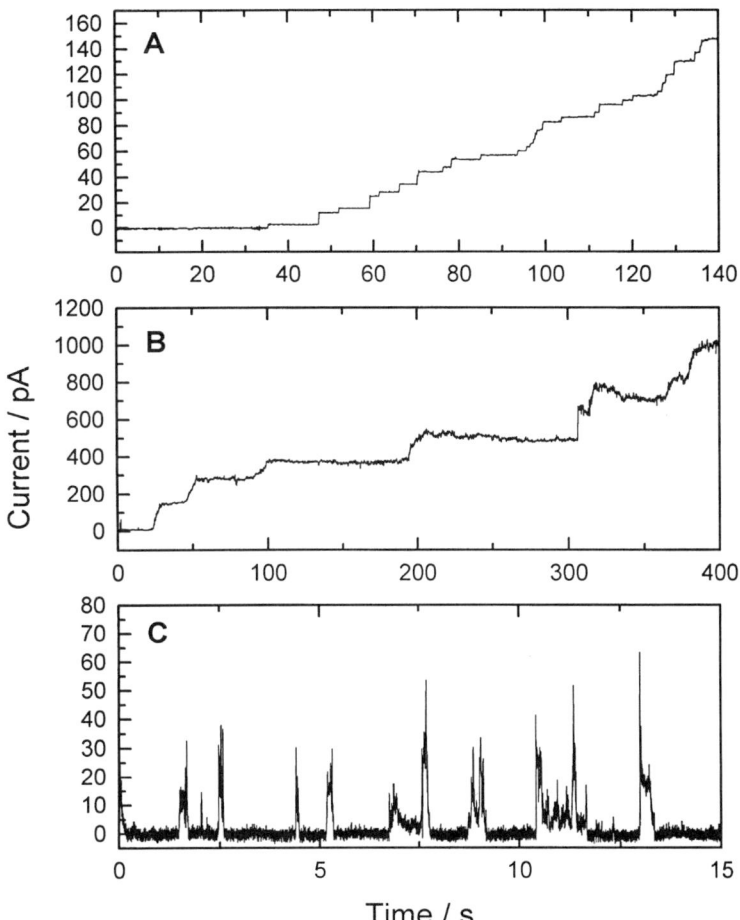

Fig. 4. **(A)** Stepwise current increase with time after the addition of 0.3 ng/mL of porin from *Paracoccus denitrificans* to the PL side of an asymmetric LPS/PL membrane. Each step represents the incorporation of one porin trimer. Bathing solution: 100 mM KCl, 10 mM MgCl$_2$, 5 mM HEPES; pH 7; T = 37°C; clamp voltage = 10 mV (LPS side). **(B)** Current increase with time after the addition of 0.1% whole human serum to the LPS side of an asymmetric LPS/PL membrane. Each step (burst) represents the formation of one complement pore. Bathing solution: 100 mM KCl, 5 mM MgCl$_2$, 5 mM HEPES; pH 7; T = 37°C; clamp voltage = 20 mV (LPS side). **(C)** Short-lived current fluctuations due to the formation of transient membrane lesions directly after the addition of 1 µM PMB to the LPS side of an asymmetric LPS/PL membrane. Bathing solution: 100 mM KCl, 5 mM MgCl$_2$, 5 mM HEPES; pH 7; T = 37°C; clamp voltage = 20 mV (LPS side). LPS used: that from the deep rough mutant strains R595 of *Salmonella enterica* sv. Minnesota (A, B) and F515 of *Escherichia coli* (C), respectively.

connector with the output of the DAC of the computer system. A triangular output of low frequency (0.2 min^{-1}) with an amplitude of 1.8 V (stim. scaling = 0.1) will expose the membrane to a voltage up to ±180 mV, which for most porins should induce channel closing.

3.6.2. Interaction of Serum Proteins with the Membrane, Here: the Complement System

To study the interaction between serum proteins and the membrane, human serum can directly be added to the bathing solution on the LPS side (*14,15*). Pore formation by proteins of the complement cascade, which is directly activated by the glycolipid surface, leads to an increase of membrane current (**Fig. 4B**). From the differently structured current steps, two pore geometries can be deduced, the circular pore and the leaky patch. Use of sera depleted in certain complement components allows the determination of the complement activation pathway.

3.6.3. Interaction of Drugs with Differently Composed Membranes, Here: Polymyxin B

The addition of membrane active drugs to the bathing solution may induce current changes. In **Fig. 4C**, short-lived current fluctuations soon after the addition of the polycationic antibiotic polymyxin B (PMB) are shown. From the amplitudes of the current fluctuations, the sizes of the membrane lesions can be estimated. For membranes made from LPS from PMB-sensitive strains (relatively high negative surface charge density) the lesions are large enough to allow the passage of PMB molecules across the outer membrane, whereas they are too small when glycolipids from PMB-resistant strains (relatively low negative surface charge density) are used (*16,17*).

3.6.4. Determination of the Intrinsic Membrane Potential

To obtain information on the transmembrane potential profile, the procedure of membrane formation requires slight changes. In this case, prior to membrane formation, ion carriers need to be present in the bathing solution. To this end, the K$^+$ carrier nonactin is added at a final concentration of 10 μM to a 100 mM KCl bathing solution. Current/voltage (I/U) traces are then recorded as described for the investigation of porin channel closing. The I/U curves are evaluated according to a protocol described by Schoch et al. (5). Briefly, the membrane current I as function of the voltage U applied with the voltage clamp is given by

$$I = K \cdot \frac{\Delta\Phi + (n_2 - n_1) \cdot U}{n_2 - n_1} \cdot \frac{\exp(a \cdot U) - 1}{\exp(a \cdot (\Delta\Phi + n_2 \cdot U)) - \exp(a \cdot n_1 \cdot U)} \qquad (1)$$

where a = $(Z\,e_0)/(k\,T)$, with Z = valence of the ions in the bathing solution, e_0 = electron charge, k = Boltzmann constant, and T = temperature of the bathing solution. K is a constant for each membrane (depending, among other parameters, on its area and thickness), n_1 and n_2 are the edges of the potential walls for the two leaflets, and $\Delta\Phi$ is the potential difference between these edges. The three parameters describe the shape of the trapezoidal energy barrier and are determined from the experimental curves by computer fitting of the above equation.

As the carrier is present in the bathing solution before membrane formation, a noncapacitative membrane current can be observed depending on carrier concentration, which can prevent the determination of membrane capacitance. In this case, from the I/U curves (which, according to Eq. 1, may not be linear) information on correct membrane formation can also be obtained.

4. Notes

In the previous sections, we described a method for the reconstitution of the lipid matrix of the outer membrane of Gram-negative bacteria as it is applied in our laboratory in such a way as to make this technique easily adaptable in other laboratories and also for matrices with other lipid composition/distribution. Here, we want to refer to important alternatives to this protocol and point out certain problems that might occur during the experiments.

1. Adjustment of pH: In the literature various other buffers are listed, for example, 10 mM NaHCO$_3$ (18), 2.5 mM 2-(N-morpholino)ethanesulfonic acid (19), and 10 mM Tris(hydroxymethyl)-aminomethane (20,21) for different pH ranges.
2. Mechanical and electrical setup: Various alternatives with respect to amplifiers and filtering devices exist. We have tried to reduce high-frequency electrical noise and to establish high sample rates. In case you build your own setup, you should thoroughly balance your needs with the respective possibilities of the devices you plan to use. In this context, the electrically controlled heating plate should be mentioned. As an alternative to an electrical heating foil, water-thermostatted heating plates may be used. The water flow, however, tends to induce mechanical vibrations that interfere with the membrane current (your membrane behaves like a very sensitive microphone!). To avoid electrical interferences, the heating foil is operated by a carefully smoothed direct current. Heating to temperatures >50°C may cause problems owing to a fluidization of the silicon paste, which on the one hand may contaminate the lipid monolayer and, on the other hand, lead to an incomplete sealing of the septa.
3. Electrodes: We have had good experience using solely Ag/AgCl electrodes; nevertheless, in the literature some experiments have been described (e.g., at low ionic strength or if channels are influenced by K$^+$ or Cl$^-$ ions) in which in addition, agar bridges are necessary. An incomplete chlorination of the electrodes may cause a shift in the electrode potential superimposing the applied voltage

and, thus, distort the results. To check electrode quality, the zero-current potential—which should not be higher than 2 mV—should be determined during the experiment. For additional details, please refer to *(22)*.

4. Septa: The production of clean, plane, and leakproof septa is a prerequisite for membrane formation. The sintering procedure requires some experience, in particular concerning the sintering time. Septa, which are still crystal clear and can easily be disassembled after sintering, have not been heated long enough. Septa, which stick to the aluminum plates, however, have been heated too long. The margins of the heating time has to be determined within a few seconds.

   As bilayer thickness is three orders of magnitude lower than the thickness of the septa, the quality of the rim of the aperture is very important. Best results are obtained if several sparks are used to form an aperture of appropriate diameter; otherwise the intensity of the spark must be so high that a rim of molten material surrounds the aperture. Alternatively, the aperture can be punched with a sharpened syringe needle *(23)*. According to our experience, however, this procedure is even more complicated.

5. Capacitance calibration: In our experimental setup, the proportionality constant was about 300 mF/A. For apertures with a diameter in the range of 100–150 µm, membrane capacitances in the range of 90–120 pF have to be expected.

6. Membrane formation: A prerequisite for the formation of membranes is that the acyl chains of the membrane forming lipids are in the liquid crystalline (α) and not in the gel (β) phase *(24)* at the temperature of membrane formation. Furthermore, the tendency of the lipids to form nonlamellar structures usually decreases membrane stability and may even prevent membrane formation. Instead of forming the monolayers by spreading lipid solutions on top of the bathing solutions, the monolayers can be generated from liposomes, proteoliposomes, or a mixture of isolated membrane fragments and liposomes added to the bathing solution *(25,26)*.

   As most of the glycolipids are negatively charged, divalent cations are known to stabilize the membranes *(16,24)*. Therefore, the presence of 5 mM MgCl$_2$ in the bathing solution makes membrane formation much easier and increases the life of membrane.

Acknowledgments

The establishment of the technique and the acquisition of the cited scientific data were supported by the German Minister of Education, Science, Research, and Technology (Grant 01 KI 9851, Project A6) and the Deutsche Forschungsgemeinschaft (SFB 470, Project B5).

References

1. Nikaido, H. and Vaara, M. (1985) Molecular basis of bacterial outer membrane permeability. *Microbiol. Rev.* **49,** 132.
2. Rietschel, E. T., Brade, H., Holst, O., Brade, L., Müller-Loennies, S., Mamat, U., Zähringer, U., Beckmann, F., Seydel, U., Brandenburg, K., Ulmer, A. J., Mattern, T.,

Heine, H., Schletter, J., Loppnow, H., Schönbeck, U., Flad, H.D., Hauschildt, S., Schade, U. F., Di Padova, F., Kusumoto, S., and Schumann, R. R. (1996) Bacterial endotoxin: chemical constitution, biological recognition, host response, and immunlogical detoxification, in *Pathology of Septic Shock* (Rietschel, E. T. and Wagner, H., eds.), Springer Verlag, Berlin, pp. 39–81.

3. Osborn, M. J., Gander, J. E., Parisi, E., and Carson, J. (1972) Mechanism and assembly of the outer membrane of *Salmonella typhimurium. J. Biol. Chem.* **247,** 3962–3972.

4. McLaughlin, S. (1989) The electrostatic properties of membranes. *Annu. Rev. Biophys. Biophys. Chem.* **18,** 113–136.

5. Schoch, P., Sargent, D. F., and Schwyzer, R. (1979) Capacitance and conductance as tools for the measurement of asymmetric surface potentials and energy barriers of lipid bilayer membranes. *J. Membr. Biol.* **46,** 71–89.

6. Mueller, P., Rudin, D. O., and Tien, H. T. (1962) Reconstitution of cell membrane structure *in vitro* and its transformation into an excitable system. *Nature* **194,** 979–981.

7. Tien, H. T. (1974) *Bilayer Lipid Membranes (BLM): Theory and Practice.* Marcel Dekker, New York.

8. Montal, M. and Mueller, P. (1972) Formation of bimolecular membranes from lipid monolayers and a study of their electrical properties. *Proc. Natl. Acad. Sci. USA* **69,** 3561–3566.

9. Schindler, H. (1980) Formation of planar bilayers from artificial or native membrane vesicles. *FEBS Lett.* **122,** 77–79.

10. White, S. H. (1986) The physical nature of planar bilayer membranes, in *Ion Channel Reconstitution* (Miller, C., ed.), Plenum, New York, pp. 335.

11. Galanos, C., Lüderitz, O., and Westphal, O. (1969) A new method for the extraction of R lipopolysaccharides. *Eur. J. Biochem.* **9,** 245–249.

12. Galanos, C. and Lüderitz, O. (1975) Electrodialysis of lipopolysaccharides and their conversion to uniform salt forms. *Eur. J. Biochem.* **54,** 603–610.

13. Wiese, A., Schröder, G., Brandenburg, K., Hirsch, A., Welte, W., and Seydel, U. (1994) Influence of the lipid matrix on incorporation and function of LPS-free porin from *Paracoccus denitrificans. Biochim. Biophys. Acta* **1190,** 231–242.

14. Wiese, A., Reiners, J. O., Brandenburg, K., Kawahara, K., Zähringer, U., and Seydel, U. (1996) Planar asymmetric lipid bilayers of glycosphingolipid or lipopolysaccharide on one side and phospholipids on the other: membrane potential, porin function, and complement activation. *Biophys. J.* **70,** 321329.

15. Schröder, G., Brandenburg, K., Brade, L., and Seydel, U. (1990) Pore formation by complement in the outer membrane of Gram-negative bacteria studied with asymmetric planar lipopolysaccharide/phospholipid bilayers. *J. Membr. Biol.* **118,** 161–170.

16. Schröder, G., Brandenburg, K., and Seydel, U. (1992) Polymyxin B induces transient permeability fluctuations in asymmetric planar lipopolysaccharide/phospholipid bilayers. *Biochemistry* **31,** 631–638.

17. Wiese, A., Münstermann, M., Gutsmann, T., Lindner, B., Kawahara, K., Zähringer, U., and Seydel, U. (1998) Molecular mechanisms of polymyxin

B–membrane interactions: direct correlation between surface charge density and self-promoted transport. *J. Membr. Biol.* **162,** 127–138.

18. Brunen, M. and Engelhardt, H. (1995) Significance of positively charged amino acids for the function of the *Acidovorax delafieldii* porin Omp34. *FEMS Microbiol. Lett.* **126,** 127–132.

19. Ishii, J. and Nakae, T. (1996) Specific interaction of the protein-D2 porin of *Pseudomonas aeruginosa* with antibiotics. *FEMS Microbiol. Lett.* **136,** 85–90.

20. Dargent, B., Hofmann, W., Pattus, F., and Rosenbusch, J. P. (1986) The selectivity filter of voltage-dependent channels formed by phosphoporin (PhoE protein) from *E. coli. EMBO J.* **5,** 773–778.

21. Brunen, M. and Engelhardt, H. (1993) Asymmetry of orientation and voltage gating of the *Acidovorax delafieldii* porin Omp34 in lipid bilayers. *Eur. J. Biochem.* **212,** 129–135.

22. Koryta, J. (1991) *Ions, Electrodes and Membranes.* John Wiley & Sons, Chichester.

23. Schindler, H. and Feher, G. (1976) Branched bimolecular lipid membranes. *Biophys. J.* **16,** 1109–1113.

24. Seydel, U., Schröder, G., and Brandenburg, K. (1989) Reconstitution of the lipid matrix of the outer membrane of Gram-negative bacteria as asymmetric planar bilayer. *J. Membr. Biol.* **109,** 95–103.

25. Schindler, H. and Rosenbusch, J. P. (1978) Matrix protein from *Escherichia coli* outer membranes forms voltage-controlled channels in lipid bilayers. *Proc. Natl. Acad. Sci. USA* **75,** 3751–3755.

26. Schindler, H. (1979) Exchange and interactions between lipid layers at the surface of a liposome solutions. *Biochim. Biophys. Acta* **555,** 316–336.

Index